I AM THE OTHER

Recent Titles in Contributions to the Study of Science Fiction and Fantasy

Science Fiction, Canonization, Marginalization, and the Academy
Gary Westfahl and George Slusser, editors

The Supernatural in Short Fiction of the Americas: The Other World in the New World
Dana Del George

The Fantastic Vampire: Studies in the Children of the Night
James Craig Holte, editor

Unearthly Visions: Approaches to Science Fiction and Fantasy Art
Gary Westfahl, George Slusser, and Kathleen Church Plummer, editors

Worlds Enough and Time: Explorations of Time in Science Fiction and Fantasy
Gary Westfahl, George Slusser, and David Leiby, editors

No Cure for the Future: Disease and Medicine in Science Fiction and Fantasy
Gary Westfahl and George Slusser, editors

Science and Social Science in Bram Stoker's Fiction
Carol A. Senf

Chaos Theory, Asimov's Foundations and Robots, and Herbert's Dune: The Fractal Aesthetic of Epic Science Fiction
Donald E. Palumbo

The Novels of Kurt Vonnegut: Imagining Being an American
Donald E. Morse

Fantastic Odysseys: Selected Essays from the Twenty-Second International Conference on the Fantastic in the Arts
Mary Pharr, editor

The Utopian Fantastic: Selected Essays from the Twentieth International Conference on the Fantastic in the Arts
Martha Bartter

War and the Works of J.R.R. Tolkien
Janet Brennan Croft

I AM THE OTHER

Literary Negotiations of Human Cloning

Maria Aline Salgueiro Seabra Ferreira

Contributions to the Study of Science Fiction and Fantasy, Number 85
Donald Palumbo, Series Adviser

Westport, Connecticut
London

Library of Congress Cataloging-in-Publication Data

Ferreira, Maria Aline Seabra.
I am the other : literary negotiations of human cloning / Maria Aline Salgueiro Seabra Ferreira.
p. cm.—(Contributions to the study of science fiction and fantasy, ISSN 0193–6875 ; no. 85)
Includes bibliographical references (p.) and index.
ISBN 0–313–32006–3 (alk. paper)
1. Science fiction, American—History and criticism. 2. Human cloning in literature. 3. Science fiction, English—History and criticism. 4. Difference (Psychology) in literature. 5. Identity (Psychology) in literature. 6. Doubles in literature. 7. Cloning in literature. I. Title. II. Series.
PS374.H83F47 2005
813'.087620936—dc22 2004044382

British Library Cataloguing in Publication Data is available.

Library of Congress Catalog Card Number: 2004044382
ISBN: 0–313–32006–3
ISSN: 0193–6875

First published in 2005

Praeger Publishers, 88 Post Road West, Westport, CT 06881
An imprint of Greenwood Publishing Group, Inc.
www.praeger.com

Printed in the United States of America

The paper used in this book complies with the Permanent Paper Standard issued by the National Information Standards Organization (Z39.48–1984).

10 9 8 7 6 5 4 3 2 1

Copyright Acknowledgments

The author and publisher gratefully acknowledge permission for use of the following material:

Parts of Chapter 4 are reprinted from *Proceedings of the Fourteenth International Conference on Literature and Psychoanalysis*, edited by Frederico Pereira (Lisboa: Instituto Superior de Psicologia Aplicada, 2000), 191–203.

Most of Chapter 5 appeared in *Biotechnological and Medical Themes in Science Fiction*, edited by Domna Pastourmatzi (Thessaloniki: University Studio Press, 2002), 186–207.

An earlier version of Chapter 6 is reprinted from *Science Fiction and Organization*, edited by Warren Smith, Matthew Higgins, Martin Parker, and Geoff Lightfoot (London and New York: Routledge, 2001), 73–89.

Some parts of Chapter 7 first appeared in the *Journal of the Association for Research on Mothering*, Vol. 4, Number 1 (Spring/Summer 2002): 113–130.

To my parents,
without whom I would not have gone to London in the first place

To Nicholas and Kate,
who patiently accepted my absences and put up with long working hours at the computer, when I could have been playing with them

To David,
who always supported me and believed I could do it

Contents

Acknowledgments

I owe thanks to many people who discussed parts of my argument with me and generously shared their ideas with me. I particularly wish to thank Prof Donald Palumbo for his enthusiasm and encouragement, and Prof Lyman Tower Sargent who always showed interest in my work.

Introduction

> À la réflexion, je crois que, de toute façon, du clonage il y en a eu, il y en a et il en aura. Les législations n'empêcheront pas le clonage.
> (Jacques Derrida and Elisabeth Roudinesco, *De quoi demain . . . Dialogue*, 93)[1]

My aim in this book is to investigate the topic of human cloning from the point of view of literary history and criticism.[2] I wish to suggest that many of the ramifications implicit in the subject of cloning were already present in, and indeed have informed the work of, numerous authors from well before the issue became of general concern. It is thus the areas dealt with by those writers that are contiguous or overlap with the concerns adjacent to cloning, taken in its wide sense of reduplication and copying, that I wish to explore here, as well as works dealing explicitly with cloning.

The fantasy of human cloning has become one of the most important myths of our time. New technologies inevitably bring about a crisis in the cultural and social scene, a climactic turning point that in the case of our contemporary Biotechnological Age can be characterized as the advent of the "posthuman," the genetically altered, the technologically enhanced, a zone where human cloning can be inscribed. The implications of this dream of engendering life bur circumventing the need for joint male and female intervention (including the uterus, with the development of ectogenetic chambers) are serious, far reaching, and at present difficult to gauge in their multifarious ramifications.

Cloned humans will inevitably challenge commonly held perceptions of our own bodies and our relations to other bodies. Cloned people can aptly be called posthuman in the sense that they have been "artificially" created by means of asexual reproduction. In many ways they incarnate, as Donna Haraway, that eloquent thinker of the posthuman condition, phrases it, "what it means to be embodied in high-tech worlds" ("A Cyborg Manifesto: Science, Technology and Socialist-Feminism in the Late Twentieth Century," 173). N. Katherine Hayles has remarked that "for some time now there has been a rumor going around that the age of the human has given way to the posthuman. Not that humans have died out, but that the human as a concept has been succeeded by its evolutionary heir" ("The Life Cycle of Cyborgs: Writing the Posthuman," 157).[3] This is similarly emphasized by Judith Halberstam and Ira Livingston, who argue that "the posthuman condition is upon us" and diagnose in "posthuman" narratives a "nostalgia for a humanist philosophy of self and other, human and alien" (*Posthuman Bodies*, vii), a nostalgia I also see at play in some of the works I will be examining here, such as Pamela Sargent's *Cloned Lives* (1976) and Kate Wilhelm's *Where Late the Sweet Birds Sang* (1976).

The stories that circulate in a given culture are essential to start understanding and coming to terms with fresh models of subjectivity, which will necessarily develop as a response to modern technologies. In order to reconfigure these novel potential social and cultural scenarios, a whole new set of ideas will be required. In this book I investigate and attempt to theorize some of the complex scenarios posited by the emergence of human cloning, the advent of which will inevitably entail a pressing need for redefinitions of social roles within the new, posthuman society we already inhabit. As Scott Bukatman notes, this "provocative set of new posthuman contours . . . necessarily contain[s] the germ of new political and philosophical orientations to accompany their new spatial and bodily configurations" (*Terminal Identity: The Virtual Subject in Postmodern Science Fiction*, 325). Pamela Sargent herself has remarked that "the current pace of biological research suggests that we may soon have to alter our most fundamental beliefs about human life, human values, and human nature" (Introduction to *Bio-Futures*, xi–xii), new social contours that she further reflects on and vividly dramatizes in her fiction.

I see the posthuman era dominated by the identity crisis of new, genetically engineered people, principally clones. Indeed, the proliferation in the twentieth century of cyborg fantasies, the development of genetic engineering, and the great impact of biology in the explanation of the workings of the human body have made of this apparently "natural" object, the human organism, a privileged crossroads of interdisciplinary enquiry. In the sense that clones can be seen as "artificially" created, in a laboratory environment, as the offspring of technology, they can thus be said to inflect to a certain extent a cyborgian ontology, such as Donna Haraway's influential definition of a cyborgian politics.

In this scenario, where cloned people may potentially become an integral part of our society, several pressing questions need to be asked: Will there still be a form of "Oedipus complex," defined by Freud as the founding stone and deep structuring device of societal coexistence? Will Oedipus survive human cloning? How will the functions of authority be restructured in a post-Oedipus society with a no-longer-clearly-located paternal authority figure? How will these new post-Freudian familial configurations be redefined? Moreover, will Lacan's "Symbolic Order" still carry the same weight in this putative postpatriarchal, posthuman society? Will the advent of human cloning mean an even greater exacerbation of narcissism in an era already haunted by the figure of Narcissus?

Jacques Derrida participates in the debate about human cloning, noting the pull and allure it exerts on people's imagination, as well as its potential role in reshaping the social fabric. As Derrida maintains, "les 'progrès' de la génétique libèrent ou accélèrent notre imagination—réjouie, terrifiée, ou les deux fois, devant toutes sortes de choses que je ne dirais pas inconnues, surtout de l'inconscient, mais non encore 'enregistrées' par ce qu'on pourrait appeler, au sens large, l'état civil" (*De Quoi Demain: Dialogue*, 69).[4] The fantasy of human cloning can thus be seen as focusing in a consummate way the widespread millenial anxieties that permeate contemporary literature, popular culture, science, and medicine. The proliferation of discourses of science and politics applied to the technologized body have accelerated at a vertiginous pace, mostly impelled by advances in biomedicine. If cyborgs are inevitably with us, so now are clones.

Narratives grappling imaginatively with the implications of cloning have been appearing over a period that began many years before the issues sifted through and infiltrated the public imagination. It is precisely this potential for apprehending crosscurrents of thought and impending technological novelties, which have the capacity to revolutionize the world, that such narratives have seized on. In the case of cloning, this function of prognostication renders the fiction dealing with the prospect of human cloning as so many cautionary tales, often suggesting a sensible and prudent approach, but also frequently conjuring up far-fetched scenarios that only with great difficulty can ever come to pass—and many years into the future at that. Indeed, a great deal of the literature dealing with human cloning can be described as evincing an apocalyptic sensibility, depicting highly improbable and strained alternative worlds, as well as conveying an ominous foreboding of disastrous events that may lead to or facilitate the emergence of cloning, seen as a solution for population and environmental problems in certain societies doomed by catastrophic occurrences. It seems thus greatly appropriate that human cloning, an eminently millenial and apocalyptic theme, has significantly come to general attention almost at the end of the millenium, suggestive of portentous changes to come.

Our contemporary world is fascinated with the fantasy of human cloning, which since the birth of Dolly, the cloned sheep, on July 5, 1996, but only publicly announced on February 27, 1997, in the Roslin Institute, in Scotland, by Dr Ian Wilmut and his colleagues, has fired the popular imagination and gradually permeated popular discourse, proving to be an endless source of inspiration for the media, advertisements, literature, and film.[5] The popular press is now teeming with books dealing with narratives of cloning, and biothrillers have become widely sought after and popular.

From Aldous Huxley's *Brave New World* (1932)[6] to Nancy Freedman's *Joshua Son of None* (1973) and Ira Levin's *The Boys from Brazil* (1976), Evelyn Lief's *The Clone Rebellion* (1980), and Michael Marshall Smith's *Spares* (1998), fictional representations of human cloning have been predominantly negative,[7] arousing feelings of deep-seated horror in many readers.[8] These and many other books and films have been largely responsible for spreading frightening visions of armies of clones taking over the Earth. Indeed, misconceptions[9] about cloning abound,[10] producing heavily distorted pictures in the popular imagination of what might be an impending reality.

Why does the concept of human cloning provoke such deep and often contradictory feelings?[11] Where does the common reaction of aversion and horror,[12] but also of guilty fascination, elicited by the very mention of cloning stem from? There are, of course, many possible explanations. From the fear of losing one's identity and uniqueness, becoming one in an endless series of duplications, to such widespread misapprehensions as the conjecture that a clone would lack a soul or the horrifying spectacle of standardized humanity moving, robot-like, as a sole organism without free will or volition, in a parody of human standardization—all these visions are dramatized in these narratives.[13] This is the kind of scenario developed in Aldous Huxley's *Brave New World* (1932), a book that has once again captured media attention since the cloning of Dolly the sheep by Dr Ian Wilmut and his colleagues was broadcast, giving rise to generalized fears that human cloning might also be imminent. Among these fears and manifold anxieties is the fact that it touches on the root of humankind's deepest ontological sense of being, giving rise to profound and almost unnameable ontological concerns. Where do we come from? Who are we? Where are we going? Cloning taps into nothing less than such crucial questions, which will constitute a recurring thematic concern throughout this book.

The word *clone*, which derives from the Greek word for twig, was coined in 1903,[14] describing genetically identical groups of plants.[15] The word first appears in science fiction in 1915, in a collection of short stories, *Master Tales of Mystery by the World's Most Famous Authors of Today*, edited by Francis Joseph Reynolds.[16] J.B.S. Haldane, the famous British biologist, for his part, was among the first to employ the word clone in a speech delivered at a scientific symposium, "Biological Possibilities for the Human Species in the Next Ten Thousand Years," when he wondered about what might happen "if we cloned people with 'attested ability'" (352).[17] Although his forecasts have proved wide of the mark in terms of the specific years in which they were to take place, Haldane's prognostications about future developments in various fields, from botany to zoology to reproductive technologies, are nowadays the object of renewed ethical debate.[18]

Aldous Huxley's *Brave New World* (1932), for its part, was arguably the book that called wide attention to and sounded alarm bells about some of the dangers of this new reproductive technology. When it appeared, it horrified its readers with its haunting visions of dehumanized, robotlike human beings who were described as lacking will, as numbered copies serially produced like objects in a factory line. However, one year before Aldous Huxley's book was published, his older brother Julian Huxley, an evolutionary biologist, had published *What Dare I Think?* (1931) in which many of the radical concepts dramatized in *Brave New World* were already mooted, such as genetic manipulation and improvement of the human species,[19] a goal that is considered eminently desirable.[20]

As Jon Turney notes, in his history and overview of the development of science and genetics in popular culture, "Meanwhile, biology proper occasioned little comment. . . . *Brave New World* and, for that matter, *Frankenstein* and *Moreau* were still being read, and plagiarised. But there was no real advance in fictional ideas to compare with Huxley's book" (*Frankenstein's Footsteps: Science, Genetics and Popular Culture*, 133). In 1959 the translation into English of *Can Man Be Modified? Predictions of Our Biological Future*, a book by Jean Rostand, a French experimental biologist, published three years earlier, reopened the debate that had raged in the 1920s about life sciences and their potential to change the future of humankind, which had centered mostly around J.B.S. Haldane, Julian Huxley, and Bertrand Russell.[21]

Rostand's book supplied an overview of some of the most important scientific developments in the field of molecular biology to date. The possibility of artificial parthenogenesis could lead, in Rostand's view, to new reproductive patterns. Extrapolating from the experiments on nuclear transplantation carried out by Briggs and King in 1955, which Jon Turney sees as the "foundation of modern research on asexual reproduction by cloning" (*Frankenstein's Footsteps: Science, Genetics and Popular Culture*, 140), Rostand prognosticates that "this new technique of generation would in theory enable us to create as many identical individuals as might be desired. A living creature would be printed in hundreds, in thousands of copies, *all of them real twins*. This would, in short, be *human propagation by cuttings*" (*Can Man Be Modified? Predictions of Our Biological Future*, 14; emphasis original).

Jon Turney considers Rostand's book "a link between the time of Loeb and Carrel and the biological revolution of the 1960s" (*Frankenstein's Footsteps: Science, Genetics and Popular Culture*, 142) in terms of its influence on many other books that popularize science. Indeed, as Turney emphasizes, in the 1960s, the decade after Rostand's book appeared in English, the decade of the "biological Revolution" (143), the "public face of biology was transformed" (143), a change that derived from the predominance of molecular biology as the science that provided the way to explain living things. As Turney elucidates, the "centrality of DNA, the notion that life might be created by synthesising DNA and its forms altered by altering DNA—were major motifs in the discursive literature about molecular biology which grew up from the beginning of the 1960s" (145). Moreover, throughout the 1960s, the key ideas developed by scientists in these areas were quickly taken up by journalists and authors who specialized in explaining science to the public.

Gordon Rattray Taylor's *The Biological Time Bomb* (1968), once highly popular, considered many of the social ramifications of the implementation of new biotechnological techniques[22] and had a profound impact in its time on many science fiction writers, such as Ursula K. Le Guin, who explicitly cites it as the driving influence and inspiration that led her to write her short story "Nine Lives" ("On Theme," 205). In the anthology of stories dealing with biotechnological themes that she edited, *Bio-Futures* (1976),[23] Pamela Sargent similarly mentions Gordon Rattray Taylor's *The Biological Time Bomb*, describing it as "as speculative as some science fiction" (xxxv).

The idea of cloning was first brought to more widespread public attention around thirty years ago, largely through the writings of Nobel Laureate geneticist Joshua Lederberg, who argued for the eugenic benefits of human cloning in an article in *The American Naturalist* (1966).[24] In 1967 he wrote a regular column in the *Washington Post* in which he considered the possibility of human cloning, which gave rise to a public debate. Leon R. Kass, a biochemist, philosopher, and ethicist at the University of Chicago, was one who participated in this discussion, arguing, as he explains, "against Lederberg's amoral treatment of this morally weighty subject and insisting on the urgency of confronting a series of questions and objections, culminating in the suggestion that 'the programmed reproduction of man will, in fact, dehumanize him'" ("Human Cloning Should Be Banned," 27).

Princeton theologian Paul Ramsey, in turn, similarly took part in the debate on cloning sparked by Lederberg's article, sounding a strong warning against new reproductive technologies, particularly cloning, in his *Fabricated Man: The Ethics of Genetic Control* (1970).[25] In two articles he wrote for the *Journal of the American Medical Association*, entitled "Shall We Reproduce?" published in June 1972, Ramsey cautioned that "human procreation has already been replaced by the idea of 'manufacturing' our progeny. Unless and until that concept is reversed, mankind's movement toward Aldous Huxley's Hatcheries must surely prove irreversible."[26]

For his part, futurist Alvin Toffler, in *Future Shock* (1970), speculates on some of the consequences of the development of molecular biology, taking up cloning again for special attention. As Toffler muses, cloning "would make it possible for people to see themselves anew, to fill the world with copies of themselves" (183),[27] a scenario that would give rise to serious moral, ethical, and political issues that, as Toffler puts it, would "simply boggle the mind" (184).

James D. Watson, the Nobel laureate scientist who, together with F.H.C. Crick, discovered the structure of DNA in 1953, declared before Congress in 1971 that he was convinced scientists might soon develop the necessary technique to clone human beings. In his influential article "Moving toward Clonal Man" in *Atlantic*, Watson cautioned against this possibility, evoking the work of British researchers Patrick Steptoe and Robert Edwards in in vitro fertilization as potentially leading to human cloning.[28]

In 1972, William Gaylin, president of the Hudson Institute,[29] wrote an article for the *New York Times Magazine*, "Frankenstein Myth Becomes a Reality," dealing with human cloning. Engaging with Kass and Ramsey's work, Gaylin, drawing on *Frankenstein* as a metaphor suggestive of human cloning, addresses some of the implications evoked by that potentiality. As Gaylin asserts, cloning "commands our attention more because it dramatises the developing issues in bioethics than because of its potential threat to our way of life. Many biologists, ethicists and social scientists see it not as a pressing problem but as a metaphoric device serving to focus attention on identical problems that arise from less dramatic forms of genetic engineering,"[30] which include in vitro fertilization and artificial insemination.

This increase in attention confirms Turney's observations that whereas discussion about the possibility of human cloning began with Aldous Huxley's *Brave New World*, it "really emerged as a recurrent motif in debate in the late 1960s and early 1970s" (*Frankenstein's Footsteps: Science, Genetics and Popular Culture*, 211). Indeed, when the concept of cloning began to circulate in scientific circles, gradually sifting through to the popular press, David Rorvik, a reputed writer in the field of human biology, wrote a best-selling book drawing on precisely that image, *In His Image: The Cloning of a Man* (1978), a sensationalist book that caused widespread interest and polemic in the United States.[31] As the narrator declares, "Cloning was, to put it mildly, a touchy subject in the scientific community—not too far removed from Frankenstein's monster, at least in some people's minds. In my own writing on the subject I had tended to see its potentially positive side. I was intrigued by it—but no more so than had been many reputable scientists, including some Nobel Prize winners" (7). In *In His Image: The Cloning of a Man* the narrator, a journalist who specializes in medical themes and reproductive technologies, is asked to help a wealthy man in his ambition to have a clone son. Blurring the boundaries between fact and fiction, Rorvik claimed the story was true, calling copious attention to the book and the possibility of human cloning but also eliciting a barrage of criticism from the scientific world for irresponsibly insisting that a human being had indeed been created with recourse to cloning technology.

The plot revolves around a wealthy man's narcissistic ambition to have a cloned son. Recruiting a selected number of scientists in an undisclosed tropical location, he asks the narrator to help him, although at first the narrator is very reluctant. A local woman is carefully chosen to carry the child, a love story eventually developing between her and the child's father. The baby is born in due course, a healthy child, and at the end of the book, when one year has passed, the couple and the baby are together and well. The novel is explicitly concerned not only with a number of ethical issues relating to cloning but also with issues that hinge around the exploitation of poorer people and the morality attached to it.

In this book Rorvik anticipates and engages with most of the ethical, philosophical, and social issues that have shaped the public debate since the cloning of Dolly the sheep was announced, a debate that has certainly not died down, quite the contrary.[32] At the time of writing, the Raelians, a sect that asserts we are all clones, having been created by extraterrestrial beings, claimed that the first cloned baby, meaningfully called Eva, was born on December 26, 2002, in the United States, and a few days later they claimed another cloned baby was born to a lesbian couple in The Netherlands.[33]

"PROBLEMATIC SELVES AND UNEXPECTED OTHERS"

Rosi Braidotti contends that one needs to turn to genres like science fiction "to find fitting cultural illustrations of the changes and transformations that are taking place at present" (*Metamorphoses: Towards a*

Materialist Theory of Becoming, 182), a perspective with which I fully concur, although the forward-looking and speculative function of science fiction also needs to be stressed. In our increasingly technological times, it is fundamental to consider in imaginative form the many possible arrangements and ramifications that scientific and technological advances might bring about in the near future. Science fiction can thus be seen as potentially functioning as both a diagnostic and a prospective tool in tandem with its admonitory but also premonitory, forward-looking and often technologically friendly narratives. Indeed, looking forward, in a Blochian sense,[34] is an integral part of science fiction narratives, which can be said to enact and create the Blochian *Novum*, an aspect similarly stressed by Carl Freedman, who describes the *Novum* as "such a *radical* novelty as to reconstitute the entire surrounding world and thus, in a sense, to create (though certainly not ex nihilo) a new world" (*Critical Theory and Science Fiction*, 69).[35] This repeated critical stress on futurity as a central structuring element of science fiction leads me to reflect, as many other commentators have done, on the celerity with which current medical and technological advances implemented in those narratives seem to be accomplished and gradually become part of our daily life, science fiction expeditiously becoming science fact.[36] This is certainly also the case with the idea of human cloning, a widespread science fiction trope that has always congregated a great repository of human fantasies, nowadays almost science fact.

Donna Haraway's insights into those she calls "inappropriate/d others" are particularly relevant for our examination of cloned people. According to Haraway,

> Science fiction is generically concerned with the interpenetration of boundaries between problematic selves and unexpected others and with the exploration of possible worlds in a context structured by transnational technoscience. The emerging social subjects called "inappropriate/d others" inhabit such worlds. SF—science fiction, speculative futures, science fantasy, speculative fiction—is an especially apt sign under which to conduct an inquiry into the artifactual as a reproductive technology that might issue in something other than the sacred image of the same, something inappropriate, unfitting, and so, maybe, inappropriated. ("The Promises of Monsters: A Regenerative Politics for Inappropriate/d Others," 320)

This seems to me a concise and illuminating description of the most important aims of SF and also of the main territory it moves in, as well as the generic terrain this book will be concerned with.

In Chapter 1, "'The Hell of the Same': From Plato to Baudrillard," I deal mostly with the theoretical framework I bring to bear on my examination of clone narratives. I look at Jean Baudrillard's discussions of cloning in a society characterized by the proliferation of images and simulacra, as well as Gilles Deleuze's provocative investigation of the simulacrum in philosophy, both analyses informed by a critical reading of Plato's considerations about copies and simulacra. Walter Benjamin's essay "The Work of Art in the Age of Mechanical Reproduction" and its implications for the debate about cloning is also taken into consideration. Freud's and Lacan's theories concerning narcissism and the death drive, insofar as they can be seen as applying to the issue of cloning, provide relevant points of intersection with the arguments I examine and develop.

Chapter 2, "'This Sex Which Is One': Writing Men Out in Selected Science Fiction Texts," looks at the premises behind such matriarchal Utopias as Mary Bradley Lane's *Mizora* (1890), Charlotte Perkins Gilman's *Herland* (1915), and Suzy McKee Charnas's *Motherlines* (1978), which place a strong emphasis on the importance of mother–daughter relations and the perpetuation and transmission of a feminine lineage. I wish to analyze and reverse the traditional assumption that woman is the other of man by inverting the terms and thinking of woman as the norm, whereas men are thus considered as her others. In order to reflect critically on these issues I draw on Luce Irigaray's perceptive theorizing of a "world for women" (*An Ethics of Sexual Difference*, 109), which she considers as "something that at the same time has never existed and which is already present, although repressed, latent, potential" (109). From "eternal mediators for the incarnation of the body and the world of man," women, according to Irigaray, "seem never to have produced the singularity of their own body and world," a world that the texts under examination go a long way toward creating. Irigaray goes on to reflect on the "originality of a sameness" (109) that would be one of the main characteristics of this world, where women would be the "same of the same" and men would become, pursuing this logic, the "other of the same," that is, the other of woman, man as woman imagines him, objectified.

My purpose here, then, is to reflect on the "dream of sameness" (*Speculum of the Other Woman*, 248), which Irigaray diagnoses as grounding patriarchal prejudices, and see how it works in utopian worlds without men created by women. In this context I want to extend Rosi Braidotti's remark that the "contemporary world is fascinated with parthenogenesis" (*Nomadic Subjects: Embodiment and Sexual Difference in Contemporary Feminist Theory*, 65) by suggesting that it is particularly fascinated with cloning, a version of parthenogenesis. All the works under scrutiny are structured around a type of parthenogenetic reproduction that circumvents the need for men. Women in these fictions can be described, by analogy with Irigaray's terminology, as "This Sex Which Is One." For Lacan, on the other hand, "there is no Other of the Other" (*Écrits: A Selection*, 311), that is, woman has no other, a proposition critically examined in the works I look at. These texts give dramatic illustration to the vexed question of whether women need an "Other" or whether they can effectively define themselves without resorting to a masculine framework, but also without perpetuating the patriarchal bias of instituting a role-reversal society, which would "reduce all others to the economy of the Same" (Luce Irigaray, *This Sex Which Is Not One*, 74), in a problematic reversal of androcentric practices of "othering" and "saming."

In Chapter 3, "The Zeus Syndrome: Womb Envy and Male Pregnancy," I examine the fantasy of male pregnancy and womb envy, translated into jealousy of woman's reproductive capabilities. The desire to bypass woman's body in order to create other human beings has a long ancestry. From Greek myths to medieval alchemists, this wish has been forcefully represented, finding powerful symbolic expression in Mary Shelley's *Frankenstein* (1818) and H. G. Wells's *The Island of Dr Moreau* (1896), to cite only the most salient instances. Here I concentrate on some more recent examples, which I want to connect with the powerful fantasy of human cloning, intimately correlated with the aspiration of self-generation. I analyze Sven Delblanc's *Homunculus: A Magic Tale* (1969), Peter Ackroyd's *The House of Dr Dee* (1993), and Lisa Tuttle's "World of Strangers" (1998). These texts are examined mainly through the lens of Bruno Bettelheim's *Symbolic Wounds: Puberty Rites and the Envious Male* (1962), Julia Kristeva's reflections on the potential consequences of new reproductive scenarios, and Donna Haraway's cyborg myth and biopolitics.

Chapter 4, "The Eternity of the Same: Human Cloning and Its Discontents," looks at the ways in which the fantasy of human cloning is dramatized in Pamela Sargent's *Cloned Lives* (1976), Kate Wilhelm's *Where Late the Sweet Birds Sang* (1974), Damon Knight's "Mary" (1964), and Ursula K. Le Guin's "Nine Lives" (1969) and examines from a psychoanalytic point of view some of the inevitable reconfigurations and retheorizations that a society that includes cloned persons will necessarily entail. Questions of identity, the relation of the cloned person to his or her genetic parent, family ties and dynamics, all need to be rethought. These narratives are crucially structured around concepts of wholeness and lack, of individuality and standardization, of a nostalgic yearning for a society before asexual reproduction and genetic engineering take hold and become the norm. Cloning will inevitably reshape many of society's deepest underlying myths. How do incest taboos apply to a family of cloned brothers and sisters? What happens when one's "double" effectively materializes, as in cloning? Does cloning abolish the subject?

My purpose in Chapter 5, "'The Malediction of the Clones': Huxley, Haldane, and Mitchison," is to examine Naomi Mitchison's *Solution*

Three (1975) and J.B.S. Haldane's *The Man with Two Memories* (1976), on the one hand, as related visions of future worlds coming out of a shared scientific and cultural background and, on the other hand, as critical responses to Aldous Huxley's *Brave New World*. Both *Solution Three* and *The Man with Two Memories* provide thoughtful alternative scenarios to the profoundly totalitarian and repressive societal patterns put forward in *Brave New World*, as well as to its wholly negative view of human cloning at the service of a dictatorial and eugenicist social policy. I also read *Solution Three* as a feminist response to *Brave New World*, both novels giving voice to contemporary widespread fears and misgivings at the prospect of human cloning.

My reading is principally informed by a Baudrillardian framework. Baudrillard's polemical analysis of human cloning could almost be said to have taken as one of its prime examples *Brave New World*, a book that dramatizes all the negative concepts attached to cloning without considering any of its many redeeming features, as Mitchison's *Solution Three* does. Indeed, *Solution Three* can be said to constitute a nuanced corrective to many of the grim visions dramatized in Huxley's novel, putting forward alternative scenarios that are gradually subject to change when modifications in the almost exclusive reproductive practice of cloning are deemed beneficial to society at large. Mitchison appears thus to provide, in *Solution Three*, a feminist revision of male-practiced science and a fruitful reflection on "this horrid idea," as she describes the tyranny of the almost exclusive practice of cloning and the consequent drastic reduction in the gene pool, both human and vegetal.[37] Both Aldous Huxley's *Brave New World* and Naomi Mitchison's *Solution Three* can be seen as vital contributions to the debate carried out about new reproductive technologies throughout much of the twentieth century, as well as women's positions in relation to these and the many ramifications and consequences their widespread use would bring about.

Chapter 6, "Cloning and Biopower: Joanna Russ and Fay Weldon," looks at Joanna Russ's *The Female Man* (1975) and Fay Weldon's *The Cloning of Joanna May* (1989), two books I see as engaged in a critical dialogue. Male patriarchal values and prerogatives are interrogated and contrasted, often satirically, with women's subordination and feelings of disempowerment over their bodies. Indeed, Weldon's *The Cloning of Joanna May* constitutes a powerful denunciation of the consequences of the male domination and monopoly of control over reproductive power, providing a forceful illustration of Evelyn Fox Keller's argument in "From Secrets of Life to Secrets of Death," according to which domination of technology by men can be interpreted as a male procreative fantasy gone wrong. Keller suggests that when creation is disassociated from the maternal, as in Frankenstein's paradigmatic story, it constitutes a dangerous permutation of the maternal urge. This impulse can be seen at work in Carl May's highly hazardous incursions into the arena of reproductive technologies, which he believes he can dominate just as he controls his employees and his business corporations.

In Chapter 7, "The Sexual Politics of Human Cloning: Mothering and Its Vicissitudes," I argue that one of the most important consequences of the implementation of human cloning would in fact be the potential empowerment it could give women, who would be able to reproduce without depending on male sperm, beacuse they could use one of their own cells (or another woman's or man's). Indeed, the enabling feeling of independence that the possibility of giving birth without male agency would potentially give women would ultimately transform in a radical way the dynamics of the politics of gender. The very act of giving birth, moreover, and pushing the whole scenario we are developing here a few steps further, could eventually be made obsolete by artificial wombs, so that ectogenesis might be chosen by a great number of women. According to this unfolding story line, then, the female body, traditionally seen in Western culture as a vessel, a receptacle for the penis and later the baby, would gradually lose these connotations and acquire new meanings and new symbolisms. The advent of human cloning, with the prospects outlined, could mean the inception of a whole new social order where women would no longer be regarded as castrated, as lacking something,[38] according to Freud's theory that regards women as lesser men, confined to their role as child bearers and, in that sense, the potential objects of male womb envy. The new reproductive technologies would spell out a fundamental change in the societal outlook for men and women and their respective roles, with women potentially enjoying a much greater freedom of choice in view of the novel alternatives offered to them. Science can thus be said to play a deeply liberating role as far as women's reproductive possibilities are concerned, granting them, by the same token, a measure of sexual equality never before attained, given the specificity of their childbearing attributes,[39] an argument prophetically put forward by Simone de Beauvoir in *The Second Sex* (1949) and Shulamith Firestone in *The Dialectic of Sex* (1972). Human cloning might consecrate and concretize the fantasy not only of a motherless birth, thus conferring on the father the possibility of bypassing the (m)other and fulfilling the dream of self-reproduction, but also of a fatherless birth, circumventing the obligatory agency of the father and allowing the mother to give birth to her own (genetic) child or somebody else's, or even, in a not too distant future, to completely sidestep the womb, with the development of artificial wombs. The sexual power politics would thus inevitably undergo an unfathomable transformation that would unavoidably extend to all spheres of human life, gradually causing brand new social configurations to arise.

UNDER THE SIGN OF THE CLONE

Crucially for my argument throughout the book, Haraway sustains that cyborg imagery exhorts us not to dismiss the study of technology and artifice, discarding it for a naturalistic, essentialist model. Her words carry a crucial message that I want to expand on here:

> A socialist–feminist science will have to be developed in the process of constructing different lives in interaction with the world. Only material struggle can end the logic of domination. . . . I do not know what life science would be like if the historical structure of our lives minimized domination. I do know that the history of biology convinces me that basic knowledge would reflect and reproduce the new world, just as it has participated in maintaining an old one. (*Simians, Cyborgs, and Women*, 68)

Haraway's call for a much more significant representation of women in the effective shaping and evolution of the sciences of life, so that the balance of domination can be evened out and the new medical practices might not reinscribe patriarchal ascendancy and control, has to be heeded and put into practice so that any future society does not perpetuate deeply questionable political power asymmetries.

Indeed, Donna Haraway's cyborg theories provide a far-reaching contribution to the cloning debate. Haraway's question of whether there could be "a family of figures who would populate our imagination of these postcolonial, postmodern worlds that would not be quite as imperializing in terms of a single figuration of identity" ("Cyborgs at Large: Interview with Donna Haraway," 13) seems to me centrally pertinent and needs to be probed further. If for Haraway cyborgs are clearly the answer, in the society of the near future these could be reshaped under the sign of the clone. Indeed, the proliferation of discourses of science and politics applied to the technologized body have accelerated at a vertiginous pace, mostly impelled by advances in biomedicine. It seems to me that in our contemporary world, increasingly shaped by biopower relations, human cloning will eventually soon be with us.

In 1967, Joshua Lederberg speculated that "it is an interesting exercise in social science fiction to contemplate the changes that might come about from the generation of a few identical twins of existing personalities."[40] This is precisely what cloning narratives have been doing, conjecturing what those changes might be and considering the manifold ramifications and implications of those alterations in the near future. This book will look at the ways those transformations have been articulated, examine the main representations those conjectures have taken, as well as offer ways of theorizing some of those radical modifications that are certain to have impact on the way we consider human nature itself and the very world we live in.

NOTES

1. "When we think about it, I believe there always was cloning, and there will always be. Legislation will not prevent cloning" (my translation).

2. I will only tangentially be concerned here with the ethical issues raised by the prospect of human cloning. Medical details are obviously beyond my expertise and the scope of this study.

3. One of the many criticisms directed against cloning has to do with the fact that many consider it antievolutionary; that is, it will alter the natural course of the evolutionary process.

4. "'Progress' in genetics liberates or accelerates our imagination, happy and terrified at the same time, when confronted with all sorts of things that I would not describe as unknown, particularly in relation to the unconscious, but not yet 'registered' by what could be called, in general terms, civilian society" (my translation).

5. I will deal here predominantly with fictional works. Filmic negotiations of cloning will be examined in a future project.

6. As Jon Turney notes about *Brave New World,* "It is still hard to do justice to the cumulative impact of the book. It looms large over a whole area of public debate, shaping reactions to and evaluations of a wide range of biomedical research" (*Frankenstein's Footsteps: Science, Genetics and Popular Culture* [New Haven and London: Yale University Press, 1998], 115).

7. Although representations of human cloning in popular fiction tend to be heavily negative, there are several examples of alternative fictional future societies where cloned people predominate and where they are considered on the same footing as those born in the traditional way. Among these could be cited Naomi Mitchison's *Solution Three* (1976), J.B.S. Haldane's *The Man With Two Memories* (1976), and David Brin's *Glory Season* (1993). For an overview of some of the best-known examples of negative representations of human cloning in fiction and films, see Gregory E. Pence, *Who's Afraid of Human Cloning?* (1998), Chapter 4. Pence cites C. J. Cherry's classic trilogy *Cyteen* (1988), Kate Wilhelm's *Where Late the Sweet Birds Sang* (1976), and Fay Weldon's *The Cloning of Joanna May* (1989). As filmic examples, Pence cites *Battlestar Galactica* (Richard A. Colla, 1978) with its six cloned sisters, selected episodes from *X-Files*, a series obsessed with cloning, Woody Allen's *Sleeper* (1973), *The Stepford Wives* (Bryan Forbes, 1974), Franklin J. Schaffner's *The Boys from Brazil* (1977), based on Ira Levin's novel of the same name, Ridley Scott's *Blade Runner* (1982), based on Philip K. Dick's novel *Do Androids Dream of Electric Sheep?* (1968), and *Multiplicity* (Harold Ramis, 1996).

8. In his book *Biology and Christian Ethics,* Stephen R. L. Clark cautions against potential abuse of new reproductive technologies, warning against

the danger that brain-dead human babies might be cloned to supply "spare body parts" for adults who might need them.

9. For a wider discussion of the most common misconceptions associated with human cloning, see Gregory E. Pence, *Who's Afraid of Human Cloning?* (1998), Chapter 4.

10. A recent example of these aberrant and mistaken views of human cloning is illustrated in *The Sixth Day* (Roger Spottiswoode, 2000), where people who are supposedly cloned reappear almost instantly in what look like the same bodies. This vision corresponds to a further stage in cloning techniques that could conceivably come to be feasible in a much more distant future, when there would be spare cloned bodies waiting for a head or brain transplant that would give a new life chance to the original people from whom they were cloned. In *The Sixth Day*, however, the bodies are revealed to be mechanical, mixtures of human and machine, which adds to the uneasy feelings of horror and aversion they give rise to. This kind of scenario would be utterly impossible at our present state of technological development, where if cloning were implemented the resulting embryo would grow in a womb and be born like any other normal baby. Films like *The Sixth Day*, then, only contribute to the generalized misunderstandings and expressions of bewilderment that surround visions of human cloning. Similar feelings of bemusement and apprehension are given sharp focus in Steven Spielberg's *Artificial Intelligence* (2001), which plays, in a much subtler way, with cyborg fantasies and the dread of overtaking the boundaries between human and machine and of there being no way back. David, the protagonist, embodies precisely the wish to be human, of going from his mechanical state to a fully human one, a thematic concern that is never really addressed in *The Sixth Day*.

11. Cloning can also, from a certain point of view, be seen as forming part of the Gothic strand that seems to be so powerful in the contemporary Western world. The uncanny horror that informs the popular manifestations of the Gothic, mainly in films and mass fiction, functions to a certain extent in people's minds as an antidote to the displays of abject terror that permeate society. They are very much the manifest, visible aspect of what one suspects is the layer of evil that lies just below the surface and is likely to erupt at any moment, given such conspicuous representations in, for example, the *Alien* tetralogy and the films of David Lynch. These haunting images of terror may be seen as an analogue to the surrealistic paintings that purported to bring to light the repressed contents of the unconscious and its more murderous supressed desires. The idea of cloning brings up a cluster of deep-seated fears and uncanny feelings of dread and anxiety that sum up what might be described as the Gothic atmosphere of our times. Indeed, there is something inherently Gothic and uncanny in the idea of human cloning and the vision of a series of duplicated human beings side by side, as in Warhol's multiple depictions of soup cans or famous public figures. Ours is to a great extent a culture permeated by Gothic features, fears, and images, of which the idea of cloning can be seen as one particular strain, inscribed within the more generalized dread and misgiving with which some scientific advances are regarded, conjuring up the figure of Frankenstein and his creature.

12. Dr Ian Wilmut, the creator of Dolly, the first cloned sheep, poses a similar question: "Is the present aversion that many people feel—including us [Wilmut and the coauthors of the book]—simply a fear of novelty? Or is there really a qualitative difference between the technology of cloning and the reproductive technologies that are already commonplace, and seem on the whole to be acceptable, even when they seem exotic?" (*The Second Creation*, 298).

13. For a discussion of the "wisdom of repugnance" elicited by the idea of human cloning, see Leon R. Kass, "Human Cloning Should Be Banned," 32–33.

14. See Clara Pinto Correia, *Clonai e Multiplicai-vos*, 56. Lee M. Silver, for his part, notes how the word *clone* "appeared in the language of science at the beginning of the twentieth century to describe 'groups of plants that are propagated by the use of any form of vegetative parts.' Since that time, *cloning* has been used to describe the process by which a cell, or groups of cells, from one individual organism is used to derive an entirely new organism, which, according to the definition, is a 'clone' of the original. When multiple individuals are cloned from a single ancestor, they are all considered to be 'members of a clone'" (*Remaking Eden: Cloning and Beyond in a Brave New World*, 93–94).

15. *The Compact Oxford English Dictionary* describes a clone, in terms of its biological meaning, as "any group of cells or organisms produced asexually from a single sexually produced ancestor" (271), whereas as far as its figurative meaning is concerned a clone refers to "a person or animal that develops from a single somatic cell of its parent and is genetically identical to that parent" (271).

16. See Clara Pinto Correia, *Clonai e Multiplicai-vos*, 57. According to Everett Bleiler, the first of Clement Fézandie's "Doctor Hachensaw's Stories," which appeared in the May 1921 issue of Hugo Gernsback's *Science and Invention*, "offers a very clear statement of cloning and genetic engineering" (*Science-Fiction: The Early Years*, 241). Brian Stableford describes Fézandie's story as dealing with the "creation of superior farm animals and interspecific hybrids as well as the possibility of bringing a human embryo to term in a bovine uterus" ("Biotechnology and Utopia," 192). Pamela Sargent, in her anthology *Bio-Futures* (1976), describes Fritz Lieber's "Yesterday House" (1952) as an early clone tale, in spite of the fact that the word is never mentioned (xxxiii). For an overview of literature dealing with clones, see Janeen Webb's "Bone of My Bones, Flesh of My Flesh: A Brief History of the Clone in Science Fiction" (2002). See also Brian Stableford's entry on clones in John Clute and Peter Nicholls, eds., *The Encyclopedia of Science Fiction* (1995), 236–237.

17. According to Gina Kolata, "the idea of cloning dated back to 1938, with Hans Speman's proposal for a 'fantastical experiment'" (*Clone: The Road to Dolly and the Path Ahead*, 71), although Speman did not use the word clone. For an overview of the cloning debates, which started with experiments conducted on frogs by Robert Briggs, Thomas King, and later John Gurdon, see Kolata, *Clone: The Road to Dolly and the Path Ahead*, Chapter 4.

18. For a discussion of Haldane's book *The Man with Two Memories* and his visionary forecasts, see Chapter 5.

19. American biologist Herbert Spencer Jennings, in his book *Prometheus, or Biology and the Advancement of Man* (1925), for example, argues that the eu-

genic project is a lost cause, for it is impossible to predict in what ways certain genes will be combined in sexual reproduction. Interestingly, though, Jennings speculates that if identical genetic copies of a person could be produced (as with cloning, as we would say nowadays), then "man would have his fate in his own hands. He could multiply the desirable combination until the entire population consisted of that type" (quoted in Daniel J. Kevles, *In the Name of Eugenics*, 146–147, which also contains an overview of the development of eugenicist ideas). See also Chapter 5 of this book.

20. In his novel *Elementary Particles* (2000), Michel Houellebecq engages with the future posthuman contours of humanity, eventually substituted by a new "intelligent species made by man 'in his own image'" (262), mainly through cloning, taking us as far as the creation of the first being of this new species, which occurred on March 27, 2029. The character who developed the techniques and the theoretical framework that argued for the necessity of such a revolutionary change, Michel Djerzinski, reads about the trajectory of Julian and Aldous Huxley and their respective work: "In 1946, just after the war, Julian Huxley was appointed director-general of UNESCO, which had just been founded. Aldous Huxley had just published *Brave New World Revisited*, in which he tried to portray the first novel as a social satire. Years later, Aldous would become a pillar of the hippie experiment. . . . Aldous Huxley is probably one of the most influential thinkers of the century" (132). Clearly the work of Julian and Aldous Huxley provided crucial inspiration for the directions Djerzinski's thought and research later took. Indeed, as the narrator explains, the practical consequences of Djerzinski's work "were dizzying: any genetic code, however complex, could be noted in a standard, structurally stable form, isolated from disturbances or mutations. This meant that every cell contained within it the possibility of being infinitely copied. Every animal species, however highly evolved, could be transformed into a similar species reproduced by cloning, and immortal" (258).

21. In *The Scientific Outlook* (1931) Bertrand Russell expressed the view that "men will acquire power to alter themselves, and will use this power" (169), which will inevitably lead to the fact that man [*sic*] "will tend more and more to view himself as a manufactured product" (169). Quoted in Jon Turney, *Frankenstein's Footsteps: Science, Genetics and Popular Culture*, 102.

22. As Jon Turney asserts, Taylor's *The Biological Time Bomb* was seen "at the time as something of a landmark in public debate. [Taylor] produced a text which, rather like Haldane's *Daedalus*, both exemplified the reasons why biology came to be a matter of public concern, and indicated some of the directions in which that concern would develop" (*Frankenstein's Footsteps: Science, Genetics and Popular Culture*, 155).

23. Pamela Sargent's opinion about the possible implementation of human cloning is that it "can be right in one case and wrong in another" (*Bio-Futures*, xvi).

24. See Gina Kolata, *Clone: The Road to Dolly and the Path Ahead*, 73. See also Leon R. Kass's detailed account of the public debate that followed the publication of that article, as well as Kass's own rebuttal of Lederberg's views ("Human Cloning Should Be Banned," 26–48).

25. As Daniel Callahan explains, referring to the emergence of writings on cloning in the 1960s and 1970s, "cloning became one of the symbolic issues of

what was, at the time, called 'the new biology.' . . . Over a period of five years or so in the early 1970s a number of articles and book chapters on the ethical issues appeared, discussing cloning in its own right and cloning as a token of the radical genetic possibilities" ("Cloning: Then and Now," 309).

26. *Journal of the American Medical Association* 220, 11 (1972): 1980.

27. Toffler goes on to note that cloning "would, among other things, provide us with solid empirical evidence to help us resolve, once and for all, the ancient controversy over 'nature v nurture' or 'heredity v environment.' The solution of this problem, through the determination of the role played by each, would be one of the great milestones of human intellectual development" (183), a recurring theme in many clone narratives. Toffler cites Joshua Lederberg's prognostications as far as the advent of cloning is concerned as anywhere between then (1970) and fifteen years from then (184).

28. Referring to Steptoe and Edward's achievements in the field of in vitro fertilization, Turney similarly notes how their work is "embedded in a series of layers [including one of] speculation about the prospects for human cloning" (180).

29. The Hudson Institute was one of the two most important centers for bioethics in the United States, as Jon Turney explains (*Frankenstein's Footsteps: Science, Genetics and Popular Culture*, 179).

30. Quoted in Jon Turney, *Frankenstein's Footsteps: Science, Genetics and Popular Culture*, 180.

31. For a thorough overview of the debate raised by Rorvik's book, see Gina Kolata, *Clone: The Road to Dolly and the Path Ahead*, Chapter 5.

32. Lee M. Silver notes how by the "early 1980s, the notion of cloning had become entrenched in popular culture, appearing again and again in movies, television shows, and science fiction novels. And it entered the inanimate world as well, with clones of computers and even perfumes. Clones were seen as almost, but not quite, perfect copies of the original, usually cheaper and assumed to be not as 'sharp' in some way" (*Remaking Eden: Cloning and Beyond in a Brave New World*, 97).

33. In the meantime, another birth was announced in Japan. None of these babies have been confirmed as clones, no tests having been allowed to be carried out on them or their mothers by scientists who do not belong to the Raelians.

34. See Ernst Bloch's *The Spirit of Utopia* and *The Principle of Hope*.

35. Engaging with Ernst Bloch, Carl Freedman maintains that "the dynamics of science fiction can on one level be identified with the hope principle itself [since the] cognitive rationality . . . of science fiction allows utopia to emerge as more fully itself, genuinely critical and transformative" (*Critical Theory and Science Fiction*, 69). Continuing to draw on Bloch, Freedman further states that "the defining features of science fiction are located on the In-Front-Of-Us, at the level of the Not-Yet-Being, and in the dimension of utopian futurity" (70).

36. Marleen S. Barr also calls attention to "the rapidity with which the science fictional element emerges from science fiction's pages and becomes human cognition" ("'We're At the Start of a New Ball Game and That's Why We're All Real Nervous': Or, Cloning—Technological Cognition Reflects Estrangement from Women," 193).

37. In the book's dedication to Jim Watson.

38. In *Imaginary Bodies: Ethics, Power and Corporeality*, Moira Gatens describes the post-Oedipal female body image as "that of a partial, passive object—a castrated body that requires first a man and then a baby to 'complete' it. Put bluntly, women's bodies are not seen to have integrity, they are socially constructed as partial and lacking . . . they are not thought of as whole beings" (41). With the gradual waning of the Oedipal scenarios in societies shaped by new reproductive technologies, women would no longer be described as inferior versions of men, needing the latter to complete them.

39. I elaborate on this issue and discuss the vexed sexual politics at work in the marketplace governing access to women's bodies and women's agencies as far as decisions over their reproductive organs are concerned, as well as political resolutions within the medical establishment. See Chapters 6 and 7.

40. The *Washington Post*, September 30, 1967. Quoted in Gina Kolata, *Clone: The Road to Dolly and the Path Ahead*, 73.

1

"The Hell of the Same": From Plato to Baudrillard

> Likeness does not make things "one" as much as unlikeness makes them "other."
>
> Montaigne, *Essays*, III: 13
>
> À cet égard [the question of the identity of clones], je pense que Freud aurat été passionné par les problèmes actuels.[1]
>
> Elizabeth Roudinesco, *De quoi demain . . . Dialogue*, 93

In this chapter I attempt a synthesis of some of the most important reflections proposed by the cultural theorists I consider the most relevant as far as the many isssues surrounding the possibility of human cloning are concerned. I will be engaging with the ideas put forward by Jean Baudrillard, Gilles Deleuze, Slavoj Zizek, and Jacques Derrida with respect to human cloning, while also considering the unavoidable Freudian and Lacanian contributions to the formation of identity and the subject, and how they might impact on cloned persons.

Reflecting on the concept of human cloning, in the wake of reading Gordon Rattray Taylor's *The Biological Time Bomb* (1968),[2] Ursula K. Le Guin asserts that "the duplication of anything complex enough to have personality would involve the whole issue of what personality is—the question of individuality, of identity, of selfhood" ("On Theme," 205). These are precisely the crucial questions that stand at the very core of the concept of human cloning, along with other fundamental issues raised by cloning, which include human freedom itself and the very limits of knowledge. I also examine the fantasy of human cloning as it can be inscribed in the context of a discussion of our image-dominated society, where copies, duplications, and simulacra increasingly constitute our contemporary cultural landscape.

Human beings have always felt a strong fascination with copies, duplications, and doubles as extensions of themselves and hence as narcissistic replicating mirrors. Cloning here touches on the profoundest sources of life, on deeply rooted archetypes and ontological structures. Indeed, the endless attraction, laced with profound anxieties, that the idea of cloning gives rise to stems to a great extent from the human longing to touch the origins of life, the sources of creation, with which cloning and its technology of necessity resonate. This obsession with copies appears to be especially intense in contemporary society, in which the culture of imitation, images, and simulacra can be seen as contiguous with the attraction exerted by the idea of human clones. Related to the strong appeal of the concept of cloning are such manifestations as twins, doubles, automata, mannequins, cyborgs, and a whole array of beings that purport to imitate the human form, from golems to marionettes. Indeed, the literature dealing with imitation human beings goes back many centuries, finding some of its foremost manifestations in myths and legends.[3]

The drive to clone human beings can arguably be inscribed in the general *Zeitgeist* of postmodernity, of the "culture of the copy" (*The Culture of the Copy: Striking Likenesses, Unreasonable Facsimiles*, 351),[4] to borrow Hillel Schwartz's terminology, in a culture controlled by "simulation" in Jean Baudrillard's account, and by an "impassioned cult of similarity," according to Walter Benjamin, ("The Image of Proust," 206), but also in a society ruled by biotechnology, which can be described as its dominant myth, with DNA as one of its main governing metaphors. As Hillel Schwartz states, in a somewhat polemical vein,

> Copying makes us what we are. Our bodies take shape from the transcription of protein templates, our language from the mimicry of privileged sounds, our crafts from the repetition of prototypes. Cultures cohere in the faithful transmission of rituals and rules of conduct. To copy cell for cell, word for word, image for image, is to make the known world our own. (*The Culture of the Copy: Striking Likenesses, Unreasonable Facsimiles*, 211)

On the other hand, as Schwartz goes on to point out, copying can introduce imperfections, namely at the genetic level, deviations that raise serious questions with respect to identity, integrity of the self, and security. Moreover, as Schwartz asserts, "we are not identical, nor do we wish to think of ourselves as clones" (212). This resistance to human cloning is very widespread, finding in Baudrillard one of its most outspoken and incisive proponents. In his often sarcastic and derogatory writings addressing the subject, he gives voice to many of the implications of cloning from a philosophical, psychological, and social perspective.

Baudrillard's work is permeated with references to cloning and clones, carrying out an astute examination of what he meaningfully calls "our present clone-loving society" ("After the Orgy," 7). Baudrillard is invariably dismissive and critical of clones and cloning, in spite of the attraction they have consistently exerted on him.[5] The appeal that cloned human beings possess for him reverberates throughout his whole work and can be said to be a logical extension of his attraction toward and analysis of all sorts of replicas, copies, reduplications, reproductions, simulacra, and simulations. Indeed, Baudrillard's interest in cloning was fuelled by reading the story of a cloned boy in a fictionalized account of cloning (David Rorvick, *In His Image: The Cloning of a Man*),[6] which its author claimed to be true. As Baudrillard remarks,

> In the United States, a child was born a few months ago like a geranium: from cuttings. The first clone child (the lineage of an individual via vegetal multiplication). The first child born from a single cell of a single individual, his "father," the sole progenitor, of which he would be the exact replica, the perfect twin, the double. ("Clone Story," 95)

For Baudrillard, the main and more scandalous transgression in this outcome would have to do with the fact that only one person is necessary in this type of reproduction, "allowing one to do without the other, to go from same to same" (96), as Baudrillard puts it.[7] This scenario would potentially lead to a reformulation of the maternal and paternal roles, to the gradual disappearance of the Oedipal complex, to novel psychological traumas and family configurations, and to an intensification of narcissism, as well as a heightened concern with issues of identity in a society dominated by the stigma of copying and reproducibility.[8]

CLONES AND SIMULACRA

Jean Baudrillard describes three orders of simulacra: the "counterfeit," which he defines as the dominant form from the Renaissance to the Industrial Revolution; "production,"[9] which dominates the industrial era, and "simulation," which Baudrillard regards as "the reigning scheme of the current phase that is controlled by the code" ("The Orders of Simulacra," 83).[10] As Baudrillard explains in "The Precession of Simulacra," the way simulation works is "nuclear and genetic, and no longer specular and discursive," carrying in its wake the disappearance of "all of metaphysics" (3). As Baudrillard notes, in a characteristically pessimistic and controversial vein, with simulation there will be "no more mirror of being and appearances, of the real and its concept. No more imaginary coextensivity: rather, genetic miniaturisation is the dimension of simulation. The real is produced from miniaturised units, from matrices, memory banks and common models—and with these it can be reproduced an indefinite number of times" (3). It is precisely in the order of simulation[11] that Baudrillard polemically inscribes cloning, which he sees as inaugurating the end of the subject by means of its potential reproducibility with recourse to the genetic code. The crucial problem for Baudrillard, then, is the reduction of the human being to a formula susceptible of being replicated, thus making people, in their potential for endless duplication, comparable to objects in an industrial production system. Baudrillard's controversial vision of the disappearance or dilution of the subject in the age of cloning can be situated within his more general drive to demote the subject in favor of the object, which gradually begins to occupy a central role in a heavily consumerist society. As a counterpoint to Baudrillard's gloomy and pessimistic vision, however, Douglas Kellner puts forward a contrasting possibility that this very "profusion and hyperization of objects" in a consumer society might "produce a new subject, who might in turn produce a freer, happier, better world?" (*Jean Baudrillard: From Marxism to Postmodernism and Beyond*, 165). For Kellner, indeed, this exaltation of the object might even "augment, not diminish, subjectivity" (165). Significantly, Walter Benjamin hints at a similar scenario when he suggests that through a reorientation of desire directed toward the future the very deterioration of the object's aura can signal novel societal and libidinal configurations.[12]

While for Baudrillard capitalist societies and the phenomenon of globalization are profoundly implicated within the domain of the industrial code, reproduction, and simulacra, their logic can also be extended to the genetic code, which in his view will soon become another commodity to be exploited by capital and the mass market. In "The Orders of Simulacra," Baudrillard aptly considers DNA[13] the "prophet" (103) of our contemporary world, a world whose dominant myth may be appropriately described as biotechnology.[14] For Baudrillard, thus, we moved "from a capitalist-productivist society to a neo-capitalist cybernetic order that aims at total control. This is the mutation for which the biological theorization of the code prepares the ground" (111).[15] Much like Baudrillard, Dorothy Nelkin and Susan Lindee consider the gene as an emblematic cultural icon of our age. As they argue,

> Instead of a piece of hereditary information, it has become the key to human relationships and the basis of family cohesion. Instead of a string of purines

and pyrimidines, it has become the essence of identity and the source of social difference. Instead of an important molecule, it has become the secular equivalent of the human soul. Narratives of genetic essentialism are omnipresent in popular culture, here explaining evil and predicting destiny, there justifying institutional decisions. They reverberate in public debates about sexuality and race, in court decisions about child custody and criminal responsibility, and in ruminations about the meaning of life. (*The DNA Mystique: The Gene as a Cultural Icon*, 198)

Cloning can be seen as deeply imbricated in this line of reasoning, which sees the genetic code as standing at the heart of life and all its manifestations, as the primordial explanatory concept of identity and behavioral patterns. Considering some of the ramifications of the omnipotence and ubiquity of the genetic code in our society, namely, as far as the concepts of simulation and replication are concerned, Baudrillard maintains that "it is in effect in the genetic code that the 'genesis of simulacra' today finds its most accomplished form" ("The Orders of Simulacra," 103–104).[16] The genetic code then, according to Baudrillard, is at the basis of a revolution that will contribute to the demoting of human beings to the level of reproducible consumer objects.

AURATIC CLONES

In any discussion that deals with copies and reproductions in an age of commodity production,[17] Walter Benjamin is an indispensable reference. For Benjamin, the mechanical reproduction of an object undermines its unique aura ("The Work of Art in the Age of Mechanical Reproduction"), given that through repetition of the same they become evened out to relative unimportance and worthlessness, a perception that can help to shed light on the investigation of the reactions and profound anxieties exhibited by cloned characters in many clone narratives, such as Pamela Sargent's *Cloned Lives* (1976) and Kate Wilhelm's *Where Late the Sweet Birds Sang* (1976).

The influence of Walter Benjamin's ideas reverberates throughout Baudrillard's work, and he openly acknowledges his respect for Benjamin.[18] In *Seduction*, Baudrillard links Benjamin's comments about the work of art in the age of mechanical reproduction, which, by dint of that very system of reduplication, loses its "aura," to cloning. As Baudrillard maintains, engaging with Benjamin's ideas, the work of art, having relinquished

the unique quality of its here and now, its aesthetic form . . . is no longer destined for seduction but reproduction, and in its new destiny, takes on a *political* form. The original is lost, and only nostalgia can restore its "authenticity." The extreme form of this process is to be found in our contemporary mass media, where there never was an original, things being conceived from the start in terms of their unlimited reproducibility. (*Seduction*, 171)

As Baudrillard claims, this is precisely what would happen to human beings in the event of cloning, when the person is envisaged only as information to be decoded and processed: "Nothing then prevents its serial reproduction in the same terms Benjamin used when speaking of industrial objects or images" (*Seduction*, 171). Are clones, then, according to Baudrillard's argumentation, simulacra, that is, copies without an original, since they are potentially endlessly reproducible? Baudrillard describes cloning technology as making possible "the generation of identical *beings*, without any possible return to an original being" (171). Here, as in many other passages from his work, I have problems with Baudrillard's sensationalist pronouncements, his facile equation of clones with serial human beings and his erasure of the person whose set of genes gave origin to a given cloned person.

A related question, with which both Benjamin and Baudrillard are deeply concerned, is that of authenticity, the genuineness of a reproduction vis à vis its original. Authenticity is indeed a vexed word. In Western society, where the copy might come to surpass the original, the concept of authenticity comes under heavy strain, a complex problem faced by many cloned characters in the narratives we will be concerned with here, such as the Swenson clones in Pamela Sargent's *Cloned Lives*, Mark in Kate Wilhelm's *Where Late the Sweet Birds Sang*, or Mary in Damon Knight's "Mary," tales that centrally revolve around the concept of origins and fakes, as well as genealogy and generations. As Walter Benjamin declares, "The authenticity of a thing is the essence of all that is transmissible from its beginning, ranging from its substantive duration to its testimony to the history which it has experienced. Since the historical testimony rests on the authenticity, the former, too, is jeopardized by reproduction when substantive duration ceases to matter. And what is really jeopardized when the historical testimony is affected is the authority of the aura" ("The Work of Art in the Age of Mechanical Reproduction," 223).[19]

In postmodern thought, however, the notion of the original has lost much of its conceptual weight, and original and reproductions can be said to occupy very similar ground as far as the notion of relative value is concerned. If, as Benjamin claims, on the one hand, the "presence of the original is the prerequisite to the concept of authenticity" ("The Work of Art in the Age of Mechanical Reproduction," 222),[20] on the other, with the advent of technical reproduction, the aura of the original will inevitably diminish, a process with an importance, according to Benjamin, that extends beyond the sphere of art. As Benjamin remarks, "the technique of reproduction detaches the reproduced object from the domain of tradition. By making many reproductions it substitutes a plurality of copies for a unique existence" (223), an insight that can be usefully extrapolated to the question of cloning. Indeed, in Benjamin's account, as the reproduction confronts the observer in a particular historical time and situation, it "reactivates the object reproduced" (222), leading to a rupture with tradition, which is the hallmark of our contemporary world and which Benjamin considers as intimately connected with mass movements. Reflecting on the social bases of the contemporary withering of the aura, which for Benjamin is closely bound up with the increasingly greater importance of the masses in our modern world, he notes the growth of the impulse to "get hold of an object at very close range by means of its likeness, its reproduction" (225), a phenomenon accompanied by the decay of the aura of those objects, which Benjamin reads as the emblem of a viewpoint according to which a generalized sense of the equality of things has come to apply, through the process of reproduction, even to a unique object. The aura that pertained only to unique objects, then, can be said to have diminished but also to have dispersed and been disseminated, attaching to mechanically reproduced objects almost as much as to "original" ones.[21]

Many of these insights can be profitably used to reflect on the question of human cloning. Indeed, extrapolating from Benjamin's view that "even the most perfect reproduction of a work of art is lacking in one element: its presence in time and space, its unique existence at the place where it happens to be" ("The Work of Art in the Age of Mechanical Reproduction," 222), a similar argument can be made for the uniqueness of clones, who will always be separate, as far as the time and circumstances in which they are brought up, from those of their genetic mothers or fathers.[22] Given that environment is unreproducible, genetic makeup alone will not be enough to produce identical human beings, since social and cultural surroundings obviously play a fundamental role in a child's development.

Hillel Schwartz, considering the question of standardization brought about by reproduction and the related issue of authenticity, maintains that "the best we can manage in a world of simulacra is a visage unavoidably generic" (*The Culture of the Copy: Striking Likenesses, Unreasonable Facsimiles*, 89). On the other hand, as Schwartz further observes, in the kind of society where "the culture of the copy is pervasive [the] upshot is that we must reconstruct, not abandon, an ideal of authenticity in our lives. Whatever we come up with, authenticity can no longer be rooted in singularity, in what the Greeks called *idion*, or private Person" (17).

Copies, reproductions, and simulacra, then, have gradually acquired an increased relative value and significance in comparison with originals, whose aura continues to decay in a society whose cultural productions are increasingly dominated by mimesis and a reduction in the authority of authenticity, a concept that is sharply problematized in numerous narratives dealing with cloning.[23]

Scott Durham, in his examination of the role of the simulacrum in the present-day world, for which he considers "it has become the emblem" (*Phantom Communities: The Simulacrum and the Limits of Postmodernism*, 4), explains how the version of simulacrum predominant in contemporary investigations of mass culture envisions the simulacrum as the copy of a copy "which produces an effect of identity without being grounded in an original" (7). Indeed, this concept of the simulacrum can already be found in Plato, for whom the good copy or icon is contrasted with the false copy or simulacrum,[24] which for Plato is a copy of a copy, or, as Rodolphe Gasché points out, "a double of a double, which itself signifies an original" (*The Tain of the Mirror*, 226),[25] an insight that would then make human clones commensurate to other human beings. If we extrapolate from these ideas to the concept of human cloning, would a clone fall into the category of a good copy or would he or she be considered as a false copy or simulacrum, according to the terms laid out by Plato?

In "The Double Session," Jacques Derrida addresses the vexed questions of appearance, resemblance, and duplication in ways that can be applicable to our discussion. Polemically, indeed, Derrida suggests that "the image can precede the model, that the double can come before the simple" (190), a version of Baudrillard's concept of simulation. Drawing on Plato's theory of mimesis, Derrida considers there is an internal division within mimesis,

> a self-duplication of repetition itself, *ad infinitum*, since this movement feeds its own proliferation. Perhaps, then, there is always more than one kind of *mimesis*; and perhaps it is in the strange mirror that reflects but also displaces and distorts one *mimesis* into the other, as though it were itself destined to mime or mask *itself*, that history—the history of literature—is lodged, along with the whole of its interpretation. Everything would then be played out in the paradoxes of the *supplementary double*: the paradoxes of something that, added to the simple and the single, replaces and mimes them, both like and unlike. (*Dissemination*, 191; emphasis mine)

Borrowing Derrida's terminology, then, one can see prefigured here the plight of the clones as "supplementary doubles," for, as Derrida further declares, there is "the 1 and the 2, the simple and the double. The double comes *after* the simple; it multiplies it as a *follow-up*. It follows . . . that the image *supervenes* upon reality, . . . the imitation upon the thing, the imitator upon the imitated" (*Dissemination*, 191). Derrida carries on to examine the aporias at the heart of Platonism and the scene of representation, especially as far as art is concerned, for in this domain, due to the emphasis placed on artistic creations to avoid slavish imitation of what is portrayed, "art can create or produce works that are more valuable than what they imitate. . . . The extra-value or the extra-being makes art a richer kind of nature, freer, more pleasant, more creative: more natural" (191). If in traditional philosophical discourse the original precedes the imitation, an order of appearance which Derrida describes as *"the order of all appearance"* (178), but which also spells the "closure of metaphysics" (180), Derrida also postulates a reversal of conventional mimetology, in which clones and simulacra can be inscribed. According to John Sallis, "mimesis both furthers and hinders the disclosure of the thing itself, disclosing the thing by resembling it but obscuring it by substituting a double in place of it" ("Doublings," 122). Indeed, according to this mechanism, the original is obscured by its image or representation, in a fusion between original and image. As Derrida states, in words that can be applied to the situation in which cloned people might find themselves, as reflections of an original being:

> There is no longer a simple origin. For what is reflected is split *in itself* and not only as an addition to itself of its image. The reflection, the image, the double, splits what it doubles. . . . What can look at itself is not one; and the law of the addition of the origin to its representation, of the thing to its image, is that one plus one makes at least three. The historical usurpation and theoretical oddity that install the image within the rights of reality are determined as the *forgetting* of a simple origin. (*Of Grammatology*, 36–37)[26]

If, on the one hand, with cloning, through the very possibility of endless reproducibility, the original might be displaced to such an extent as to be forgotten, on the other, in other cases, the genetic origin is the focus and impetus for the cloning of a given subject, who may later on take the place of the original, thus again stressing the inversion of Platonism addressed by Derrida, and also by Gilles Deleuze.

DELEUZIAN SIMULACRA

There is, however, another notion of simulacrum germane to our discussion of clones and (evil) doubles, a concept developed notably in the writings of Gilles Deleuze, where the idea of simulacrum takes up a daemonic guise, evoking the metamorphic and affirmative connotations of the "false" repetition of an image. As Deleuze declares in his reading of Nietzsche's discussion of the "power of the false," "For the artist, *appearance* no longer means negation of the real in this world but this kind of selection, correction, redoubling, and affirmation. Then truth perhaps takes on a new sense. Truth is appearance" (*Nietzsche and Philosophy*, 103). Appearance, then, is valorized in relation to the notion of some immanent essence that is intrinsically constitutive of human beings, which might be called the soul or the spirit and which opponents of human cloning controversially claim might be destroyed by cloning and absent in clones.

In "The Simulacrum and Ancient Philosophy" Deleuze contends that Plato considers not only that the simulacrum is not merely a false copy but also that it interrogates the very concepts of copy. According to Deleuze's interpretation of the Platonic allegory of the cave, "*copies* are secondary possessors. They are well-founded pretenders, guaranteed by resemblance; *simulacra* are like false pretenders, built upon a dissimilarity, implying an essential perversion or a deviation. It is in this sense that Plato divides in two the domain of image-idols: on one hand there are *copies-icons*, on the other there are *simulacra-phantasms*" (256). According to Deleuze, then, the Platonic project consists of choosing among these pretenders, by selecting between good copies and simulacra, which always harbor a measure of dubious dissimilarity. In this light, Plato is supposedly bent on guaranteeing the victory of the copies over simulacra, keeping these repressed and submerged. In this context, Deleuze emphasizes the "demonic character of the simulacrum" (258),[27] the possibility that these simulacra might take over, while provocatively sustaining that "we have become simulacra" (257). To reverse Platonism, Deleuze maintains, "means denying the primacy of original over copy, of model over image" as well as "glorifying the reign of simulacra and reflections" (*Difference and Repetition*, 64). Reflecting on what he considers as the anti-Platonism at the heart of Platonism, Deleuze suggests that at the end of the *Sophist* we can "glimpse the possibility of the *triumph of the simulacra*" (128; emphasis mine). In the consideration of this anti-Platonism, then, Deleuze points out how "in the infinite movement of degraded likeness from copy to copy, we reach a point at which everything changes nature, at which copies themselves flip over into simulacra and at which, finally, resemblance or spiritual imitation gives way to repetition" (128). We are then irremediably steeped in the reign of simulation, irreparably transformed into simulacra. In addition, drawing on Pierre Klossowski's theories, Deleuze, referring to the Nietzschean concept of the eternal return, notes that "taken in its strict sense, eternal return means that each thing exists only in returning, copy of an infinity of copies, which allows neither original nor origin to subsist" (67). Extrapolating to the fantasy of cloning, we might say that while, on the one hand, a clone can emblematize a narcissistic projection (an idealized copy of oneself), on the other, it inevitably subjects the concept of original to interrogation, depreciating it in the process instead of making it appear more valuable. Indeed, it is in this context, where the copy can no longer be considered as an inferior imitation of the original, that clones could be situated within a Nietzschean and Deleuzian logic. Both Baudrillard and Deleuze thus carry out a problematization of the Platonic concepts of copies and simulacra in terms that are useful in the context of my discussion of clones as simulacra.

THE TRIUMPH OF THE FALSE PRETENDER

Deleuze, in tune with Nietzsche, who described the task of his philosophy as the attempt to "reverse Platonism" ("The Simulacrum and Ancient Philosophy," 262), asserts the rights of simulacra to rise over icons and copies. As Deleuze affirms, the "simulacrum is not a degraded copy. It harbors a positive power which denies *the original and the copy, the model and the reproduction*" (262).[28] This reversal of Platonism, then, can be seen as amounting to "the triumph of the false pretender" (262). However, as Deleuze is quick to point out, "the false pretender cannot be called false in relation to a presupposed model of truth, no more than simulation can be called an appearance or an illusion" (262–263).

In this context, and in an analogous vein, Baudrillard posits in "The Clone or the Degree Xerox of the Species" that "the clone can be seen—this is the positive side of repetition—as the parody of the original, as its ironic, grotesque version" (201). Thus, while for Deleuze the simulacrum harbors a pointed transgressive potential, Baudrillard, in somewhat less radical fashion, considers it as just another version of the original, albeit a distorted one, based on pastiche and mockery, demoted and demoting (although carrying a similarly subversive capacity). Original and copy, original and simulacrum, could be considered thus as equiprimordial, with no hierarchy attached to them.

Is a clone, then, according to this line of argumentation, a "false pretender"? To judge from the negative assessment generated by most narratives about clones, that would seem to be the case. There are, however, exceptions, as in Naomi Mitchison's *Solution Three* (1975), J.B.S. Haldane's *The Man with Two Memories* (1976), and David Brin's *Glory Season* (1993), which envisage future societies where clones are an integral part of the population and fully accepted as such, a subject that will be further developed in Chapter 5.

Fredric Jameson's notion of pastiche can be instructive here, for Jameson, engaging with Baudrillard's ideas, considers pastiche as a neutral form of parody, as "blank" parody in his "Postmodernism and

Consumer Society" (114). In his essay "Postmodernism, or The Cultural Logic of Late Capitalism," he rephrases his definition of postmodern pastiche to include the adjective "blind." Jameson sees pastiche gradually replacing parody in our postmoderm world and describes it as "the imitation of a peculiar mask" (65), as "blank parody, a statue with blind eyeballs" (65), an image that could aptly illustrate the groups of interchangeable clones in such clone narratives as Kate Wilhelm's *Where Late the Sweet Birds Sang*, Damon Knight's "Mary" (1964), or Ursula K. Le Guin's "Nine Lives" (1969). As if engaged in a critical dialogue with Baudrillard, Jameson also speaks about the "death" of the subject brought about by the gradual disappearance of norms ("Postmodernism and Consumer Society," 114), a concept that can be pertinently linked to Jameson's notion of the neutral, blank nature of pastiche. In the sense that Baudrillard would conceive of a clone as a diluted version of the original, pastiche seems to refer appropriately to a cloned person in a Baudrillardian framework.

THE REVENGE OF THE SIMULACRUM

As part of this debate, another paradoxical question arises: If the cloned person can be described as a simulacrum, a copy of a copy that has lost sight of an original, then this copy can be said to subvert the authority and legitimacy of the original. In that sense, the whole notion of human cloning as a narcissistic project is undermined by precisely the emergence of this potentially threatening figure that will in fact make of one's very image the model for a repetition matrix. Human clones would thus inevitably force us to rethink the secondary status attributed to the copy in relation to the original, for this copy could be as perfect as or more perfect than the original, given the potential addition of health-promoting genetic material or removal of harmful genes.

Clones could thus come to be perceived as a menace in the sense that they might become the very embodiment of a more perfect "you," while encapsulating the capacity of infinite reproduction of the same. In our society of the simulacrum, and in a Deleuzian vein, it could be argued that the clone as simulacrum incarnates the possibility of always becoming other, of the production of endless repetitions of the same and of the recreation of a putative original, in a ceaseless engendering of masks and performances that generate the effect of an original. By stressing the verisimilitude of the simulacrum, Deleuze claims that it is as real as the original and extends that notion to rethink the world and life in general as always already shaped by simulation, with its capacity for production and emphasis on the idea of creation and becoming.

In addition, and further pursuing this logic of the potential power of the simulacra to take over, one of the numerous scenarios that the implementation of human cloning could bring about in terms of new family and social configurations, would be, as Baudrillard suggests, that the "disqualified original" would take "revenge on his clone" ("The Clone or the Degree Xerox of the Species," 201). As Baudrillard pointedly asks, "For what becomes of the human being when he [*sic*] is pushed out by his own clone and rendered useless? A reserve? A relic? A fetish? An art object? There is no immediate end in sight to the conflict between the original and its double, the clash between the real and the virtual" (201–202). Here again we have a case where the creature turns against his or her creator, as with Frankenstein's monster, a recurring theme in clone narratives, which can aptly be described, in Baudrillardian terms, as "the Revenge of the Simulacrum" ("The Orders of Simulacra," 95). This retaliation of the manufactured being also takes place within a Western discourse that articulates a widespread fear and mistrust of science and its complicity in tinkering with birth processes, as well as the abuses on the part of scientists of that knowledge to exert power to achieve often questionable ends without taking account of the troubling consequences for the people involved, not to mention shaping reproductive politics in ways that seem unethical to most observers.

From a cloned person's point of view, it may be speculated that feelings of jealousy and insecurity would be endemic and have wide repercussions in terms of family dynamics, as is the case in, for instance, Lisa Tuttle's "World of Strangers," Felicia Ackerman's "Flourish Your Heart in This World," Martha C. Nussbaum's "Little C," Ian R. MacLeod's "Past Magic," Nancy Kress's "To Cuddle Amy," and Naomi Mitchison's "Mary and Joe," short stories that illustrate the complex psychological symptoms and traumas that may arise with new family reconfigurations, which will be discussed in Chapter 7.

From the cell donor's perspective, however, in view of the genetic material that he or she has provided, there is the danger that the resulting clone will mostly be defined according to the donor's terms in a reductive way that will deprive the donor of the sense of a unique and distinct personality.[29] On the other hand, in his or her similarity or close sameness to the genetic donor, the clone could also be seen as countering the fear of the other, the irruption of the putatively monstrous other in our lives, that concern so powerfully expressed in Mary Shelley's *Frankenstein*. The clone would thus provide a foil, an antidote to what Žižek, in *Enjoy Your Symptom* (1992), described as the contemporary fascination with the "Thing, a foreign body within the social texture" with all its marks of "latent corruption" (123).

DOUBLES AND CLONES

While clones have often been portrayed in fiction and film as simulacra,[30] evil or otherwise, another common trope attached to clones is that of the evil double, a feature that can be perceived as going back to antiquity, appearing to resemble an original by reproducing its exterior but neglecting the essential, immanent features of that same original. The two emblematic and closely intertwined figures of Narcissus and the Double can be seen as crucially implicated in any discussion about human cloning. Indeed, the contemporary fascination with duplication, duality, resemblance, and immortality can be said to be the millenial equivalent to the romantic attraction to the double, the dual, the alter ego.[31]

Otto Rank, in his pioneering study of the double in mythology and literature,[32] notes the many links between those two figures, stressing the narcissistic import of the double in both its positive connotations and its more threatening avatars. As Rank maintains, "the double, who personifies narcissistic self-love, becomes an unequivocal rival in sexual love; or else, originally created as a wish-defense against a dreaded eternal destruction, he reappears in superstition as the messenger of death" (*The Double: A Psychoanalytic Study*, 86).[33]

In "The Uncanny" (1919) Freud smilarly argues that the double functioned, in its original formation, as "an insurance against the destruction of the ego" (356),[34] as "a preservation against extinction" (356).[35] Like Rank, Freud goes on to establish a connection between the double and "the primary narcissism which dominates the mind of the child and of primitive man" (357). As Freud further states, "when this stage has been surmounted, the 'double' reverses its aspect. From having been an assurance of immortality, it becomes the uncanny harbinger of death" (357).[36] Indeed, the idea of human clones or doubles is considered frightening, disturbing, and uncanny by many people, as Freud similarly stresses, considering that the feeling of uncanniness attached to the double derives from the fact of the double's "being a creation dating back to a very early mental stage, long since surmounted—a stage, incidentally, at which it wore a more friendly aspect" (358).[37] In "The Uncanny" Freud also considers the idea or actual occurrence of the repetition of the same thing, as in the fantasy of human cloning, as a source of impressions of uncanniness. The something familiar that has been repressed and resurfaces at some stage, as is the case with the emergence of the double, is also applicable to the resurgence of the clone as the person's alter ego, or ideal ego, in Freudian terms.[38] Indeed, as Freud maintains, in words that supply a useful context and justification for the appearance of the imaginary double or clone, "it may be true that the uncanny [*unheimlich*] is something which is secretly familiar [*heimlich-heimisch*], which has undergone repression and then returned from it" (368).

As will become apparent, feelings of uncanniness are constantly attached to human clones in clone narratives, a reservoir of disquiet that many of these clone characters attempt to dispel by trying to build a life predicated, as much as possible, on normality, which will be extremely difficult to sustain because they will always be looked at with feelings of suspicion and a certain awe. In "The Uncanny" Freud provides a possible explanation for this sense of disquiet many people experience as far as the idea of cloning is concerned, when he notes that an uncanny effect is "often and easily produced when the distinction between imagination and reality is effaced, as when something that we have hitherto regarded as imaginary appears before us in reality" (367), as might be the case with the emergence of one's double or clone, envisaged in the fantasy of coming across someone who is identical to oneself. Indeed, encountering one's other self is necessarily a profoundly disturbing, haunting, reflexive, thought-provoking experience, visually evoked long before contemporary technologies in Dante Gabriel Rossetti's painting *How They Met Themselves* (1851–1860), with its connotations of auguries of an announced death. But meeting your own younger and older selves can be described as potentially even more unsettling. Confrontation with the other who is yourself can be seen as a profoundly life-changing and life-expanding experience, permeated with uncanny resonances.[39]

According to Marina Warner, "The uncanny double has expressed, since the late eighteenth century, modern intimations of inner demons, of being multiple rather than integrated. The shadow of the doppelgänger above all reveals that the threat to personhood comes from bodily manipulation and psychological multiplicity, the monstrous threat of the 'many-in-the-one'" (*Fantastic Metamorphoses, Other Worlds: Ways of Telling the Self*, 165). For Warner, however, the double "also solicits hopes and dreams for yourself, of a possible becoming different while remaining the same person, of escaping the bounds of self" (165). Engaging with the idea of representation as "a form of doubling," potentially able to "make something come alive, *apparently*," Warner further remarks that the "figure of the other you inside you threatens to escape . . . and become another, usurping your being as someone else" (165). As Warner goes on to observe, "The idea of the clone is probably most frightening because even if it looks and acts like a copy, it cannot and will not be one" (165), a deep apprehension illustrated in many stories dealing with clones who act like evil incarnations of the individual.

As Michel Foucault states, referring to the appearance of "man and his doubles" in modern thought, in words that can also be seen as a prefiguration of the emergence of those other doubles, the clones, "The unthought (whatever name we give it) is not lodged in man like a shrivelled-up nature or a stratified history; it is, in relation to man, the Other: the Other that is not only a brother but a twin, born not of man, nor in man, but beside him and at the same time, in an identical newness, in an unavoidable duality" (*The Order of Things: An Archeology of the Human Sciences*, 326). This other, then, in present-day terms, can aptly be configured as the clone.

THE ETERNITY OF THE SAME (BAUDRILLARD, "CLONE STORY," 95)

In "The Hell of the Same," Baudrillard likewise addresses this theme of the double, which constitutes a fundamental recurrent motif in his work, linking the double with the human fantasy of encountering precisely one's double or clone.[40] As Baudrillard claims, "Everyone may dream—and everyone no doubt does dream all his life long—of a perfect duplicate, or perfect multiple copies, of his own being; but the strength of such copies lies precisely in their dream quality, and is lost as soon as any attempt is made to force dream into reality" (113).[41] In a reversal of this scenario, Baudrillard sees cloning as abolishing confrontation with one's double and thus with death. In "Clone Story" he posits the double as "an imaginary figure, which, just like the soul, the shadow, the mirror image,[42] haunts the subject like his other . . . like a subtle and always averted death. This is not always the case, however: when the double materializes, when it becomes visible, it signifies imminent death" (95).[43] For Baudrillard, "it belonged to our era to wish to exorcize this phantasm like the others, that is to say, to want to realize, materialize it in flesh and bone and, in a completely contrary way, to change the game of the double from a subtle exchange of death with the Other into *the eternity of the Same*" (95; emphasis mine). That is to say, the human clone constitutes, for Baudrillard, the materialization of the fantasy of one's double "by genetic means" ("Clone Story," 97), while at the same stroke eliminating some of the uncanny (*unheimlich*) connotations of that vision[44] and abolishing its perceived contiguity with death. One's clone thus becomes an assurance of life, of immortality, although simultaneously a threat to one's sense of singularity, of personhood.[45] In a related vein, John Lash notes how as "self-reflective human beings, we are all twins caught in a love/hate relationship with ourselves. There is no way out but there is a perennial temptation to go 'through the looking-glass' and merge with the other as reflected self" (*Twins and the Double*, 44).[46]

Indeed, an interrelated theme pertaining to the idea of the double and, by extension, of cloning is the projection onto the double or clone of the potentialities that another lease of life might provide, imaginatively brought about by the connection with those figures. In his analysis of the question of the double in Dostoevsky, Bakhtin addresses precisely this issue, while simultaneously problematizing it, when he argues that through the double "the possibilities of another man and another life are revealed. . . . The dialogical attitude of man to himself . . . contributes to the destruction of his integrity and finalizedness" (*Problems of Dostoevsky's Poetics*, 96).

As Baudrillard asks in his more recent reflections on cloning, "Is there not a terror of and a nostalgia for this double and, to go further, for the whole multiplicity of *semblables* from whom we have divided ourselves in the course of evolution? Do we not, after all, deeply regret our individuation?" (*The Vital Illusion*, 13–14). Rosemary Jackson similarly comments on the psychological drama that involves the negotiations of duality in terms of the acquisition and development of a sense of identity, considering stories of the double as "graphic depictions of the alienation which is involved in becoming 'human' at all: they protest against and then reenact that drama of insertion into human culture which is the time when, with the acquisition of identity, our many protean selves, our undifferentiated elements, are 'unified' and stabilized as 'one' character—the ego, the I, the self, indivisible and integral, upon which society depends" ("Narcissism and Beyond: A Psychoanalytic Reading of *Frankenstein* and Fantasies of the Double," 44). This drama of the unification of undifferentiated elements—which, according to Lacan's theory of the mirror phase, will coalesce into a unified body picture at some stage between six and eighteen months of age when the infant perceives clearly his or her body as a whole—can be seen as being given dramatic illustration when it does not follow the normal pattern of entry into the symbolic realm. In such a scenario many elements associated with the imaginary are retained in some clone narratives where the cloned beings are depicted as parts of an aggregate, as fragments of a bigger unit, the group of clones, unable to function without the support of their respective brothers or sisters.

Indeed, at the heart of the idea of human cloning is the question of identity itself,[47] of the formation of the ego, as is also the case in narratives of the double. Rosemary Jackson further notes that "fantasies of dualism have more to do with a quest for wholeness and integration than with mere moral division" ("Narcissism and Beyond: A Psychoanalytic Reading of *Frankenstein* and Fantasies of the Double," 45), a quest that also lies at the core of many clone narratives. According to Joseph Francavilla, "the double threatens the extinction of differences between oneself and all others, which means that the double jeopardizes individual identity (defined by such differences) by threatening usurpation of, possession of, substitution for, or the obliteration of, the self" ("The Concept of the Divided Self in Harlan Ellison's 'I Have No Mouth and I Must Scream' and 'Shatterday,'" 111),[48] as clones are suspected of potentially doing to other identical beings. It is precisely this kind of scenario that is illustrated in numerous clone narratives, some of which I will examine in the following chapters.

> In a way, there is no Ego without its clone double.
>
> Slavoj Žižek, "Of Cells and Selves," 315

In his essay "Of Cells and Selves" Slavoj Zizek reflects at length on the concept of human cloning and asks many probing questions that revolve around the interrelated issues of human freedom in the face of the possibility of genetic engineering, genetic determinism versus environmental and social constructionism, the uncanniness of the double, as well as the philosophical and ethical arguments that address the desirability or otherwise of a prohibition of cloning. In this essay Zizek addresses the theme of the double through a Lacanian lens, observing, in relation to the "anxiety people experience apropos of being cloned [that] this fear of encountering an exact double already presupposes that I experience myself as the absolute singularity of a subject, so that the imposed sameness with another subject will consist of an imposed imaginary and/or symbolic identity, of an imposed 'self-image,' not of the direct genetic sameness-in-the-real. . . . In clear contrast to it, a true clone would feel no anxiety at being the same as his [*sic*] genetic double, since he would lack the very sense of absolute singularity which constitutes the self-referential identity of a subject" (315). I have some problems with this view, for a cloned human being would be as much a singular individual as any other, a point Zizek himself makes earlier in the same essay, when he stresses the "uniqueness" (314) of clones.[49] Furthermore, Zizek is imputing feelings to a clone that can be read as suggesting, at least in part, that a clone is not in fact fully human,[50] a doubt that persists in his mention of true clones, a terminology I have trouble understanding. What is a "true clone?" Are there false clones? Only in a symbolic sense, such as the false pretenders Deleuze refers to in his discussion of the Platonic copies and simulacra.

Evoking the vexed concept of the soul in connection with the related topics of doubles and clones, Zizek provides a Lacanian reading of the anxiety of meeting one's double or clone. As Zizek points out, "the true point of anxiety is . . . not that of doubling the body, but that of doubling my unique Soul. What makes me creep at the very thought of encountering my double is not that he looks exactly like me, but that, in his personality, he 'is' another me, that he clones the very uniqueness of my personality" ("Of Cells and Selves," 316).[51] He further proposes to locate "in my double, in the encountered object that 'is' myself, the Lacanian *objet petit a*: what makes the double so uncanny, what distinguishes it from other inner-worldly objects is not simply its resemblance to me, but the fact that it gives body to 'that which is in myself more than myself,' to the inaccessible/unfathomable object that 'I am,' to that which I forever *lack* in the reality of my self-experience" (316).

These are, of course, very important points, fundamental for any discussion of the philosophico-ethical issues pertaining to human cloning, but it seems to me they need to be somewhat qualified. I believe Zizek is in part contradicting himself, or at least falling prey to the force of paradox at work in this discussion. Does the double inexorably clone "the very uniqueness of my personality?"[52] Isn't that very uniqueness, in fact, unreproducible, impossible to copy, given the interplay between genes and the environment, as Zizek himself maintains earlier in the same essay? Indeed, according to Zizek, "even if science defines and starts to manipulate the human genome, this will not enable it to dominate and manipulate man's [*sic*] subjectivity: what makes me 'unique' is neither my genetic formula (genome) nor the way my dispositions developed owing to the influence of the environment, but the unique self-relationship which emerges out of the interaction between the two" ("Of Cells and Selves," 313–314). Significantly, Dr Ian Wilmut, the creator of Dolly, the first cloned sheep using a somatic cell from an adult animal, makes a very similar point when he emphasizes, in connection with the question whether clones are identical, that "genes are not as constant as we imagine [and] genes operate in constant dialogue with their surroundings" (*The Second Creation: The Age of Biological Control by the Scientists Who Cloned Dolly*, 300).[53]

As Zizek sustains, engaging with Lacan, "the way I 'see myself,' . . . the fantasy which provides the ultimate co-ordinates of my being is neither in the genes nor imposed by the environment, but the unique way each subject *relates to him or herself*, 'chooses him or herself,' in relationship to his or her environs, as well as to (what he or she perceives as) his 'nature'" ("Of Cells and Selves," 314). Thus, Zizek goes on to postulate "a third mediating agency," which he calls "the subject" although it "has no positive substantial Being, since, in a way, its status is purely 'performative'" (314). For Zizek, then, "I emerge through the interaction between my biological bodily base and my environs—but what both my environs and my bodily base are is always 'mediated' by my activity" (314). This emphasis on the performative is carried over to Žižek's views about clones. As he argues, "In the case of clones (or, already today, of identical twins), what accounts for their difference, for the *uniqueness* of each of the two, is not simply that they were exposed to different environments, but the way each of the two formed a *unique* structure of self-reference out of the interaction between their genetic substance and their environment" (314; emphasis mine). So, according to Žižek, each clone is, after all, indisputably a unique human being,[54] a point left somewhat blurred in Žižek's discussion of clones.

NARCISSUS AND THE DOUBLE

While ours is eminently a culture of the copy, of reproducibility, of dualism,[55] it is also very centrally a culture grounded on the cult of identity and, increasingly, of narcissism. The concept of human cloning can also be said to be inextricably linked with the notion of narcissism. In the Western world, the figure of Narcissus has been the most enduring emblem for male self-reflection. Lacan has described our time as the "era of the ego,"[56] characterized by a pervasive narcissism, while Julia Kristeva establishes the ubiquity of the notion of narcissism in her discussion of the myth of Narcissus from Ovid to contemporary times,[57] considering Narcissus "a modern character" (*Tales of Love*, 121), very close to us. In similar fashion, according to Gray Kochhar-Lindgren, "The figure of Narcissus haunts the twentieth century" (*Narcissus Transformed: The Textual Subject in Psychoanalysis and Literature*, 1), while "Narcissus brings with him a concern for *origins*, indeed a myth of origins that is a fantasy of autogenesis" (20; emphasis mine).[58] Such a fantasy can crucially be seen as one of the main driving impulses for human cloning, a dream of self-reproduction that will produce a putative copy of the self. The figures of Narcissus and his double, his watery reflection, can thus be said to be emblematic of the contemporary psychological landscape in the Western world while simultaneously prefiguring a possible representation of the fantasy of human cloning, as well as some of its psychological impact.

In his "On Narcissism: An Introduction" (1914), Freud describes a group of people who "in their later choice of love-object . . . have taken as a model not their mother but their own selves. They are plainly seeking *themselves* as a love-object, and are exhibiting a type of object-choice which must be termed 'narcissistic'" (81). Freud postulates a "primary narcissism in everyone" (82). As he explains, "The development of the ego consists in a departure from primary narcissism and gives rise to a vigorous attempt to recover that state" (95). When, however, the person does not move beyond primary narcissism to love of others, his or her development remains arrested at that stage, which then comes to be regarded as pathological. This view, I suggest, can be used to cast light on several of the texts I will be examining in this book that revolve around the profoundly narcissistic urge of certain characters to clone themselves, as well as others, to satisfy their selfish and megalomaniac desires, as in Huxley's *Brave New World* (1932), Fay Weldon's *The Cloning of Joanna May* (1989), and Anna Wilson's *Hatching Stones* (1991).

The fantasy of human cloning can also be said, in Freudian terms, to be closely linked with Eros, the life-drive, the impulse to reproduce one's very image as faithfully as possible, to perpetuate one's undiluted genes, to carry on living in the body of another just like oneself. Indeed, closely linked with Eros and the pleasure principle, narcissism is a powerful explanatory concept of the drive toward cloning both oneself and others. In "On Narcissism," Freud notes, "If we look at the attitude of affectionate parents towards their children, we have to recognize that it is a revival and reproduction of their own narcissism, which they have since abandoned" (84), something the fantasy of human cloning could bring back in an exacerbated form.

Baudrillard considers the fantasy of cloning among many other things, as we have seen, as a "monstrous parody of the myth of Narcissus" ("The 'Ludic' and Cold Seduction," 168), for the source of this narcissism, according to Baudrillard, is "no longer a mirror but a formula" (168). Cloning, then, can be seen as a "projection and internment in the mirror of the genetic code" (167), words that hint at a profound narcissistic confinement from whose centripetal force it is hard to escape. Indeed, in "Clone Story," Baudrillard further meditates on this issue, observing that

> one can say that the genetic code, where the whole of a being is supposedly condensed because all the "information" of the being would be imprisoned there (there lies the incredible violence of genetic manipulation) is an artifact, an operational prosthesis, an abstract matrix, from which will be able to emerge, no longer even through reproduction, but through pure and simple *renewal*, identical beings assigned to the same controls. (98–99)

In this light, he suggests that "there is no better prosthesis than D.N.A., no finer narcissistic extension than that new image bestowed on modern human beings in place of their specular image: their molecular formula. Here is where one will find one's 'truth'" ("The 'Ludic' and Cold Seduction," 167). The concept of "truth," then, can be said to have shifted to "appearance," on the one hand, while on the other, and paradoxically, to have gone deeper into the heart of the cell, to the DNA helix, folded around itself as if mimicking another narcissistic *mise-en-abîme*. The genetic code, then, becomes the new image of "truth" in the world we live in, shaped by biotechnology. Images thus emerge as a crucial concept in a discussion of the philosophical and cultural implications associated with the idea of human cloning.

Baudrillard argues that the creation of a biological clone will severely undermine the capacity of "playing with one's own image and, thereby, playing with one's death" ("The 'Ludic' and Cold Seduction," 168). In analogous fashion, Kochhar-Lindgren remarks that with Narcissus "there is always reflection and self-reflexivity," as well as "the anxiety of death" (*Narcissus Transformed: The Textual Subject in Psychoanalysis and Literature*, 3), which the appearance of one's clone can avert, although as Kochhar-Lindgren also suggests, "the threat of absolute solipsism marks the presence of Narcissus" (6). It is precisely the danger of this absolute solipsism that the presence of one's clone(s) can ward off, although simultaneously it can cause the risk of absorption into a group continuum with a consequent loss of individuality, a recurrent thematic concern in clone narratives.

Sabine Melchior-Bonnet, for her part, notes how "the true sin of Narcissus lay in the preference he granted to his own singularity. . . . The mirror, instead of offering him the clarity of reciprocal admiration, reflected his somber double with a single desire: a bubble of solipsistic confinement" (*The Mirror: A History*, 148),[59] so endemic in our Western society, obsessed as it is with appearances, where massification can turn human beings into almost interchangeable elements in an endless series. Hence the lure of Narcissus for our contemporary world,[60] in which societal configurations, increasingly stressing muted solitude in the midst of the masses as a widespread way of life, enclose the subject in a potentially reductive solipsistic sphere.[61] In a sense, the resistance to standardization can be read as implicated in a positive narcissistic drive, which the concretization of the fantasy of a clone could be seen as the materialization of.

This obsession with external features, described by Sabine Melchior-Bonnet as "the overinvestment of the mirror image" (*The Mirror: A History*, 273), can be seen as contributing to a dilution and demoting of the subject, as well as an increasing quest for that most elusive concept, identity. Melchior-Bonnet goes on to problematize this issue in words relevant for this debate. As she points out, and in tune with Baudrillard's analysis,

> With modern electronic means of reproduction, capable of showing what was heretofore invisible . . . the image has acquired new powers. The infinite possibilities of duplication and of the re-creation of synthetic images drains the subject of his [*sic*] alterity and his [*sic*] mystery, but the ingenuity of the machine frees the image from its devalued status as a mere copy, and reproduction thus ceases being the simple tracing of the real. (*The Mirror: A History*, 273)

Bluntly, "The individual is transformed into an image" (273). Appearance, then, has come to play a fundamental role in our clone-fascinated society; it has been elevated to the status of "truth," while copies and simulacra have been equally valorized, partaking of a similar philosophical and sociological nexus.

There is, however, another important dimension to the myth of Narcissus that can be profitably connected both with the problematic role of the image in our society and with clones. It can be summed up as the impulse to know ourselves, particularly through those contested figures. Melchior-Bonnet comments, "It is precisely because there is resemblance or likeness that there is the possibility of knowing oneself" (*The Mirror: A History*, 112).[62] However, as Gérard Genette argues in his essay "The Narcissus Complex," considering the interrelated topics of the construction of the self and identity through his or her images:

> In itself, the reflection is an equivocal theme: the reflection is a *double*, that is to say at the same time an *other* and a *same*. This ambivalence provokes in baroque thought an inversion of significations which makes identity fantastic (*I am an other*) and otherness reassuring (*There is another world, but it is similar to this one*). (*Figures I*, 21–22)[63]

Following on from this perspective, a clone (double, twin) can be seen, on the one hand, as providing a unique opportunity of observing oneself, of accessing an image of oneself otherwise unavailable, since we can only ever see ourselves by means of reflections, or through the eyes of others. On the other hand, any reflection is always an illusion, a double, a clone; our idea and image of ourselves are always already mediated.[64] Tellingly, however, were we to come across our clone or double, he or she would occupy exactly the same place as our own image in a mirror: Our right eye would be the mirror equivalent of our reflection's left eye, and the same would occur with a live identical being. As Gérard Genette notes in "The Narcissus Complex," through the mirror image "the Self is confirmed, but under the species of the Other: the mirror image is a perfect symbol of alienation" (*Figures I*, 21–22).[65]

Indeed, behind the interrelated tropes of Narcissus and the double stands the motif of the mirror, with its multiple connotations of reflection, duality, duplication, (re)creation of identity, vanity.[66] As Rosemary Jackson observes, "By presenting images of the self in another space (both familiar and unfamiliar), the mirror provides versions of self transformed into another, become something or someone else" (*Fantasy: The Literature of Subversion*, 87).[67] In effect, it presents us with an image of our double. Indeed, seeing oneself again, not in the mirror but as another person, as in the fantasy of cloning, could from one point of view constitute a very affirmative sign of recognition, of validation of the self, while from another it becomes an unbearable sign of anxiety, a threat of the extinction of one's unique body and personality, as if these had been uncannily stolen.

Like the double, the clone can be seen as the mirror image onto which one can project either dreams and wishes unfulfilled in one's lifetime or even socially unacceptable desires. Laplanche and Pontalis explain the mechanism of this kind of projection by describing it as "the operation whereby qualities, feelings, wishes or even 'objects,' which the subject refuses to recognize or rejects in himself are expelled from the self and located in another person or thing. Projection so understood is a defence of very primitive origin" (*The Language of Psychoanalysis*, 359). This projection, indeed, can be translated into a wish to return to a pre-Oedipal state of greater fusion with the mother, of undifferentiation, which the fantasy of human cloning can also be seen as potentially suggesting, as simultaneously a recovery and reversal of the mirror stage.

Also, like the double the clone can be regarded as symbolizing a person's relationship with him- or herself, in a kind of narcissistic *mise-en-abîme* which also stands for narcissistic self-love. Narcissus's words can thus be seen as illustrating not only the drama of self-knowledge but also prefigurations of the encounter between a person and his or her double or clone: "I am myself the boy I see! I know it. My own reflection does not deceive me" (Ovid, *The Metamorphoses*, III: 86). The Narcissus myth is crucially about the power and elusiveness of the image, of reflection. Julia Kristeva significantly diagnoses the contemporary sickness that afflicts many inhabitants of big cities in this turn into the new millenium, which she associates precisely with the role played by images and reflections as far as ego formation and life in society are concerned, in terms of the "abolition of psychic space" (*Tales of Love*, 373). Kristeva then goes on to associate this psychological condition with the plight of Narcissus, whom she describes as "in want of light as much as of a spring allowing him to capture his true image, Narcissus drowning in a cascade of false images (from social roles to the *media*), hence deprived of substance or place" (373). Can that spring be symbolically provided by the clone as reflection of the self, as the exteriorization of narcissistic drives to self replication? Tellingly, Kristeva considers that many of the psychological problems experienced by these modern characters have to do with difficulties in being able to "elaborate primary narcissism" (374), to love themselves, a condition that can be potentially alleviated by the projection of wishes and mental pictures onto, as well as identification with, an idealized double or clone figure. Significantly in the context of the topic of this book, Kristeva considers that those characters, who embody the predicament of Narcissus, can be seen as extraterrestrials[68] in need of love. Kristeva sees this Narcissus figure as an exile, an "uneasy child, all scratched up, . . . without a precise body image, having lost his specificity, an alien in a world of desire and power" (382–383), yearning to recreate love. As Kristeva insists, the "ET's are more and more numerous. We are all ET's" (383). In a sense, we are all to a certain extent clones too, going through a process of mental cloning, translated into the adoption of similar and standardized thought conventions, tastes, and behavior. We are each other's mirrors, as well as doubles or clones.

OF MIRRORS, IMAGES, AND MIRAGES

The fantasy of human cloning can also be read, through the lens of Lacan's mirror stage, as the deep-seated wish to recover the illusion of (re)possessing one's image as a whole, without fragmentation, a return to that inaugural moment in front of the mirror when the child first sees its body as a whole, a unity, a jubilant instant that constitutes a rite of initiation into the symbolic order, introducing the subject to his or her own image, always already, however, a delusional mirage.[69] By seeing ourselves mirrored in our clones, so the fantasy would go, we would be able to not only metaphorically rejuvenate, go back to our childhood, but also and putatively, by reliving that moment, achieve greater insight into ourselves, or at least have the illusion of getting to know ourselves better through the act of contemplating this other "me" directly, without the mediation of any reflecting surface. At the same time, this fantasy could also be said to work toward the recuperation of what Lacan termed the Imaginary and Kristeva described as the semiotic, a stage before the entrance into the symbolic order of the father and described by Mary Jacobus as "regressive, narcissistic, mother-identified" (*First Things: The Maternal Imaginary in Literature, Art, and Psychoanalysis*, iv). Ruth Parkin-Gounelas succinctly summarizes the experience of the six- to eighteen-month-old child going though the "mirror stage" according to Lacan:

> In gazing in the mirror, the infant experiences two contradictory responses, not necessarily simultaneously. On the one hand, it fulfills that perpetual human desire for identification and fusion when it "recognizes" itself with a "flutter of jubilant activity" (*Écrits* 1). Yet on the other, it learns that this image, which offers a promise of wholeness and (self-) identity, is in fact a mirage, a mere reflecting surface which disguises the fragmentation of the infant's felt experience. (*Literature and Psychoanalysis: Intertextual Readings*, 6)

Parkin-Gounelas goes on to assert that "this realization that the image is a fake, that its representation does not fit, is the founding experience of *alienation* on which all subjectivity is based. . . . For the child, like Narcissus, 'loses' itself at the very point of self-recognition, which is in fact a misrecognition" (*Literature and Psychoanalysis: Intertextual Readings*, 6; emphasis mine).[70] Engaging again with the concepts of fakes, simulacra, and counterfeiters, Julia Kristeva inscribes the figure of

Narcissus within this nexus of ideas, arguing that Narcissus killed himself because he eventually understood that he had fallen in love with a fake. In Kristeva's words, "The moral condemnation that concludes the Ovidian myth thus reveals the concomitance of *narcissism* and the *fake*" (*Tales of Love*, 126). Kristeva goes on to analyze the appearance of Narcissus in Dante's Canto XXX in a context that deals with "forgers, counterfeiters, and falsifiers" (127).

Explaining the process of ego formation, Lacan distinguishes between the ego and the *alter ego* in words that may throw light on the psychological conflicts attendant upon the idea of human cloning:

> The human ego is the other and . . . in the beginning the subject is closer to the form of the other than to the emergence of his own tendency. He is originally an inchoate collection of desires—there you have the true sense of the expression *fragmented* body—and the initial synthesis of the *ego* is essentially an *alter ego*, it is *alienated*. The desiring human subject is constructed around a center which is the other insofar as he gives the subject his unity, and the first encounter with the object is with the object as object of the other's desire. (*Seminar, livre III: Les psychoses, 1955–1956*, 39; emphasis mine).[71]

If we take this alter ego as an ideal version of oneself, one could, admittedly stretching Lacan's point, associate this ideal ego with the clone, who can function as a double, a figure onto which a subject's desire for perfection can be projected.

So for the ego to take shape it needs another, the ego ideal. According to Freud, in the "Twenty-Sixth Lecture," when it comes to libidinal choice of an object, a process can be detected "where the subject's own ego is replaced by another one that is as similar as possible" (88).[72] In this scenario, "this ideal ego is now the target of the self-love which was enjoyed in childhood by the actual ego. The subject's narcissism makes its appearance displaced onto this new ideal ego, which, like the infantile ego, finds itself possessed of every perfection that is of value" (94.). Cloning can thus be seen as literally recreating the symbolic figure of the ego ideal, who would resemble the subject as closely as possible.

Indeed, according to Rosemary Jackson, many fantasies of dualism can be seen as dramatizations of the conflict between a state of primary narcissism, where the self constitutes his or her own model of perfection, and an ideal ego, in relation to whom desire is always thwarted. Jackson tells us that many of these fantasies illustrate "a return to a state of undifferentiation, to a condition *preceding* the mirror stage and its creation of dualism. For prior to this construction, in a state of primary narcissism, the child is its own ideal, and experiences no discrepancy between self (as perceiving subject) and other (as perceived object)" (*Fantasy: The Literature of Subversion*, 89). Within this context, the fantasy of human cloning can be seen as dramatizing this wish for a return to a state of primary narcissism, to reverse the ego formation movement and to recreate a stage of putative original unity, prior to the insertion of the subject into the symbolic order and the inevitable division this construction of the subject entails. However, as Jackson pertinently notes, since this realm of the imaginary is devoid of language, it is not possible for human beings to go back to it without being depleted of their language; and those who try a return to this kind of undifferentiation, then, can never achieve it.[73] As Jackson points out, "Most versions of the double, for example, terminate with the madness, suicide, or death of the individual subject: 'self' cannot be united with 'other' without ceasing to be" (91), an insight that sheds light on, for example, Kaph's predicament after his clone brothers and sisters are killed in Ursula K. Le Guin's "Nine Lives," which can be seen as emblematized in the Narcissus myth, for, as Francette Pacteau notes, "Narcissus died of love, trapped in the body of his specular rival" (*The Symptom of Beauty*, 194).

The image, then, that necessarily always mediates our perception of ourselves, becomes our perpetual other, as Lacan argues (*Écrits*, 5). However, the yearning for unity, wholeness, and identification with that very image continues to pursue and afflict one throughout life.[74] For Parkin-Gounelas, "Lacan sees this dependence on the image as a trap, luring the ego into attempt at self-replication and sameness which are perpetually frustrated. Future identification will always be haunted by this Imaginary wholeness, projected onto others" (*Literature and Psychoanalysis: Intertextual Readings*, 7). As Francette Pacteau notes, this image that Lacan talks about

> implies a corporeal form which is perceived as coherent and fixed—in a way that the child's own disjointed and turbulent phenomenological body is not. The first outline of an ego is thus an ideal image which the existential subject is unable to sustain in actuality. No sooner than it is found, this ideal image is lost: not only does it not tally with the actuality of the subject, but it is given, in the first place, "as a *Gestalt*, in an exteriority . . . ," that is, as already "other." (*The Symptom of Beauty*, 105)

In a Lacanian context, one might speculate whether clones will be able to form a whole picture of their bodies, as a specular image of a unified body picture, rather than a "*corps morcelé*," since they might perceive themselves as the same or similar to others and, because of that, feel threatened and experience lack. According to this order of ideas, then, they might be said to symbolically remain trapped in a version of the Lacanian imaginary, prior to the mirror stage, their bodies metaphorically scattered in bits and pieces among the bodies of the other clones, their brothers and sisters, since they are the same, in a complex play of reflections. The fantasy of cloning can thus be read as a reenactment of the entry into the imaginary before any feeling of lack intrudes.

The dream that with cloning one would have a second chance of seeing and perceiving oneself, as it were, without the mediation of a mirror (for one can never regard oneself directly without the help of a mirroring device) by gazing at someone who looks just like one (albeit younger)—and thus achieving that impossible sense of a whole body image supposedly experienced by the child in the Lacanian mirror stage—would of necessity remain a dream, a mirage, since the clone is not you yourself, he or she is a separate human being who would, at best, function as a physical mirror and thus repeat the founding experience of alienation suffered by the infant during the Lacanian mirror phase. Indeed, as Roland Barthes maintains, "*You are the only one who can never see yourself except as an image; you never see your eyes unless they are dulled by the gaze they rest upon the mirror or the lens*" (*Roland Barthes by Roland Barthes*, 36).

In related fashion, pursuing his reflections on the connections between doubles and clones and drawing on the Lacanian concept of the mirror phase, Žižek asserts that "I constitute myself as Ego only by recognizing myself in the mirror-image—that is, by encountering my virtual double, with whom I then engage in an ambiguous love–hate relationship (loving him because he is like me, hating him for the very same reason, because he threatens to occupy my place). So, in a way, there is no Ego without its clone double" ("Of Cells and Selves," 315). The latter can thus be seen as functioning as the Ego ideal, the idealized "object" that gives credence and a reassuring feeling of unity to one's reflected image.

In "Clone Story" Baudrillard addresses the related questions of the Lacanian mirror stage and how the advent of human cloning might change this fundamental phase in a child's development, as well as the role of the mirror image in achieving individuation:

> The subject is . . . gone, since identical duplication puts an end to his division. The mirror stage is abolished in cloning, or rather it is parodied therein in a monstrous fashion. Cloning also retains nothing, and for the same reason, of the immemorial and narcissistic dream of the subject's projection into his ideal alter ego, since this projection still passes through an image: the one in the mirror, in which the subject is alienated in order to find himself again, or the one, seductive and mortal, in which the subject sees himself in order to die there. None of this occurs in cloning. No more medium, no more image—any more than an industrial object is the mirror of the identical one that succeeds it in the series. One is never the ideal or mortal mirage of the other, they can

only be added to each other, and if they can only be added, it means that they are not sexually engendered and know nothing of death. (97)

I have to take issue again with Baudrillard's polemical approach. In derogatory fashion, Baudrillard is describing a clone as an industrial object in a series, failing to consider that *one* cloned person has nothing to do with multiple copies or duplications (only in the sense of potential reproducibility, a scenario which, in the age of cloning, would apply to everybody). In addition, as I have been suggesting, the cloned person may arguably function as a projection of precisely that ideal alter ego to the subject, while—from the point of view of the cloned person—he or she would presumably follow a normal process of ego formation (unless, indeed, he or she was a part of a number of identical clones, a situation that would pose different problems and lead to distinct paths of ego formation and individuation).

Rimbaud also gives expression to the splitting of the self when he famously declares "Je est un autre" (I is an other) in a letter to Paul Demeny from May 15, 1871. Rimbaud is here referring to the creative process he witnesses within himself, when he is the spectator of the development of his own thinking and artistic conception, from an almost detached perspective, as it were, experiencing himself as double, as two distinct characters, a scene of representation that could be seen as a foreshadowing of a person and his or her clone, an other who might enact his or her wishes, ambitions, projects. In "Aggressivity and Psychoanalysis," the essay following "The Mirror Stage," Lacan meaningfully refers to Rimbaud (23). According to Lacan, human beings, following on from the mirror stage, will suffer from what he describes as "paranoiac knowledge" (17), the conviction that we are all uncannily haunted by the feeling that there is always already an "other" who governs or influences our behavior. In *Strangers to Ourselves*, Kristeva similarly engages with Rimbaud's resonant words, associating this other to my "self" not only with the "psychotic ghost that haunts poetry" (13) but also with the foreigner. For Kristeva, "being alienated from myself, as painful as that may be, provides me with that exquisite distance within which perverse pleasure begins, as well as the possibility of my imagining and thinking, the impetus of my culture. Split identity, kaleidoscope of identities: can we be a saga for ourselves without being considered mad or fake?" (13–14). The double or the clone can be regarded as manifestations of what Kristeva describes as "our disturbing otherness . . . that threat, that apprehension generated by the projective apparition of the other at the heart of what we persist in maintaining as a proper, solid 'us'" (192).

Indeed, and further pursuing this line of enquiry, it could be argued that the fantasy of cloning constitutes a powerful foil to the equally powerful fantasy of the body in pieces, the fragmented body. There is, however, an interesting paradox at work here, from a Baudrillardian perspective: If, on the one hand, the fantasy of cloning can be seen as projecting the dream of the body made whole again, on the other hand, as Baudrillard asserts in relation to cloning and the domination of the genetic code, "If all information can be found in each of its parts, the whole loses its meaning" ("Clone Story," 97), a state of affairs that also means for Baudrillard "the end of the body, of this singularity called body, whose secret is precisely that it cannot be segmented into additional cells, that it is an indivisible configuration" ("Clone Story," 97–98).

Furthermore, and circling back to the question of images, while on the one hand multiplication or duplication can be seen to demote and dilute the subject, on the other, with the valorization of the copy to the level of the original, of appearance to the status of truth, clones are, from this perspective, equal to other human beings, not inferior or subhuman, as Baudrillard would have it. Inasmuch as Baudrillard inscribes clones within a nexus of simulacra who lack an ultimate referent, or a non-Platonic idea of simulacrum, as serial human beings presumably would embody, he continues to equate clones with secondary or minor human beings, a point of view I take issue with, as already stated.

PRIMAL FANTASIES, OEDIPUS, AND CLONING

Freud identified three primal fantasies: the primal scene proper, in which the child supposedly watches his or her parents in the act of copulation; that of seduction; and that of castration.[75] As Hal Foster elucidates, Freud considered them fundamental "because it is through these fantasies that the child teases out the basic riddles of *origins*: in the fantasy of seduction the origin of sexuality, in the primal scene the origin of the individual, and in the fantasy of castration the origin of sexual difference" (*Compulsive Beauty*, 58; emphasis mine).[76] With cloning, however, the primal scene of origins, the fantasy of watching parental sex, would become superfluous, unnecessary for the purposes of procreation. In addition, the advent of cloning and its more or less widespread use could contribute, in the long run, to the gradual disappearance of the Oedipal and Electra complexes.[77] Cloning would do away with the need for both father and mother, since only one cell from one person is necessary. In that sense, both Victor Frankenstein's creation of a being without a biological mother and the parthenogenesis scenario, where women reproduce without a biological father, as in for instance Charlotte Perkins Gilman's *Herland* (1915), could become feasible with obvious implications as far as the sexual politics of reproduction are concerned.[78] Baudrillard addresses this issue in "Clone

Story," where he comments with respect to cloning and in words that shed light on the disappearance of the Oedipal plot:

> Cloning radically abolishes the Mother, but also the Father, the intertwining of their genes, the imbrication of their differences, but above all, the joint act that is procreation. The cloner does not beget himself: he sprouts from each of his segments. One can speculate on the wealth of each of these vegetal branchings that in effect resolve all oedipal sexuality in the service of "nonhuman" sex, of sex through immediate contiguity and reduction—it is still the case that it is no longer a question of the fantasy of auto-genesis. The Father and the Mother have disappeared, not in the service of an aleatory liberty of the subject, but in the service of a matrix called code. No more mother, no more father: a matrix. And it is the matrix, that of the genetic code, that now infinitely "gives birth" based on a functional mode purged of all aleatory sexuality. (96)[79]

But doesn't the very word "matrix" inevitably suggest the maternal feminine that, in one of its versions, cloning would remove? On the other hand, it could be argued that with human cloning the role of the mother would be even stronger, since women would have no need of men to reproduce. The same would apply to the idea of the father, provided there were artificial wombs,[80] so the figures of mother and father would remain, albeit separate. Along with this, it can be argued that the Oedipal complex is crucially about narcissism, about the wish to be the only one to be loved by the mother, to the exclusion of the father. In this context, then, it is centrally a narcissistic project. Significantly, Baudrillard, who considers the clone the "materialization of the genetic formula in human form [posits a] digital Narcissus instead of a triangular Oedipus" ("The 'Ludic' and Cold Seduction," 173). Narcissus is thus never far from the fantasy of human cloning, whereas the plight of Oedipus might never have taken place in the age of human cloning, nor need it ever occur again in a society where clones may come to predominate, with a concomitant rewriting of psychological structures and traumas.

A NEW MYTH OF ORIGINS

Lacan's psychoanalytic theories are centrally structured around concepts of wholeness and lack, issues that are also crucially implicated in the context of an investigation of the question of human cloning. Indeed, the concept of "lack," in genetic and psychological terms, can be seen as determining, as far as clones' perceptions of themselves are concerned, namely, the potential feeling of apprehension provoked by the fact that a clone will have the genetic information of only one parent, a mother or a father. The fear of a lack or certainty about one's genetic origins, which can be said to be a primordial human anxiety (what if the stories one has always been told about one's origins turn out to be false?) can be seen as applying in pertinent respects to serially produced people, if that scenario ever materializes, but even, to a certain extent, to a human clone who will have the genetic code of only a mother or a father, not the combination of both. If the drive to write, to produce narratives, can be said to always be in some respects crucially about a quest for origins, a search for father and mother but also for one's children, with the implementation of cloning the centrality of this thematic concern at the very core of writing might become even more acute. If all texts can be described as, to some extent, governed by Oedipal structures, in Sophocles' play itself, as Terry Eagleton remarks, "narrative hierarchies of cause and effect, parent and child, self and other, past and present, are radically undermined" (*Walter Benjamin, or, Towards a Revolutionary Criticism*, 67).

Here we might also recall William Wordsworth's famous line, "The Child is Father of the Man," which with cloning might literally come to pass. This reversal of the normative biological narrative might mean, then, that the child's desire to become its own progenitor, to be "self-originating," as Eagleton explains, "at the point of Oedipal crisis," when it "rejects the emplotments of genealogy" and "strikes against the authority of origin," this "impossible conundrum" could actually come true (*Walter Benjamin: Or Towards a Revolutionary Criticism*, 67). On the other hand, as Eagleton points out, should this nonfeasible and unrealizable conundrum be realized, it would "spell the death of all narrative—literally so, since five-year-olds cannot fertilize their mothers or be fertilized by their fathers, the narrative of human genealogy would grind to a halt, and with it the production of narrative discourse" (67). With cloning, however, these scenarios could literally be possible, without meaning the end of narrative. On the contrary, they could even provide a fresh impetus to the revision of the grand narratives, creating new myths and novel genealogies.

Unsurprisingly, many clone narratives are haunted by Oedipal questions, by the search for a father, for origins and a sense of identity conferred by those origins which, when missing, can lead to an ontological vacuum, as is the case of Bernard in *Brave New World* and of Mark in Wilhelm's *Where Late the Sweet Birds Sang*.

"Who was I? What was I? Whence did I come? What was my destination?" (*Frankenstein*, 128). These are the searching questions Frankenstein's creature asks, which remain unanswered by his creator. In a similar way, in a society where clones would have been turned into mere ciphers, cogs in an industrial conveyer belt, mass produced and conditioned, the very idea of origins, of a family, of the maternal or paternal genes, would have disappeared, conceivably as taboo, as anachronistic, or as an aberration, as in Aldous Huxley's *Brave New World*. While neither Frankenstein's monster nor the cloned people in *Brave New World* can be said to have a notion of "origins" in the traditional sense of the word, the difference is that the creature created by Frankenstein yearns for a father, a family, some reassurance as to where he comes from, whereas those questions are of no relevance whatsoever for the clones in Huxley's novel, with the exception of a few like Bernard, who deviate from the norm and whose punishment for this deviance will be to be sent into exile to a remote island.

Also unlike Frankenstein's creature, whose body, made of many body parts, can be said to encapsulate in a graphic way the Lacanian fantasy of "le corps morcelé," the body in bits and pieces, before the mirror stage, a clone can, by contrast, be seen as the perfect representation of the unified body, in the sense that it derives from only one body, a fantasy of wholeness with a new myth of origins. Indeed, for Lacan, a sense of primordial chaos and discord, of "le corps morcelé," is gradually substituted by an impression of corporeal unity. This presymbolic fantasy, so graphically illustrated in the body of Frankenstein's creature, literally made up of bits and pieces from other bodies, is described by Lacan as "a jig-saw puzzle, with the separate parts of a man or an animal in disorderly array" ("Some Reflections on the Ego," 13). The unified body picture that will eventually arise from the mirror stage is, however, always provisional and needs to be constantly reinforced, a function that the dream of a clone, a renewed version of oneself with a strongly unitarian body image, can potentially provide.[81]

THE FAUSTIAN PACT AND THE TEMPTATIONS OF SCIENCE

If scientists engaged in new reproductive technologies, and more specifically cloning, have often been regarded as contemporary versions of Victor Frankenstein, they can also be aptly compared, in a sense, to Faust, who falls prey to temptation in an attempt to satiate his search for knowledge, entering into a pact with the devil that will lead to his downfall. Several questions need to be asked at this stage: Are there and should there be limits to knowledge? Who is entitled to impose those limitations if, indeed, they are perceived as necessary?[82] How should ethical questions be dealt with, and who should make sure that the putative resolutions that are passed will then be implemented?[83] While I don't believe we can impose limits to knowledge, I think we can and should make every effort to tease out the moral, ethical, personal, political, and social implications of the impact of any particular area of knowledge, in this case that of human cloning. After all, no legislative environment allows unfettered experimentation; all polities are aware of the temptation of the Faustian bargain.

Slavoj Žižek, in his essay "Of Cells and Selves," poses a very similar question in the context of his examination of issues related to human cloning, inscribed in a meditation on the limits of knowledge, genetic engineering, and the putative freedom of the subject, sealed as it were in his or her genome. As Žižek observes, the concept of human cloning "confronts us with the most fundamental ethico-ontological alternative" (307), going on to forcefully assert,

> Either our genome does determine us, and then we are just "biological machines," and all the talk about prohibiting cloning and genetic manipulation is just a desperate strategy for avoiding the inevitable, for sustaining the false appearance of our freedom by constraining our scientific knowledge and technological capacities based on it; or our genome does not determine us thoroughly, in which case, again, there is no real cause for alarm, since the manipulation of our genetic code does not really affect the core of our personal identity. (308)[84]

Zizek then rejects the type of conservative understanding according to which only by restricting scientific research and imposing limits on what can and should be known can human freedom and dignity be spared. Jacques Derrida also, reflecting on human cloning, believes,

> Le fantasme de fabrication d'un être human (clonage reproductif) relève du scientisme, d'un imaginaire de la science don't il n'y a pas lieu d'avoir peur dans les circonstances actuelles. Même si techniquement une telle reproduction est possible, le statut du clone ne sera pas celui que l'on imagine aujourd'hui, précisément parce que, pour exister, un clone devra être un sujet et trouver une identité singulière. À cet égard, je pense que Freud aurat été passionné par les problèmes actuels. (*De quoi demain . . . Dialogue*, 93)[85]

Freud would unquestionably be fascinated with the idea of the clone, since it taps onto so many of the anxieties and repressed concerns with one's identity, as well as the fear of losing it to someone else, narcissism, the formation of the ego, and the discontents of contemporary civilization, which may have contributed to the attraction exerted by the dream of cloning and the misconception of potential immortality associated with it.

Just another biological machine

Slavoj Žižek, "Of Cells and Selves," 307

In addressing some of the philosophico-ethical problems posed by biogenetic cloning, namely the question of the prohibition (or not) of human cloning, Zizek gives the example of the Catholic church in order to analyze the arguments adduced as far as potential changes in personality are concerned. From some angles, it seems that these changes would deprive the individual of his or her unique dignity. Zizek takes the Catholic church's conservative response to cloning as illustrative of what he describes as the "paradox of prohibiting the impossible" (306):

> If the church effectively believes in the immortality of the human soul, in the uniqueness of human personality, in how I am not just the result of the interaction between my genetic code and my environs, then why oppose cloning and genetic manipulations? Are those Christians who oppose cloning not yet again involved in the game of prohibiting the impossible—biogenetic manipulation cannot touch the core of human personality, so we should prohibit it? In other words, is it not that these Christian opponents of cloning themselves secretly believe in the power of scientific manipulation, in its capacity to stir up the very core of our personality? (306–307)

For Zizek, this dilemma has centrally to do with the notion that, for many people, the use of genetic manipulation inevitably suggests the disturbing vision of the human being as a biological machine in a potential series, thus depriving human beings of their singular dignity. However, as Zizek retorts, "Why should we not endorse genetic manipulation and simultaneously insist that human persons are free responsible agents, since we accept the proviso that these manipulations do not really affect the core of our Soul?" (307).[86]

Zizek considers thus that what is at stake in cloning "is the very core of 'human freedom'" ("Of Cells and Selves," 307). He then goes on to engage with Jürgen Habermas's reflections on human cloning, in an essay published in the *Süddeutsche Zeitung*, where Habermas supported the prohibition of cloning, contending that cloning can be compared, to a certain degree, with slavery. Zizek summarizes and glosses Habermas's argument:

> In short, Habermas's argument is that what makes cloning ethically problematic is the very fact that my genetic base—which hitherto depended on the blind chance of biological inheritance—is at least partially determined by the conscious and purposeful (i.e., *free*) decision and intervention of another person: what makes me unfree, what deprives me of a part of my freedom, is, paradoxically, the very fact that what was hitherto left to chance (i.e., to blind natural *necessity*) becomes dependent on the *free* decision of another person. (307; emphasis original)

Zizek is unhappy with this, and I also see some problems with Habermas's line of argumentation: How can we say that a person's decision is "free" when it is almost certainly structured by economic and social reasons? And why is that person's decision, leaving aside those imperatives, more "free" than the decision made by two people to have a baby?

As Žižek points out, does Habermas's solution "not imply that a certain minimum of ignorance is the condition of our freedom?" ("Of Cells and Selves," 308). Are we just "biological machines" (308), fully determined by our genome, or are there other factors that intimately interact with our genes, in which case genetic manipulation would not substantially affect the core of human identity? As Zizek forcefully argues in a later essay, "Bring Me My Philips Mental Jacket," "contrary to Habermas, we should take the objectivisation of the genome fully on board. Reducing my being to the genome forces me to traverse the phantasmal stuff of which my ego is made, and only in this way can my subjectivity properly emerge" (5). It is precisely the problematization of these questions in fictional form that I will investigate in the next chapters.

BEYOND SEX AND DEATH: CLONING AND ITS DISCONTENTS

Implicit in a reflection on human cloning is a meditation on the fantasy of immortality, a fallacious dream, since one's clone is not oneself, a popular misconception that is the source of intense confusion, on the one hand, but also the "pleasures" of death, on the other. However, with the advent of cloning the inexorability of death as we know it might cease to be the necessary ending to life on Earth, for in the distant future, technologies[87] could be developed to allow the cloning of a rejuvenated body in artificial wombs, so that one's brain could then be transplanted to one's new body.[88] In this sense, but also in other scenarios the popular imagination has conjured up, the possibility of cloning humans would necessarily extend into new, unexamined areas the theorizing of the death drive posited by Freud. Freud's famous contention that "the aim of all life is death" ("Beyond the Pleasure Principle," 46) might stand in need of rephrasing, given the advent of the new technology of cloning.

Freud considers Eros and Thanatos as the two main drives controlling human beings and their behavior, an aspect that Lacan took up and continued to engage with throughout his work. The life-drive, a form of the pleasure principle, which can be seen as one of the driving forces behind the impulse to develop human cloning, would appear to be diametrically opposed to the death-drive, one of the most complex of Freud's concepts.[89] This death wish can be described as a longing for a state of undifferentiation or entropy and is closely related to the instinct Freud described in "Beyond the Pleasure Principle" as the primary human drive, that toward a condition of inorganicism. As Freud postulates, the pleasure principle is related to "the most universal endeavour of all living substance—namely to return to the quiescence of the inorganic world" (336). As Rosemary Jackson explains, "Freud sees it as the most radical form of the pleasure principle, a longing for Nirvana, where all tensions are reduced. This condition he termed a state of *entropy*, and the desire for undifferentiation he termed an *entropic* pull, opposing entropy to energy, to the erotic, aggressive drives of any organism" (*Fantasy: The Literature of Subversion*, 73). While human beings can be said, from a Freudian perspective, to long for this state of inorganicism where life's tribulations are ended, which functions as a redemptive form of pleasure principle that touches on the death drive, coloring the latter with a redeeming and consolatory quality, the fantasy of human cloning can be seen as providing a kind of metaphorical immortality, an imaginative escape and extension to the closure spelled out by death, whose unavoidability can thus symbolically be regarded as potentially circumvented.

Engaging with Freud's and Walter Benjamin's theories, Terry Eagleton brings them together in a thought-provoking way. Noticing Benjamin's considerations about commodity fetishism "as succumbing to the sex appeal of the inorganic" (*Walter Benjamin: Or Towards a Revolutionary Criticism*, 34), Eagleton describes death as "the ultimate aura, in which the organism can at last discover secure refuge from the shocks that batter it. . . . Eros, in Freud's metapsychology, is in the service of Thanatos, or the death drive, in that it continually seeks a discharge of stimuli that anticipates the quiescence of death" (34). Eagleton gives the example of fashion, calling it, in a Benjaminian register, "the supreme cult of the commodity" (35) in the presence of which "we are in the presence of death—of a hectic repetition that gets precisely nowhere, a flashing of mirror upon mirror that believes that by thus arresting history it can avoid death, but in this orgy of matter succeeds only in being drawn more inexorably into its grasp" (35). It is in this sense of a mirror repetition, which can be seen as putting a symbolic end to death in its perpetual drive to duplication and reproduction of the same, that cloning can be crucially inscribed.

Otto Rank links the concepts of death, narcissism, the double, and the soul in ways germane to this discussion. In *The Double: A Psychoanalytic Study*, Rank argues that "only with the acknowledgement of the idea of death, and of the fear of death consequent upon threatened narcissism, does the wish for immortality as such appear . . . the primitive belief in souls is originally nothing else than a kind of belief in immortality, which energetically denies the power of death; and even today the essential content of the belief in the soul . . . has not become other, nor much more, than that" (84).[90] In words that could be applied to some of the putative symbolic and philosophic ramifications of the notion of human cloning, Rank goes on to state that "the thought of death is rendered supportable by assuring oneself of a second life, after this one, as a double" (85), as is the case with the fantasy of living a second life through one's clone.[91]

Linda Ruth Williams is another who establishes the connection between narcissism and the death drive when she notes that already in 1914, long before Freud had developed any cohesive theory about the death drive, his essay "On Narcissism" already makes that connection in embryonic form. As Williams remarks, the essay "offered a sexual model which turned erotic impulses back onto the self in a circular form of desire which needed no external object. Narcissism is important in this genealogy of the death drive in that it shows sexual drives being directed toward the self, with the ego as object, which revealed something of the drive's general will to return" (*Critical Desire: Psychoanalysis and the Literary Subject*, 157).[92]

Baudrillard also links sexuality with the death drive, as well as with cloning, when he muses, "What, if not the death drive, would push sexed beings to regress to a form of reproduction prior to sexuation" ("Clone Story," 96), since, as he goes on to ponder, "isn't this form of scissiparity, this reproduction and proliferation through pure contiguity that is for us, in the depths of our imaginary, death and the death drive—what denies sexuality and wants to annihilate it, sexuality being the carrier of life, that is to say, of a critical and mortal form of reproduction?" (96). In "The Clone or the Degree Xerox of the Species," he again takes up many of the themes he had already addressed as far as the subject is concerned, namely, sex and death. Baudrillard polemically sees the advent of human cloning as putting an end to sexual reproduction and, by extension, to sex, as well as to death. Baudrillard sees cloning as potentially replacing "sexual reproduction and the inalienable death of every individual being" (196) with "the biological monotony of the earlier state of affairs, the perpetuation of a minimal, undifferentiated life, for which we have perhaps never stopped yearning" (196–197), something that can be equated with the impulse behind what Freud termed the death drive. As Baudrillard controversially continues, "The paradox of cloning is that it will produce and reproduce beings which are still sexed, whereas the sexual function has become absolutely useless" (197), thus implicitly suggesting that there can be no sex that is not directed toward procreation, a ludicrous statement. Baudrillard's ironic and pessimistic vision of the contemporary state of humanity is summed up in his pronouncement that "immortality and perfection achieved through genetic engineering and cloning will only ever be the immortality and perfection of the average formula," calling this putative state of events "virtual eugenics" (198).

In tune with Baudrillard, Julia Kristeva believes that we are moving toward a future that will efface sexual difference, in which case, as she observes, two events will occur. The first has to do with the fact that "we're going to witness the end of a certain kind of desire and sexual pleasure. For, after all, if you level out difference, given that it's difference that's desirable and provokes sexual pleasure, you could see a kind of sexual anesthesia, and this in an incubator society where the question of reproduction will be posed by way of machines and bioscientific methods in order for the species to continue" (*Interviews*, 126). Kristeva, like Baudrillard, finds this prospect "extremely troubling, first, for the individual's psychic life whose leveling off rules out desire and pleasure, and second, for the individual's creative possibilities. What can that individual invent that's new, surprising, or evolving? By moving toward this sexual homeostasis, won't we see some sort of symbolic homeostasis and therefore very little creation?" (126–127). This is indeed what happens in Kate Wilhelm's *Where Late the Sweet Birds Sang*, where, as one of the characters, Mark, forcefully notes, the cloned people do not seem to be able to create anything, they can only imitate.

The implementation of human cloning, then, might arguably bring about not just sexual homeostasis, as Baudrillard and Kristeva suggest, but also the fantasy of a return to a state of entropy, as postulated by Freud in his theory of the death drive, or an "eternity of the same," in Baudrillard's terminology ("Clone Story," 95).

> "Am I a man or just a potential clone?"
>
> Jean Baudrillard,"Transsexuality," 24

Cloning as a metaphor suggesting doubling, copying, and reproducing has entered the popular vocabulary, particularly conspicuously in the area of advertising. As Baudrillard comments, "It is culture that clones us, and mental cloning anticipates any biological cloning" (*The Vital Illusion*, 25). Indeed, as Baudrillard derogatorily remarks, "Through school systems, media, culture, and mass information, singular beings become identical copies of one another" (25). This is the effect Baudrillard describes as "social cloning, the industrial reproduction of things and people," as well as "the cloning of human conduct and human cognition" (25), which is given such vivid dramatization in, for instance, *Brave New World*: "Standard Gammas, unvarying Deltas, uniform Epsilons. Millions of identical twins. The principle of mass production at last applied to biology" (7).

Not only do these visions conjure up fears, horrified contemplations of the future, but to a certain extent they have already begun, for Baudrillard goes on in crescendo: "Cloning, the *écriture automatique* of 'ready-made' individuals and their identification with a minimal formula (their mental, behavioral code), their unthinking inscription in the operational networks—all these things are already largely achieved. The clones are already there; the virtual beings are already there. We are all replicants!" (*The Vital Illusion*, 199), a scenario illustrated in such dramatically graphic detail in Huxley's *Brave New World*.

Is standardization, then, inexorably extending from consumer culture to a kind of class-structured human uniformity and, paradoxically, to elitist designer-humanity, where wealth and narcissism prescribe the continuation of one's genes in the form of cloned children? These are dystopian visions illustrated in many of the highly cautionary tales that constitute the object of analysis in this book well before the technology became feasible, alerting us to novel societal configurations that need to be prefigured and theorized before we are unwittingly caught up in them and they are upon us.

As I hope has become apparent in this discussion, the related concepts of narcissism, doubles, copies and simulacra, mirrors, and the dream of immortality are crucially implicated in an examination of the psychological and philosophical contours associated with the idea of human cloning. In the following chapters I go on to analyze these recurring themes, together with the intricately related topics of identity, individuality, and duplication, as well as a profound concern with origins and freedom, which are inextricably entwined with the concept of human cloning, as they appear dramatized in the texts I have selected.[93]

Naturally, these texts have not been directly informed by the type or range of the theoretical material I have introduced here. Nonetheless, they constitute complementary discourses. While the critical analyses of the implications of cloning and related issues in psychoanalytic and philosophical reflections have a sophistication that is often daunting, the mediation of the same issues in fictional texts possesses an imaginative richness that contributes to our appreciation of the dilemmas involved in ways that the former does not. Moreover, the fictional texts participate in a tradition of speculative analysis that is more practiced at following the implications of emerging technologies in directions that are surprising, disturbing, or just plain overlooked in other discourses. This speculative and imaginative intelligence is the main focus of the chapters to follow; yet in order to better configure some of the profound issues that are at stake, it was thought necessary to undertake the foregoing survey. More than that, what has been introduced here will often be referred to in the remainder of the book. By foregrounding the argument here, we are now better equipped to enter their more specific resonances, beginning appropriately enough with one powerful site of beginnings—motherhood itself.

NOTES

1. "In this respect [the question of the identity of clones], I think that Freud would have been passionately interested by these contemporary issues" (my translation).

2. Brian Stableford describes Taylor's book as a "frantically alarmist work which declared that the dawning era of biotechnology would be utterly horrific" ("Biotechnology and Utopia," 199).

3. See, for instance, Harold B. Segel, *Pinocchio's Progeny: Puppets, Marionettes, and Robots in Modernist and Avant-Garde Drama* (1995); Susan Stewart, *On Longing: Narratives of the Miniature, the Gigantic, the Souvenir, the Collection* (1993).

4. Analogously, Christian Moraru considers whether the technique of rewriting can similarly be said to "bear witness to an entire *Zeitgeist*" (*Rewriting: Postmodern Narrative and Cultural Critiques in the Age of Cloning* (8) and whether it is "a sign of postmodernity in which feverish reproduction of 'originals' ends up reconstructing—or rather 'deconstructing,' à la Foucault (*The Order of Things*, 322)—the repetitive nature of every 'origin?'" (*Rewriting: Postmodern Narrative and Cultural Critiques in the Age of Cloning*, 8).

5. Baudrillard repeatedly acknowledges that cloning "holds great fascination for us" ("Prophylaxis and Virulence," 70). The most important pieces Baudrillard wrote about the subject of cloning are "Clone Story," "The Hell of the Same" (a version of "Clone Story" with a different ending), and "The 'Ludic' and Cold Seduction," although references to cloning abound everywhere in his work.

6. Baudrillard's essay "Clone Boy" first appeared in *Traverses* 4, then in Z/G 11 (1986), an issue devoted to cloning. It was later included in *Simulacra and Simulation* as "Clone Story" and then, with a different ending, in *Seduction*.

7. In a seemingly derogatory manner Baudrillard goes on to add that "one still has to use the uterus of a woman, and a pitted ovum, but this support is ephemeral, and in any case anonymous; a female prosthesis could replace it" ("Clone Story," 6). It is as if the only scenario Baudrillard is contemplating with the implementation of human cloning would be the cloning of males.

8. For a discussion of the biological dimensions of human cloning and the redefinition of family relations it might give rise to, see Glenn McGee, *The Perfect Baby: Parenthood in the New World of Cloning and Genetics* (2000); Gregory Stock, *Redesigning Humans: Choosing Our Children's Genes* (2002); Ronald M. Green, "I, Clone" (2002).

9. As Scott Durham points out, drawing on the work of Pierre Klossowski, this second concept of the simulacrum, like that already found in Plato, "can also be traced back to antiquity, perhaps most strikingly to the attacks on pagan simulacra by Tertullian and Augustine. . . . For these Fathers of the Church, the simulacrum is not merely the copy of a copy that has ceased to resemble

its original. It is also the mask of an evil *simulator*, a diabolical actor who, by repeating a familiar image, assumes another's identity as the mask of malign intentions" (*Phantom Communities: The Simulacrum and the Limits of Postmodernism*, 9).

10. For a comparison between Plato's three orders of representation and Baudrillard's corresponding categories, see Julian Pefanis, *Heterology and the Postmodern: Bataille, Baudrillard and Lyotard*, 59–61. As Pefanis notes, Baudrillard's position seems to be closer to Nietzsche's than to Plato's, since both Nietzsche and Baudrillard consider that "the sham world is the real world, whereas the real world is the world of illusion" (60–61).

11. However, as Charles Levin points out, "If we are to believe Baudrillard, then we are already well into the fourth order of simulation, the viral or carcinogenic phase, whose narrative Baudrillard began with 'Clone Story' in the mid-1970s" (*Jean Baudrillard: A Study in Cultural Metaphysics*, 54).

12. See Walter Benjamin, *Charles Baudelaire: A Lyric Poet in the Era of High Capitalism*.

13. As John Lash in a related vein observes, "DNA, the double helix, is perhaps the ultimate Twin of our modern mythology" (*Twins and the Double*, 31).

14. Brian Stableford claims with missionary fervour that "the time has come ... for every thinking person to become a fervent and uncompromisingly extravagant propagandist for the only kind of activity that is truly, fundamentally and definitively human and the last, best hope for the future improvement of our lot: the Cinderella of speculative thought, biotechnology" ("Biotechnology and Utopia," 202).

15. Most clone narratives can be inscribed in this neocapitalist cybernetic logic of total control. Extreme examples of this logic would be Aldous Huxley's *Brave New World* and the film *Gattaca* (Andrew Nicol, 1997, US).

16. As Charles Levin perceptively notes, commenting on Baudrillard's analyses of the digital and genetic code as the grounding myths of our society, defined by the ascendancy of simulation, "It is as if the the opening moves of classical philosophy are being played out in a kind of epistemological endgame. Plato saw imitation as a destabilizing corruption of the norm, and Aristotle saw it as the central motif of learning. They have both been vindicated in the modern problematic of simulation" (*Jean Baudrillard: A Study in Cultural Metaphysics*, 53).

17. Including "production of the Other," as Baudrillard suggests and elaborates on (*Figures de l'altérité*, 169; my translation).

18. Baudrillard said of Benjamin that he "is someone whom I admire deeply" ("Revenge of the Crystal: An Interview with Jean Baudrillard by Guy Bellavance," 21).

19. Terry Eagleton, however, problematizes Benjamin's discussion of origins and auras when he states that "such persistence of the origin is ideological delusion.... Mechanical reproduction ... destroys the authority of origins, but in doing so writes large a plurality that was there all along" (*Walter Benjamin, or, Towards a Revolutionary Criticism*, 33).

20. Even as I am writing this, the mother of the first alleged clone baby, Eva, born on December 26, 2002, in the United States, is refusing to allow DNA tests to be carried out, tests that would prove or disprove the "authenticity" of Eva as her mother's clone.

21. As Terry Eagleton notes, "The mechanical reproduction of an object, which for Benjamin undermines its unique aura, promises to undo a fetishism that finds its highest form in, precisely, repetition. Jacques Lacan has reminded us that in Freud's texts repetition (*Wiederholen*) is never reproduction (*Reproduzieren*); and in Benjamin's writing they might be said to be antithetical" (*Walter Benjamin, or, Towards a Revolutionary Criticism*, 28).

22. Apart from groups of clones.

23. In connection with this, Scott Durham notes the "increasing domination of everyday experience by mass-produced simulacra, which effectively undercut in advance any notion of an original referent that would precede its reproduction" (*Phantom Communities: The Simulacrum and the Limits of Postmodernism*, 52). As Durham goes on to point out, from a Baudrillardian perspective, "the notions of 'original' and 'originality' have ceased to bear the weight of any epistemic, political, or aesthetic authority" (52). This leads into Baudrillard's reflections on human cloning, where he controversially inscribes clones within this nexus of simulacra, of copies without an original.

24. The Platonic simulacrum, as Scott Durham explains, "repeats only the external appearance of the icon without itself participating in the Idea that founds it" (*Phantom Communities: The Simulacrum and the Limits of Postmodernism*, 7).

25. In "The Double Session," Jacques Derrida problematizes the question of mimesis, engaging with Plato's hierarchies of original and copy, imitator and imitated, as well as with Mallarmé's drama *Mimique*, deconstructing the ideas of representation and binary thinking in the Western world. For a discussion of the theory of doubling and duplicity in the work of Jacques Derrida, as well as its wider implications, see Rodolphe Gasché, *The Tain of the Mirror: Derrida and the Philosophy of Reflection* (1986) and John Sallis, "Doublings" (1992).

26. Commenting on this passage, John Sallis argues that "determined as (merely) a forgetting of a simple origin, this mechanism by which the double splits, and thus redoubles, that of which it is the double—this doubling operation is also determined by Saussure as catastrophe or monstrosity" ("Doublings," 123), strong words that some observers are keen to apply indiscriminately to clones.

27. As Scott Durham observes, Deleuze suggests that "simulation not only provides a means of resistance to the dominant powers but also permits an emergent collective, through the displaced repetition of the figures of the dominant, to express its own power to create a new sensibility and form of life no longer legitimized by or grounded in the dominant truths" (*Phantom Communities: The Simulacrum and the Limits of Postmodernism*, 20). It is tempting to see in the image of this future community envisaged by Deleuze a form of society that would include clones, who might be seen as initiating the new sensibility heralded by Deleuze in the era of simulation.

28. According to Hal Foster, for Plato "images were divided between proper claimants and false claimants to the idea, between good iconic copies that resemble the idea and bad fantasmatic simulacra that insinuate it" (*Postmodern Culture*, 97). However, Foster further notes that, "as Gilles Deleuze has argued, the Platonic tradition repressed the simulacrum not simply as a false claimant, a bad copy without an original, but because it challenged the order

of original and copy, the hierarchy of idea and representation" (97), thus paving the way for what Baudrillard termed the "revenge of the simulacrum" ("The Orders of Simulacra," 95).

29. These concerns are powerfully dramatized in Eva Hoffman, *The Secret: A Fable for Our Time* (2001).

30. An example that comes readily to mind is Ridley Scott's 1982 film *Blade Runner*, based on Philip K. Dick, *Do Androids Dream of Electric Sheep?* (1968).

31. As Rosemary Jackson notes, however, "Recent studies of the double in literature have acknowledged its shift in the romantic period from a supernatural motif into an increasingly self-conscious psychological function" ("Narcissism and Beyond: A Psychoanalytic Reading of *Frankenstein* and Fantasies of the Double," 44).

32. Rank's "Der Doppelgänger" was first published in 1914 and subsequently expanded into book form in 1925.

33. The trope of the double as a rival in sexual love is dramatized in, for instance, Pamela Sargent, *Cloned Lives*, where the cloned brothers vie for the love of their sister.

34. Rosemary Jackson comments on the same issue, "The production of a reflection or of multiple selves defends against a fear of not-being" ("Narcissism and Beyond: A Psychoanalytic Reading of *Frankenstein* and Fantasies of the Double," 48).

35. For Mladen Dolar, the double always represents a figure of *jouissance* ("I Shall Be with You on Your Wedding Night," 13–14). As Ruth Parkin-Gounelas, engaging with Dolar's theory, explains the double, "On the one hand, it enjoys at our expense, commiting acts we would otherwise (or rather would only ever) dream of. On the other, it does not simply enjoy but rather *commands* enjoyment, forcing us into a position of servitude to our appetites. . . . Its uncanniness, therefore, stems from the same compulsion to repeat which the subject is powerless to resist, as many literary works demonstrate" (*Literature and Psychoanalysis: Intertextual Readings*, 110).

36. Cloning could also be seen as the making of a pact with the devil, a Faustian pact to achieve knowledge, youth, beauty, immortality, as is the case in, for instance, Oscar Wilde's *The Picture of Dorian Gray* (1891) or Emma Tennant's *Two Women of London* (1989) and *Faustine* (1992).

37. See also Karl Miller, *Doubles: Studies in Literary History*.

38. This "constant recurrence of the same thing" (Freud, "The Uncanny," 356), identified by Freud as uncanny, can also be related to Nietzsche's concept of the "eternal return." See Nietzsche, *Thus Spake Zarathustra*, for a discussion of this concept.

39. Fictional examples where this kind of encounter is staged include, for instance, Gene Wolfe, *The Fifth Head of Cerberus: Three Novellas* (1972), Lisa Tuttle, "World of Strangers" (1998), and Eva Hoffman, *The Secret: A Fable for Our Time* (2001). Henry James, *The Sense of the Past* (1917) constitutes another example of the search for and confrontation with someone from the past. In Disney's *The Kid* (1999), the main character, played by Bruce Willis, comes across a child who is himself, while later both characters see how they will look in the future, encountering their older selves.

40. As Michael Arnzen observes, Baudrillard's "'simulacrum' shares much with traditional theories of the 'double'" ("The Return of the Uncanny," 316).

41. What happens when one's "double" effectively materializes, as in cloning? In "Political Economy and Death" Baudrillard reflects at length on the complex status of the double, tracing the moment when, from the "non-alienated" (141) relation the "primitive" supposedly has with his or her double, "the primitive thought of the double as continuity and exchange is lost, and the haunting double comes to the fore as the subject's discontinuity in death and madness" (142).

42. As David Lomas in an analogous vein asserts, "Mirror reflections are a case of the more general phenomenon of the double, which frequently has associated with it an uncanny and malevolent aspect" (*The Haunted Self: Surrealism, Psychoanalysis, Subjectivity*, 96).

43. For a discussion of the shadow and the soul as tropes for the double, see Otto Rank, *The Double: A Psychoanalytic Study*.

44. The fantasy of encountering one's double constitutes both a dream and a nightmare. Dante Gabriel Rossetti gives strong visual expression to this most uncanny experience in *How They Met Themselves* (1851–1860), a small watercolor that represents the terror experienced by two lovers who meet their doubles in a dark wood, an encounter, as Jan Marsh notes, "symbolic of impending death, and an ominous nuptial image" (*Pre-Raphaelite Women: Images of Femininity in Pre-Raphaelite Art*, 141). With cloning, then, the double can be said to have returned to haunt us.

45. In "The 'Ludic' and Cold Seduction" Baudrillard considers the clone the "hypostasis of the artificial double . . . your guardian angel, the visible form of your unconscious and the flesh of your flesh" (173). The clone is thus, in Baudrillard's view, a projection of deep-rooted narcissistic dreams, a benign manifestation of the double, who is nonetheless an artificial one, generated with recourse to a genetic code.

46. John Lash goes on to maintain that "the greater challenge, perhaps, is simply to face ourselves in full acceptance of both our lovable and detestable traits" (*Twins and the Double*, 44).

47. Identity is a very troubled, complex issue that touches on all the subtopics I briefly address in this chapter, which, unfortunately, I will not be able to develop as it deserves for lack of space.

48. For Baudrillard, in contentious vein, "Identity is a dream pathetic in its absurdity. You dream of being yourself when you've nothing better to do. You dream of that when you've lost all singularity" (*Paroxysm: Interviews with Philippe Petit*, 49).

49. According to David Elliott, "It is a mistake to believe that human cloning would result in an objective loss of uniqueness/individuality for the clone. The reason is that if this claim were true, some very strong kind of biological determinism would also have to be true. For the only thing we are really considering in cloning humans is the duplication of an individual's genome (and perhaps this too is only partial). So the only way that we could duplicate a person by duplicating that person's genome would be if most, if not all, of what makes up a person comes from the genes. Although there are exceptions, most biologically inclined psychologists seem to resist this idea, and posit that environment plays some role in shaping personality" ("Uniqueness, Individuality, and Human Cloning," 111). Elliott further observes, "If cloning is morally objectionable because it would produce nonunique human beings,

then it must be the case that identical twins are either (1) less morally valuable or (2) somehow worse off than the rest of us" (114).

50. A suspicion Baudrillard often articulates, as for instance in *The Vital Illusion*, 21.

51. A point illustrated and at least partially revised in Eva Hoffman, *The Secret*.

52. For a discussion of the question whether a human clone would cause the nature of personal identity to be called into question, see Ina Roy, "Philosophical Perspectives," 61–62.

53. As Wilmut, Campbell, and Judge further develop this point, "For two reasons then—recurrent mutation and variable expression—even two *identical twins*, formed by splitting an embryo, may be genetically different, physically and perhaps functionally. The differences are liable to be slight, but they can be obvious, nevertheless" (*The Second Creation: The Age of Biological Control*, 301).

54. In "The Clone or the Degree Xerox of the Species" Baudrillard declares, provocatively, that "each of two twins, because he has a double, is ultimately just half an individual—if you clone him to infinity, his value becomes equal to zero" (199). This (unqualifiable) pronouncement is characteristic of Baudrillard's often scandalous, contentious, and offensive proclamations on the subject of human cloning.

55. As Hillel Schwartz maintains, "Doubleness has become an inescapable element of modernity; . . . for some, its very definition" (*The Culture of the Copy*, 87).

56. Quoted in Juliet Flower MacCannell's Introduction to *Thinking Bodies*, 7.

57. See Julia Kristeva, *Tales of Love*.

58. With the prospect of human cloning, a different ending to the myth of Narcissus could be imagined, one in which he did not drown in the waters of the lake, embracing Thanatos, and subsequently turn into a flower. Given the knowledge that he could clone himself, Narcissus would not presumably have felt the imperious need to kill himself out of desperation in the face of the prospect of never finding someone as beautiful as himself that he could love.

59. As Gray Kochhar-Lindgren observes, the myth of Narcissus can suggest "the inability to acknowledge the other (as the lover or as the world)" (*Narcissus Transformed: The Textual Subject in Psychoanalysis and Literature*, 3), a topic that is centrally implicated in an investigation of human cloning.

60. Gray Kochhar-Lindgren sums up this contemporary trend when he writes about the "entombing, rigidifying gaze of the selfsame that is so dear to the heart of Narcissus and to Western culture" (*Narcissus Transformed: The Textual Subject in Psychoanalysis and Literature*, 1).

61. Here we can counterargue that uniformization has always existed in every society, and arguably even more in so-called primitive ones.

62. Still with reference to the theme of Narcissus and resemblance, Melchior-Bonnet notes how "Narcissus's transgression . . . stems from the fact that he was ignorant of his soul and its resemblance to God. Flattered by his corporeal form, he neglected true beauty in order to follow a reflection, thereby condemning himself to grasp but a simulacrum that would never fill the aspirations of his soul" (*The Mirror: A History*, 112). Indeed, "Dante places Narcissus in Purgatory with the counterfeiters, guilty of being content with the false currency of the tangible appearance" (113).

63. Quoted in Joseph Kestner, "Narcissism as Symptom and Structure: The Case of Mary Shelley's *Frankenstein*," 69.

64. As Ruth Parkin-Gounelas pertinently observes, "Identity is simply a role" (*Literature and Psychoanalysis: Intertextual Readings*, 8).

65. Quoted in Joseph Kestner, "Narcissism as Symptom and Structure: The Case of Mary Shelley's *Frankenstein*," 70.

66. For Melchior-Bonnet, "Identity can . . . only be grasped through vanity" (*The Mirror: A History*, 166).

67. As Rodolphe Gasché in analogous fashion declares, reflection constitutes the only means "by which an ego can engender itself as a subject" (*The Tain of the Mirror: Derrida and the Philosophy of Reflection*, 14).

68. Kristeva's words can be seen as almost prescient of the predicament of clones in the future if the technique is implemented. Indeed, many people look upon clones erroneously as almost alien beings, fabricated in a laboratory environment—extraterrestrials to be feared.

69. Kathleen Woodward argues that "one of Lacan's most brilliant contributions to psychoanalysis is his insistence on the illusion of wholeness of our bodies" even though, as she also maintains, "old age finds an end in the fragmentation of the body, in the return to the body in parts" (*Aging and Its Discontents: Freud and Other Fictions*, 189).

70. Alienation and division, as well as projection, can thus be said to stand at the heart of ego formation. Indeed, upon realizing that the image in the water is the representation of himself, Narcissus, according to Kristeva, "discovers in sorrow and death the alienation that is the constituent of his own image" (*Tales of Love*, 121), a point also made by Rosemary Jackson, who asserts that "alienation is at the heart of identification" ("Narcissism and Beyond: A Psychoanalytic Reading of *Frankenstein* and Fantasies of the Double," 46). Christopher Horrocks similarly intervenes in this debate introducing the term "neo-individualism," which refers to social being "after the age of bourgeois individualism, which once forged subjectivity through a battle between free will and alienation" (*Baudrillard and the Millennium*, 51). In a Baudrillardian mode, Horrocks notes how "alienation is a luxury long gone, as it allowed the subject to imagine it could be other than what it was. Alienation has, however, now succumbed to 'identity logic': the millenial [*sic*] subject has become the *same as itself* and only related to others through *difference* rather than radical otherness" (51).

71. Engaging with Lacan, Juliet Flower MacCannell observes that "one pole of this split ego is located in a fragmented body 'before the mirror' . . . and its *alter Ego*—a synthetic, ideal, and whole unfractured Other, 'beyond the mirror'" (Introduction, 7). In some ways, the clone might be seen as embodying this other unfractured alter ego, or ego ideal, which, as Julia Kristeva points out, is "a revival of narcissism" (*Tales of Love*, 22).

72. In a related vein, Baudrillard, in *L'autre par lui-même—Habilitation*, ponders the dualism of reflections between the subject and the ego ideal, constituted as an object, in what might be seen as a mirror game where the subject becomes his or her own libidinal selection of object: "But there is, above all, in the subject itself, the passion to be object, to become object—enigmatic desire whose consequences we have hardly evaluated, in all domains, political, es-

thetic, sexual—lost as we are in the illusion of the subject, of its will and representation" (80; quoted in Francette Pacteau, *The Symptom of Beauty*, 184).

73. Kristeva, however, redefines this pre-Oedipal stage, prior to the mirror stage, as the "semiotic," as a linguistic and poetic phase where the child experiences close fusion with his or her mother.

74. Nietzsche, for his part, with his theory of the eternal return, opens up new possibilities for the splintered, fractured self in the figure of Narcissus, whose metamorphosis points to an endless renewal of life and to an alternative path, which consists in the reinvention of the self as other, as someone else (his or her clone?).

75. Freud used the term "primal phantasy" for the first time in his 1915 essay "A Case of Paranoia Running Counter to the Psychoanalytic Theory of the Disease," 154. In his *Introductory Lectures on Psychoanalysis* Freud argues that "among the occurrences which recur again and again in the youthful history of neurotics . . . there are a few of particular importance . . . observation of parental intercourse, seduction by an adult and threat of being castrated" (416).

76. As Hal Foster explains, they were first called scenes, and only later were they termed "fantasies, when it became clear that they need not be actual events to be psychically effective—that they are often constructed, in whole or part, after the fact, frequently with the collaboration of the analyst. Yet, though often contrived, these fantasies also tended to be uniform; in fact, the narratives appeared so fundamental that Freud deemed them phylogenetic: given schemas that we all elaborate upon" (*Compulsive Beauty*, 57–58).

77. Already in 1978, in an article called "The Waning of the Oedipus Complex," Hans Loewald reflected on the weakening of Oedipal patterns in contemporary society, as well as on the "decline of psychoanalytic interest in the oedipal phase and oedipal conflicts and the predominance of interest in research in preoedipal development, in the infant–mother dyad and issues of separation–individuation and of the self and narcissism" (386). In turn, I believe that we will witness in the twenty-first century a further weakening and gradual dissolution of Oedipal structures, while increasingly greater attention will be devoted to the examination of non-Oedipal configurations, prompted by the refashioning of the nuclear family in response to the implementation of new reproductive technologies, namely human cloning, as well as novel (de)formations that exacerbated models of consummer society will give rise to. As Ruth Salvaggio in a related vein observes, when "attention is directed away from the father and son and toward the mother" and the ideas derived from such a shift in focus "come to inform and influence psychoanalytic theory, then psychoanalysis becomes engaged in the process of producing something other than Oedipus" ("The Case of Two Cultures: Psychoanalytic Theories of Science and Literature," 61).

78. These are addressed in Chapters 2, 3, and 7.

79. As Douglas Kellner notes with reference to a clone child, summarizing the gist of Baudrillard's argument, "no longer caught up in the oedipal drama of the family triangle (or circle), [it] in effect engenders itself, rather than being a product of 'the duel' between the genetic information of the sperm and the egg and between the mother's and father's personal influence" (*Jean Baudrillard: From Marxism to Postmodernism and Beyond*, 101).

80. Here, as in many other passages where he writes about cloning, Baudrillard seems to engage in wild extrapolation and exaggeration for the sake of sensationalism.

81. David Hillman and Carla Mazzio, in the introduction to *The Body in Parts: Fantasies of Corporeality in Early Modern Europe,* draw on a reading of Lacan's mirror stage and the presymbolic chaos that characterizes the child's sensations prior to that phase to articulate the possibility that "the tension between body parts and corporeal wholes lies at the very heart of social and symbolic structuration" (xvii).

82. George Steiner, engaging with some of the issues raised by Gordon Rattray Taylor in *The Biological Time Bomb,* such as in vitro fertilization and recombinat DNA, believes that humanity has no other choice but to carry on with this kind of research, in spite of the potential dangers for society it may harbor ("Has Truth a Future?" *The Listener,* 12 January 1978, 42–46, quoted in Turney, *Frankenstein's Footsteps: Science, Genetics and Popular Culture,* 210).

83. Roger Shattuck starts his book *Forbidden Knowledge: From Prometheus to Pornography* with a probing question that many have also been asking as far as the idea of human cloning is concerned: "Are there things we should *not* know? Can anyone or any institution, in this culture of unfettered enterprise and growth, seriously propose limits on knowledge? Have we lost the capacity to perceive and honor the moral dimensions of such questions?" (1).

84. The complex question of the relative weight played by genetic makeup versus the influence exerted by the environment on the development of a human being is dramatized in, for instance, Fay Weldon's *The Cloning of Joanna May* (see Chapter 6). See also Ian Wilmut, Keith Campbell, and Colin Tudge, *The Second Creation,* especially Chapter 13.

85. "The dread attendant upon the fabrication of a human being (reproductive cloning) is associated with a certain type of popular science, a certain science imaginary which there is no reason to be afraid of in our contemporary circumstances. Even if technically such a reproduction was possible, the status of the clone will not be such as one imagines today, precisely because in order to exist a clone must be a subject and have an individual identity. In this respect, I believe Freud would have been fascinated by today's problems" (my translation).

86. Joseph Fletcher, a Roman Catholic theologian, defends an increase of human freedom precisely through control of human reproduction. As the National Bioethics Advisory Commission summarizes Fletcher's views, "He portrayed cloning of humans as one of the many present and prospective reproductive options that could be ethically justified by societal benefit. Indeed, for Fletcher, as a method of reproduction, cloning was preferable to the 'genetic roulette' of sexual reproduction. He viewed laboratory reproduction as 'radically human' because it is deliberate, designed, chosen and willed (Fletcher, 1971, 1972, 1974, 1979)" (Martha C. Nussbaum and Cass R. Sunstein, *Clones and Clones: Facts and Fantasies about Human Cloning,* 165–166).

87. In *Immortality: How Science Is Extending Your Life Span—And Changing the World,* Ben Bova explains some of the mechanisms involved in cloning that might help in prolonging life: "Almost every cell of your body holds, in its nuclear DNA, the blueprints for every part of you: your skin and your bones,

your brain and nervous system, every organ in your body. If only we could learn to use that information properly, we could grow new organs, limbs, skin, muscles, and any other part of the body when we need to. . . . Why clone another whole copy of you when what you really need is a new heart and a fresh pair of kidneys?" (161).

88. Damien Broderick mentions the possibility of cloning yourself mentally, that is, one might be able to "purchase a machine (or space in one) and 'xerox' yourself as an independent copy" (*The Last Mortal Generation: How Science Will Alter Our Lives in the 21st Century*, 215). Although this copy will not remain identical, since "the history of his or her consciousness would start to diverge from yours the moment the copy was complete . . . the two of you would be closer than twins" (215). Although this procedure would be an insurance against degeneration of the brain and the loss of one's memories, Broderick does not address the question of aging and general bodily deterioration.

89. For a discussion of the concept of the death drive in Freud, see, for example, Linda Ruth Williams, *Critical Desire: Psychoanalysis and the Literary Subject*, and Ellie Ragland, *Essays on the Pleasures of Death: From Freud to Lacan*, with a stronger emphasis on Lacan's reworking of the Freudian notion of the death drive.

90. As George Johnson remarks in "Soul Searching," "We each carry in our heads complexity beyond imagining and beyond duplication. Even a hard-core materialist might agree that, in that sense, everyone has a soul" (70).

91. As Rank further argues, "As in the threat to narcissism by sexual love, so in the threat of death does the idea of death (originally averted by the double) recur in this figure who, according to general superstition, announces death or whose injury harms the individual" (*The Double: A Psychoanalytic Study*, 85).

92. David Lomas, in a related vein, contends that the figure of Narcissus "encapsulates the seductive enticement of the death drive" (*The Haunted Self*, 181), an impulse to regress into a form of atavistic, almost vegetal repose, a state of entropy.

93. Given the enormous variety and number of novels and short stories dealing explicitly or more tangentially with the topic of cloning, I had inevitably to drastically select the examples I was going to use. My criteria had to do with how representative and topical they were, but personal preference also played a part when the coherence of the particular topic or cluster of works was not marred.

2

"This Sex Which Is One": Writing Men Out in Selected Science Fiction Texts

> Imagine parthenogenesis, virgin birth, as practiced by the greenfly, actually applied to human fertility.
>
> Shulamith Firestone, *The Dialectic of Sex: The Case for Feminist Revolution*, 182

In this chapter I wish to examine some texts that dramatize and reflect on the fantasy of parthenogenesis, which can be seen as a version of human cloning, depicting women-only worlds that constitute a powerful denunciation of women's position and roles in traditional androcentric societies, as well as a blueprint for improving human living conditions and transforming the world into a place where more egalitarian rules apply.

I confront and analyze Mary E. Bradley Lane's *Mizora* (1890), Florence Ethell Mills Young's *The War of the Sexes* (1905), Charlotte Perkins Gilman's *Herland* (1915), as well as two more recent texts, James Tiptree Jr.'s "Houston, Houston, Do You Read?" (1976) and Suzy McKee Charnas's *Motherlines* (1978), which describe fictional societies made up exclusively or predominately of women. In these narratives the traditional assumption that woman is the other of man is implicitly reversed and woman thus becomes the norm while men are constituted as her others. I will investigate this presupposition and its ramifications by exploring the main organizational and structuring principles that underlie these societies.

Alternative reproductive techniques conducive to distinct procreative and birth scenarios have always existed in myth and literature. From the biblical story of the creation of the first human beings, in Genesis, which features God the Father (Himself a perfect example of a self-engendering deity) creating Adam and Eve, bypassing any feminine intervention,[1] to the immense panoply of scenarios made potentially feasible with the advent of human cloning, visions of male reproduction circumventing women abound.[2] The opposite fantasy, that of female conception without men, appears to be just as extensive and encompassing, suggesting the existence of a deep-seated malaise with traditional gender roles allotted to women, which includes and stresses the generative capacity of women, a crucial distinguishing feature which—if it has served throughout the ages to give women a measure of importance as the "vessels" of life—has often been used to circumscribe women to the domestic sphere and thus restrict their roles outside that realm.[3]

Rosi Braidotti argues nothing less than that the "contemporary world is fascinated with parthenogenesis" (*Nomadic Subjects: Embodiment and Sexual Difference in Contemporary Feminist Theory*, 65). The fantasy of creating life on one's own, without the help of the opposite sex, has been a long-standing dream of humankind, nurtured by both women and men. *Mizora, Herland,* "Houston, Houston, Do You Read?" and *Motherlines* can be read as illustrations of the old fantasy of spontaneous generation of life, in this case the female fantasy of creating life without the help of a man. This dream found expression in parthenogenesis, or virgin birth, where women give birth without the intervention of a man, as is the case in these utopian worlds.[4]

Parthenogenesis is a fundamental theme in the texts I will be looking at here, since it would putatively allow women to dispense with men altogether in order to have female children. Indeed, a radical futuristic scenario which could be said to correspond to a logical, if apocalyptic, conclusion to the battle between the sexes[5] can be seen as taking two main forms: One is the eradication of one of the sexes, either by warfare (biological or otherwise), disease that affects one of them selectively, or the forceful, deliberate waging of war by one sex on the other; the other is the development of separatist societies.[6] These two main and at times overlapping scenarios can be seen at work in many of the novels under examination.[7]

The introduction and widespread use of parthenogenesis or human cloning could foreseeably, if exorbitantly, lead to one of these extravagant scenarios: a world peopled mostly by women or, alternatively, mostly by men.[8] Any of these radical dystopian scenarios could predictably come to pass if the technology necessary to carry out and implement the selection of the sexes and deliberate phasing out of one of them remained in the hands of scientists exclusively of one sex.

Indeed, the trope of the destruction of humanity on the face of the Earth, as a consequence of devastating diseases or ecocide, usually leaving only one person alive, has a long literary tradition, from Mary Shelley's *The Last Man* (1826) to Margaret Atwood's *Oryx and Crake* (2003).[9] One particular form it may take consists in the extermination of one of the sexes, leading to a single-sex world. Usually there is a cataclysmic event, such as a war or rampant disease that kills the majority of the population selectively, decimating only one sex. Typically, there is a man or woman left, who has the sole responsibility to perpetuate humanity as we know it. This is the case with Young's *The War of the Sexes*, where love intervenes to save the only man left in the British Isles. In Susan Ertz's *Woman Alive* (1935), in turn, all the women die of a mysterious illness except for one, who is eventually found and treated like a queen, for in her lies the only hope of survival for humankind.[10]

> Could you imagine a world of women only?
>
> Adrienne Rich, *The Dream of a Common Language: Poems 1974–1977*, 61

For Rosi Braidotti, "Parthenogenetic births are always a sign of the potentially lethal powers of the undomesticated female. This *topos* resurrects an ancient set of beliefs about the monstrous powers of the female imagination. . . . They simultaneously also express, however, men's sense of impotence and of increasing irrelevance" (*Metamorphoses: Towards a Materialist Theory of Becoming*, 196). Braidotti's words shed some light on the fantasy of women-only worlds, which constitute primarily a forceful political statement, a potent envisioning of a society where women would be defined in relation to a normative feminine standard and not according to a male one. These worlds can exist only in the realm of fantasy, of utopia, but the very impulse that led to their creation suggests a need to reexamine and reshape contemporary societal conditions.[11]

These utopian and science fiction societies I am analyzing here are centrally concerned with issues of organization,[12] with institutional arrangements and how they will in very practical terms affect women's daily lives and improve the quality of their existence by prioritizing the feminine. How a women-only society is organized as opposed to a predominantly patriarchal one is the fundamental operative question these texts thematize. The stress in these fictional women-only worlds falls decisively on organizational measures that emphasize the optimal management of resources, principally farming, as well as on community, on the sharing of property, services, and family roles. Joanna Russ usefully summarizes the fundamental points shared by most feminist utopias:[13]

They are explicit about economics and politics, fairly sexually permissive, demystifying about biology, emphatic about the necessity for female bonding, concerned with children . . . nonurban, classless, communal, relatively peaceful while allowing room for female rage and female self-defense, and serious about the emotional and physical consequences of violence. ("*Amor Vincit Foeminam*: The Battle of the Sexes in Science Fiction," 58)[14]

THE WAR OF THE SEXES

In Young's *The War of the Sexes*, set several centuries into our future, war and the use of parthenogenesis have combined to produce the extinction of men in the British Isles, leaving only one man alive, who is described as a "misogynist" (12). As the narrator reveals, "Some centuries back—the middle of the nineteenth century, most authorities put it—'Women's Rights' first came into fashion" (6). We further learn that "it was towards the end of the twentieth century that the subject of parthenogenesis, from being a matter of purely scientific discussion in the universities, developed from theory to fact; and man, himself, worked with extraordinary zeal in the matter of his own extinction" (8). As the narrator further explains,

The first to give practical proof of the working hypothesis of a theory which had been absorbing scientific interest for many years was Professor Loeb of Chicago University, who, at the close of the nineteenth century, is reported to have produced by the aid of certain chemicals the young of sea-urchins from the unfertile eggs of these hitherto unimportant marine animals.[15] Professor Loeb next experimented on starfish with the same success; the male of each kind being entirely unnecessary to the propagation of the species. By his own writings Professor Loeb shows that he entertained very little doubt that the theory would work equally well with mammalians, and that *parthenogenesis* would become general; but it was not for many centuries that this now familiar enough subject was brought to the state of perfection which we are privileged to know, and which has resulted, as before stated, in the comparative extermination of man. His complete extinction, so far as the British Isles were concerned, was brought about through war. (8–9; emphasis mine)

Indeed, there were only 432 men left in the whole of the United Kingdom, while at the end of the war there was a sole male survivor, Geoffrey Sterndale, a philosopher and scientist, described as a "misogynist" (12). However, later love intervenes, when Miss Julia Evans, a journalist whose job is to write a piece about him, and Sterndale fall in love with each other. In spite of his misogynism, Sterndale wrote two brilliant treatises on "Improved Parthenogenesis" and has been working on "improving the feminine species" (21), as Evans puts it in her interview to him. Sterndale's ironic answer further stresses his hatred of women: "There's so much room for improvement . . . that it appears to me the wisest thing to do with one's life, seeing that your sex is not likely to become extinct like mine" (21). In their time, many centuries into the future, Sterndale thinks of the women as "chemically-produced" (23), since they are "parthenogenetically produced" (33), describing them in scientific articles as the "imperfect production of scientific discovery" (33), to which Evans retorts, with great emphasis, that women are not "the production of science . . . but the natural result of evolution" (33).[16]

This kind of plot line dealing with the extermination of the male population and leaving a world peopled almost only or exclusively by women had already been used by Mary E. Bradley Lane in *Mizora* and can be said to prefigure Irene Clyde's *Beatrice the Sixteenth*, which portrays a female separatist society, as well as Charlotte Perkins Gilman's *Herland*. *Beatrice the Sixteenth* can be described as a utopian fantasy where gendered roles are rethought and almost totally abandoned. Linguistically it is also radically innovative, since, as the narrator explains, their language carries no words to distinguish male from female or husband and wife. As Angela Ingram and Daphne Patai point out, "From beginning to end the narrator eschews gendered nouns and the generic 'he' and instead refers to characters as 'figure,' 'person,' and 'personage'" ("Fantasy and Identity: The Double Life of a Victorian Sexual Radical," 266). Ingram and Patai further remark that "the narrative itself, on the other hand, is far from gender-neutral: it consistently and unabashedly values so-called feminine characteristics and, after the initial avoidance of gendered pronouns, reveals a predilection for 'she' and other feminine forms to refer to the Armerians, whose gender we actually don't know" (266).

In *Beatrice the Sixteenth*, then, Irene Clyde (who in fact is a man, Thomas Baty)[17] portrays a feminist and separatist society where the question of gender is seen as the fundamental obstacle to be transposed in order that an egalitarian modus vivendi can be effectively achieved, as well as a not male-centred sexuality. As Ingram and Patai maintain, in formulating this kind of critique of conventional gender ideologies, "Irene Clyde's work emerges as an important predecessor not only of recent feminist utopias that have challenged the notion of the generic male but also of current radical feminist perspectives, prefiguring, in particular, the work of Catharine MacKinnon and Andrea Dworkin" ("Fantasy and Identity: The Double Life of a Victorian Sexual Radical," 270). Interestingly, in *Alone in Japan: The Reminiscenses of an International Jurist Resident in Japan, 1916–1954* (1959), a memoir, the author of *Beatrice the Sixteenth* also engages with the theme of parthenogenesis, writing that "parthenogenesis is well within the range of scientific possibility" (196).

These visions of parthenogenetic reproduction, where the need for male intervention is abolished, can be seen as powerful symptoms of women's discontent with men's legal power over their wives and children, as well as deep-seated evidence of their wish to be independent from men. They can also be regarded as taking their place alongside a dominant and complexly ambiguous eugenicist discourse of control over reproduction and population. Indeed, submerged underneath a superficially meliorist rethoric arguing for the future benefit of humankind there lurks a powerful masculinist discourse, which takes many forms, such as the deliberate choice of the sex of one's child, as in A. Garland Mears's *Mercia: The Astronomer Royal: A Romance* (1895) and a little later in Charlotte Haldane's *Man's World* (1926), or the elimination of those considered unfit or physically inferior, as in Mrs George Corbett's *New Amazonia* (1889), where those women who bore "the slightest trace of disease or malformation"[18] were rigorously excluded.

A REPUBLIC OF WOMEN

In 1891, Mary A. Livermore was putting forward a dream of a future "republic of women" where there would be no war, an Amazonian blueprint for a country characterized by peace, egalitarianism, and an equitable distribution of resources. The event of the American Civil War had taught women "to draw more closely together, to keep in touch with each other in thought and purpose, to unite in an organization 'superior to any existing society'" ("Coöperative Womanhood in the State," 295). Livermore's vision consists of an association within the government of a "republic of women, duly organized and officered, . . . [which] would train women for the next great step in the evolution of humanity, when women shall sit side by side in government, and the nations shall learn war no more" (295). As Nina Auerbach remarks, Livemore's article is "a prophetic blueprint for the divided nature of the female communities two world wars would produce" (*Communities of Women: An Idea in Fiction*, 161–162).

Lane's *Mizora* and Gilman's *Herland* are intricately associated with the political, economic, and cultural situation of the last decades of the nineteenth century in America, constituting a fictional response to those conditions. Estelle Freedman describes the period of the 1870s to the 1920s as "an era of separate female organizations and institution building, the result on the one hand, of the negative push of discrimination in the public, male sphere, and on the other hand, of the positive attraction of the female world of close, personal relationships and domestic institutional structures" ("Separatism as Strategy: Female Institution Building and American Feminism, 1879–1930," 90). As Freedman further stresses, "The creation of a separate, public female sphere helped mobilize women and gained political leverage in the larger society. A separatist political strategy . . . emerged from the middle-class women's culture of the nineteenth century" (87). It is precisely this kind of separatist strategy, inscribed in women's struggle for political rights, that Lane and Gilman dramatize in their utopian societies, which have many similarities but also some important differences.

> Only lesser races produce men.
>
> Joan Slonzcewski, *A Door into Ocean*, 80

In *Mizora*, with action located in the sixth millennium, Lane eliminates men altogether. In the land of Mizora, which had fallen into chaos during the system of administration of a corrupt general (modeled on Ulysses S. Grant, U.S. president from 1869 to 1877), the government is eventually overthrown and taken over by a group of educated women who have vowed to hold power for a century and create a more equitable and prosperous society. In order to further those aims, the "Constitution of the National Government provided for the exclusion of the male sex from all affairs and privileges for a period of one hundred years. *At the end of that time not a representative of the sex was in existence*" (101).[19]

In Gilman's *Herland*, most of the country's population was decimated by war about two thousand years ago, following which a volcanic eruption buried what was left of its small army and isolated it from the outside world. The few men who were left, most of them slaves, rebelled against their masters, killing them as well as the young boys and the old women. The women were so incensed at this barbarous behavior that they "rose in sheer desperation and slew their brutal conquerors" (55). Only women were left to rebuild the country.

In both *Mizora* and *Herland* women reproduce parthenogenetically, giving birth only to daughters. Gilman's utopian vision in *Herland* was strongly influenced by the ideas of sociologist Lester Frank Ward, to whom she dedicated her book *The Man Made World: Or, Our Androcentric Culture* (1911). In her dedication to Ward, Gilman enthusiastically calls him

> a creative thinker to whose wide knowledge and power of vision we are indebted for a new grasp of the nature and processes of society, and to whom all women are especially bound in honour and gratitude for his gynaecocentric theory of life, than which nothing more important to humanity has been advanced since the theory of evolution, and nothing more important to women has ever been given to the world.[20]

As Debra Benita Shaw explains, Ward's gynaecocentric theory of evolution, which Gilman described as "the most important single percept in the history of thought" (*His Religion and Hers*, 57), has "as its premise that sexual reproduction is the next evolutionary stage to parthenogenesis" (*Women, Science and Fiction: The "Frankenstein" Inheritance*, 16), going on to maintain that Gilman's work is "largely an extension of this theory, relevant as it is to the status of women" (16).

Ward's theory, which he developed establishing correspondences with examples from zoology, is particularly attractive to women, who see in it the fundamental importance of their biological functions vindicated. For Ward,

> The female is the fertile sex, and whatever is fertile is looked upon as female. . . . It therefore does no violence to language or to science to say that life begins with the female organism and is carried on a long distance by means of females alone. In all the different forms of asexual reproduction, from fission to *parthenogenesis*, the female may in this sense be said to exist alone and perform all the functions of life including reproduction. In a word, life begins as female. (*Pure Sociology*, 313; emphasis mine)

In both *Mizora* and *Herland*, Ward's scenario is activated, leading to a radical revision and deconstruction of traditional concepts of femininity and motherhood. Motherhood is regarded as the highest service these women can engage in for the good of all. In *Herland*, they all descend from a first mother. On the other hand, much to the men's surprise, they learn that "as to everyone knowing which child belongs to which mother—why should she?" (76).[21] As one of the men, Van, reflects, Herlanders were "Mothers, not in our sense of helpless involuntary fecundity, forced to fill and overfill the land, every land, and then see their children suffer, sin, and die, fighting horribly with one another; but in the sense of Conscious Makers of People" (69).[22] In a related vein, Ellador explains to him that "the nearest approach to an aristocracy they had was to come of a line of 'Over Mothers'—those who had been so honored" (70). Indeed, in Gilman's *Herland*, as in Lane's *Mizora*, society is built upon a reestablishing of the mother–daughter bond. As Ellador vehemently points out to Van, "You see, we are *Mothers* . . . as if in that she had said it all" (*Herland*, 67).[23] Vera, in turn, remarks that "the only intense feeling that I could discover among these people was the love between parent and child" (*Mizora*, 32).

In *Mizora*, similarly, the feminine, mother, is all-important, the source of all life. As one of the inhabitants demonstrates to the narrator, Vera, the woman from Russia who happens to come across the women's country completely by chance, after having taken her to the Chemist's Laboratory, "Daughter . . . you are now looking upon the germ of *all* Life; be it animal or vegetable, a flower or a human being, it has that one common beginning. We have advanced far enough in Science to control its development. Know that the MOTHER is the only important part of all life. In the lowest organisms no other sex is apparent" (103), a view that can be said to correspond closely to Lester Frank Ward's gynaecocentric theory of life.

While *Mizora* demystifies the supposedly privileged links between women and nature, as well as with "normal" processes of procreation, *Herland* can be said to stress them and fortify them, sometimes falling prey to an essentialist position. As Naomi Jacobs points out, "Reversing the classic equation of men with culture and women with nature, Lane's Mizorans associate masculinity with the natural 'grossness' and animality that they seek to remove from every aspect of their culture" ("The Frozen Landscape in Women's Utopian and Science Fiction," 195).

Both Lane's *Mizora* and Gilman's *Herland*, however, articulate disturbing racial and eugenicist views.[24] While in Mizora all the women are blonde, white-skinned, thick-waisted, and endowed with formidable lung capacity, in Herland the drive to physical perfection means that those women who are not considered fit to be mothers are not allowed to procreate, a decision that conflicts somewhat with the sentimentalizing of motherhood that characterizes Herland. In addition, Herlanders made it their business to "breed out" (83) the "lowest types" (83) of girls. As Sommel explains to Van, "If the girl showing . . . bad qualities had still the power to appreciate social duty, we appealed to her, by that, to renounce motherhood. Some of the worst types were, fortunately, unable to reproduce. But if the fault was in a disproportionate egotism—then the girl was sure she had the right to have children, even that hers would be better than others" (83). That, as Somel concludes, was never allowed, for she was not considered "fit for that supreme task" (83). On the other hand, *Mizora*'s defense of a eugenicist program of gradually ridding the land first of men, then of women who did not exhibit the ideal type of beauty, and also of those who were not deemed to be fit, is much more rigid and applied more ruthlessly than the parallel process in Gilman's novel.

Luce Irigaray, who throughout her career has investigated the construction of women as inferior beings in patriarchal society and suggested remedial strategies to reverse that state of affairs, pointedly asks, in words that could apply to Lane and Gilman's novels, "Is this a utopia? Can a society live without sacrifices, without aggression?" ("Women, the Sacred, Money," 77), going on to suggest that "there are no other societies available to women, at least for the time being" (78). The answer *Mizora, Herland,* and "Houston, Houston, Do You Read?" provide to the second question, however, is an unequivocal yes, putting forward an imaginative blueprint for precisely such a society without sacrifices and violence.

As in *Mizora*, there are no criminals in *Herland*. One of the women points out that it is "quite six hundred years since we had had what you call a 'criminal'" (*Herland*, 83). *Herland* is characterized by "universal peace . . . good will and mutual affection" (99). Also as in *Mizora*, in *Herland* sickness was virtually unknown. Both societies had made science and education, respectively, one of their highest priorities, together with motherhood and the caring for children. Indeed, the two societies are remarkably similar.[25] There is no poverty, and the inhabitants lead a fruitful, fulfilling life in comfortable surroundings. The women in both *Mizora* and *Herland* do not believe in life after death, in eternal life. Their "religion" consists in making life as prosperous and worthy as possible for everybody without exception.

Mizora, however, is a much more radical novel than *Herland* with its gentle separatism. In the latter the three American men who show up are welcomed. Furthermore, "dual parentage" (137), "The New Motherhood" (137) that included a father, was cherished and honored. In Mizora, on the other hand, men, should they happen to appear, would be eliminated, the fate they presumably suffer in Tiptree's story "Houston, Houston, Do You Read?"

> A dream of a manless world.
>
> David Keller, "The Feminine Metamorphosis," 186

Lane's *Mizora* can profitably be placed alongside David Keller's "The Feminine Metamorphosis" (1929). In Keller's tale, American women, who during World War I had started working and enjoying financial independence, are driven back to domesticity after the war ends and the men return home. Despite being as clever, efficient, and hard-working as the men, women are refused promotion. One of these women, Martha Belzer, decides to conspire with other women to take over the country, as well as the rest of the world, by means of a drug that turns the women into men. Five thousand of these "men" achieve control over the business world, with the aim of bringing women back into the workforce and creating a more egalitarian society. Their plot, however, goes wrong when the drug, which was made from the unhealthy gonads of Chinese men, causes them to fall sick and develop a kind of insanity (probably from the syphilis-infected gonads). In addition, a private detective uncovers their secret scheme, exposes them, and criticizes them severely for reversing what he considers the natural order of things: Women should remain in the private, domestic sphere and leave the public arena for the men.

As a result, Miss Martha Beltzer and the other women develop a "dream of a manless world. . . . We do not want two sexes in this fair world of ours, not as long as one sex can run it so efficiently" (186), words reminiscent of a similar opinion expressed by the inhabitants of Lane's *Mizora*. Like the latter, the women in Keller's "The Feminine Metamorphosis" implement a parthenogenetic technique that enables them to have only female babies, thus (together with the fantasy of eradicating men from their society) paving the ground for a world without men,[26] which however seems very unlikely to emerge, for men will gradually become dominant again.[27]

MOTHERHOOD AND *JOUISSANCE*

> You see, we are *Mothers*, . . . as if in that she had said it all.
> Charlotte Perkins Gilman, *Herland*, 66

Barbara Creed notes that the maternal figure we find in the work of Freud, Lacan, and Kristeva is always constructed in relation to the father. However, as Creed goes on to argue, "If we posit a more archaic dimension to the mother—the mother as originating womb—we can at least begin to talk about the maternal figure as *outside* the patriarchal family constellation" (128). In this context, as Creed further asserts, "The mother-goddess narratives can be read as primal-scene narratives in which the mother is the sole parent. She is also the subject, not the object, of narrativity" (*The Monstrous-Feminine: Film, Feminism, Psychoanalysis*, 128). Lane's Mizorans and Gilman's Herlanders can be seen as inscribed in precisely such a context, where they are the proud subjects and agents of their own world and their own narratives.[28]

The ideology of motherhood in Gilman's *Herland*, however, seems still to be grounded, at least in part, on the dominant tenets governing traditional discourses on mothering. Indeed, *Herland* appears to suggest, problematically, that only through motherhood can women be fulfilled, thus illustrating what Patrice DiQuinzio, in *The Impossibility of Motherhood: Feminism, Individualism, and the Problem of Mothering*, defines as "essential motherhood" (xiii), that is, the role of women *qua* human beings is to be mothers. As the narrator explains, life for Herlanders is "just the long cycle of motherhood" (*Herland*, 61), an expression that suggests, on Van Jenning's part, his view of the concept of motherhood as reductive and vaguely limiting. For DiQuinzio, essential motherhood as an ideological formation spells out an imposition of motherhood as it operates in our Western culture, since it sees women's motherhood as both natural and inevitable, suggesting that those women who do not wish to be mothers are deviant or defective. As a way of negotiating what she sees as the "impossibility of motherhood" (248), that is, being a mother and an individual simultaneously, DiQuinzio proposes what she describes as a "paradoxical politics of mothering" (248), a politics that "accepts the impossibility of motherhood and the impossibility of individualist subjectivity, and that does not require for its foundation a univocal, coherent and exhaustive position on mothering" (248); one that "might instead make possible multiple and overlapping positions of resistance to individualism and essential motherhood, and might show how to achieve or constitute the possibility of movement among such positions" (248).

Many of the texts I look at in this chapter go a long way toward dramatizing some of the new contours that motherhood might take. While DiQuinzio asserts that "individualism and essential motherhood together position women in a very basic double bind: essential motherhood requires mothering of women, but it represents motherhood in a way that denies mothers' and women's individualist subjectivity" (xiii), I would like to suggest that Gilman's Herlanders can be seen as embodying precisely the ideal of the fundamental primacy of motherhood but without any diminution of women as individuals in their own right, a scenario that also applies to Mizorans.

On the other hand, and in tune with DiQuinzio's paradoxical politics of mothering, feminist critics have often pointed out the need for feminism to theorize the specificity of women's embodiment in order to negotiate the idea that, as in the concept of "essential motherhood," motherhood is natural and unavoidable. According to this definition, those women who do not show what essential motherhood sees as a mandatory feminine characteristic and function, maternal instinct, are then considered defective. This is the case in *Herland*, where the paean to motherhood is so extreme, so romanticized, that the very possibility that a woman might not wish to become a mother is not even considered. In Gilman's novel, women are the "Mother Sex" (*Herland*, 129), an expression Gilman used again in her last book, *His Religion and Hers* (1923), where she wrote, "Not all the long, loud struggle for 'women's rights,' not the varied voices of the 'feminist movement,' and, most particularly, not the behaviour of 'emancipated women,' have given us any clear idea of the power and purpose of *the mother sex*" (emphasis mine),[29] thus highlighting the contradictions inherent in her argumentation vis-à-vis the idea of motherhood. As Sandra M. Gilbert and Susan Gubar note, in words that could also apply to *Mizora*, "Gilman's hymns to maternal nurturance are so excessive as to evoke the retrograde sexual ideologies associated with the Victorian Angel in the House" ("'Fecundate! Discriminate!' Charlotte Perkins Gilman and the Theologizing of Maternity," 211).[30]

Related with this emphasis on motherhood is an accent on the centrality of birth, unequivocally eulogized in *Herland*. In *His Religion and Hers*, Gilman calls birth woman's "great function" (247) and argues that in contradistinction to what she considers the patriarchal death-oriented religions there should be offered a "birth-based religion" (46), altruistically concerned with the child that is born and not with one's fate after death, which Gilman regards as the mark of a "death-based religion . . . a posthumous egotism" (46).

Christine Battersby's argument about a new feminist metaphysics, which places woman as the norm in terms of defining woman's identity and selfhood, in *The Phenomenal Woman: Feminist Metaphysics and the Pattern of Identity*, includes "an emphasis on birth" (4) that can profitably highlight many issues in *Mizora* and *Herland*, novels that place great stress on birth as the supreme activity women could engage in. As she goes on to asssert, "The hypothetical link between 'woman' and 'birth' that matters is 'If it is a male human, it cannot give birth,' not 'If it is a female human, it can give birth'" (4). Battersby goes on to argue that "the dominant metaphysics of the West have been developed from the point of view of an identity that cannot give birth, so that birthing is treated as a deviation of the 'normal' models of identity—not integral to thinking identity itself" (4).[31] Lane and Gilman can be said to be doing precisely that, that is, creating a feminine identity that takes "birthing" as the normative mode of identity: If men do not or cannot give birth, then they are different, and their identity needs to be thought out according to value systems distinct from those of the women—which does not mean that these identitary structures will be inferior, just different.

As Christine Battersby remarks, "Philosophers have notably failed to address the ontological significance of the fact that selves are born," while acknowledging a "general inability to imaginatively grasp that the self/other relationship needs to be reworked from the perspective of birth" (*The Phenominal Woman: Feminist Metaphics and the Patterns of Identity*, 3), an argument similarly put forward by Hanna Arendt, for whom birth is a fundamental category of philosophical and political thought, describing human beings as the "natals,"[32] defined precisely by their natality.

Grace M. Jantzen also participates in this discussion, noting that "although not all women give birth, every person who has ever lived has been born, and born of a woman.[33] Natality is a fundamental human condition. It is even more basic to our existence than the fact that we will die, since death presupposes birth" (*Becoming Divine: Towards a Feminist Philosophy of Religion*, 144). As Jantzen further remarks, an imaginary of natality is "always material, embodied . . . one which recognizes our rootedness in the physical and material" (145), aspects crucially illustrated in *Mizora* and *Herland*.

Adriana Cavarero, for her part, drawing on Hanna Arendt's category of birth, which Arendt developed in relation to and against Heidegger's notion of being-towards-death, emphasizes the need to shift philosophical perspective to an acknowledgment of the importance of birth, since "all persons, female and male, are invariably born from their mother's womb" (*In Spite of Plato: A Feminist Rewriting of Ancient Philosophy*, 6). As Cavarero accentuates, "Humans always come into the world in this way, never otherwise" (6). This statement, however, seems very provisional nowadays, with the increasingly closer prospect of the development of an artificial womb where fetuses could be brought to term. The introduction of this technique would inevitably mean that many philosophical concepts would have to be revised and retheorized. It is not hard to imagine, indeed, with impending advances in the technology of artificial wombs, that the unconditional importance attributed to birthing might shift to being a mother or father, concepts, however, no longer inextricably tied to giving birth in a physical way with recourse to a woman's uterus. The all-important notion of birth from a woman's body, then, might be more transient and ephemeral than we might be able to conceive nowadays. In this sense, however, woman's only prerogative to "a power exclusively of her own," to rephrase Virginia Woolf's expression of desire for a room of one's own, will be taken away from them. These questions have to be the object of very careful negotiations in terms not only of their ethical implications but also of the power politics implicit in such foundational acts as that of a woman carrying a fetus in the womb to term or not, in the consideration of whether to leave that function to a machine, and in weighing the health gains for the woman and the baby when the former is not able to sustain a pregnancy, for example. Indeed, this scenario might be seen by some as a return to Aristotle's notion of the womb as receptacle and consequently of women as wombs, simply carriers of the fetus to term, with no other procreative role, since the fetus originated in the man's life force. Should unconditional philosophical importance then be placed on the act of birth from woman? In the context of the articulation of the kind of sexual politics envisioned by Shulamith Firestone in *The Dialectic of Sex: The Case for Feminist Revolution* (1970), a more egalitarian society might be created where women would not be treated or considered as breeding machines and, because of that, find themselves at a disadvantage in the workplace. If protective laws are not implemented, as is the case in many countries, then I do not think such emphasis should be laid on the act of birth, and future reproductive technologies may provide the means for alternative scenarios to materialize that would broaden women's choices.[34] While this symbolic of natality is very important, I wish to suggest that motherhood does not necessarily need to include birth and indeed, as many futuristic scenarios evoke, mothers who so wish might be able to have recourse to ectogenesis to bring their babies to term outside their own wombs. Indeed, birth can take many forms, and future technologies will no doubt envisage many more options.

A DESIRE OF ONE'S OWN

In *La Révolution du langage poétique*, Julia Kristeva had already perceptively asserted that it was not *woman* who was oppressed and subdued in the masculinist world but *motherhood*, and the maternal *jouissance* associated with the reproductive stages, which is denied to men. As Kristeva reasons, "If the position of women in the social code is a problem today, it does not at all rest in a mysterious question of feminine *jouissance* . . . but deeply, socially and symbolically in the question of reproduction and the *jouissance* that is articulated therein" (462). This point is also stressed by Irigaray in "The Bodily Encounter with the Mother," where she argues that "if we are to discover our female identity, I do think it important to know that, for us, there is a relationship with *jouissance* other than that which functions in accordance with the phallic model" (45). The reassertion of the mother's power and *jouissance* is one of the central thematic concerns illustrated in *Herland*.

Jouissance in *Herland* is associated with maternity. As Somel explains, "Before a child comes to one of us there is a period of utter exaltation—the whole being is uplifted and filled with a concentrated desire for that child" (71). In *Mizora*, similarly, as the Preceptress explains to Vera, "Our children come to us as welcome guests through portals of the holiest and purest affection" (130), although there is no mention of pleasure or desire. As Jean Pfaelzer explains, "By eliminating the female body as the site of sexuality, Mizora resolves sentiment's embarrassing contradiction between chastity and the esteem for motherhood. Celibacy annuls the reproductive imperative that shaped gender roles and thereby frees the Mizoran woman from romance" ("Utopians Prefer Blondes—Mary Lane's *Mizora* and the Nineteenth-Century Utopian Imagination," xxviii).[35]

Both Lane's *Mizora*[36] and Gilman's *Herland* abolish the feminine body as the site of sexuality,[37] thus at a stroke eliminating romance and the dangers of relationships characterized by power maneuvers. James Tiptree Jr.'s "Houston, Houston, Do You Read?" on the other hand, includes sexual activity among the women. In "Note on 'Houston, Houston, Do You Read?'" James Tiptree Jr. comments that "it is known that certain clones are attracted ('fated') to each other. Deep and serious and abiding love is suggested here. But it isn't 'tragic,' because, quite practically, if one member of a clone doesn't reciprocate, another, identical member may!" (374).[38] Joanna Russ, in turn, points out

Whether tentative or conclusively pessimistic, the invented, all-female worlds, with their consequent lesbianism, have another function: that of expressing the joys of female bonding, which—like freedom and access to the public world—are in short supply for many women in the real world. Sexually, this amounts to the insistence that women are erotic integers and not fractions waiting for completion. Female sexuality is seen as native and initiatory, not (as in our traditionally sexist view) reactive, passive, or potential. ("Recent Feminist Utopias," 142)

A NEW ERA

Luce Irigaray's recurrent fantasy of a "new era" (*Marine Lover of Friedrich Nietzsche*, 170), which she sees as dependent on the emergence of a maternal genealogy, on the restoration of the mother and daughter lineage (since both sons and daughters have always been, in legal terms, the father's progeny), might then materialize with human cloning, in parallel with the father–son lineage,[39] a vision also put forward in Gilman's *Herland* and *With Her in Ourland*, where the day that Ellador, Celis, and Alima get married to Van, Jeff, and Terry is described as "the dawn of a new era" (*Herland*, 119). As Somel, one of the women, explains to the three men: "You don't know how much you mean to us. It is not only Fatherhood—that marvelous dual parentage to which we were strangers—the miracle of union in life-giving—but it is Brotherhood" (119). In *Herland* and *With Her in Ourland*, then, Gilman—unlike Lane in *Mizora* or James Tiptree Jr. in "Houston, Houston, Do You Read?"—puts forward a vision of harmony and cooperation between the sexes, not of separatism.

Interestingly for my argument, Irigaray, drawing on the vocabulary of Christianity (since for her Christianity constitutes the main mythology of the West), suggests that we stand on the threshold of the "third era" (*An Ethics of Sexual Difference*, 148) that follows the era of the Father (the Old Testament) and that of the Son (the New Testament). This "third era," then, according to Irigaray, would correspond to the era of the "Bride,"[40] where the woman would have "her own territory: her birth, her genesis, her growth. With the female becoming in self and for self" (*An Ethics of Sexual Difference*, 149). Firestone's own millenarian dreams, much more radical than Irigaray's, similarly borrow tropes from the biblical narrative. As she maintains,

> The revolt against the biological family could bring on the first successful revolution, or what was thought of by the ancients as the Messianic Age. Humanity's double curse when it ate the Apple of Knowledge . . . that man would toil by the sweat of his brow in order to live, and woman would bear children in pain and travail, can now be undone through man's very efforts in toil. We now have the knowledge to create a paradise on earth anew. (*The Dialectic of Sex: The Case for Feminist Revolution*, 242)

In Firestone's conception of the future, a feminist revolution would mean

> production and reproduction of the species would both be, simultaneously, reorganized in a nonrepressive way. The birth of children to a unit which disbanded or recomposed as soon as children were physically able to be independent . . . would eliminate the power psychology, sexual repression, and cultural sublimation. Family chauvinism, class privilege based on birth, would be eliminated. (*The Dialectic of Sex: The Case for Feminist Revolution*, 241)

We might say, then, that once again Lane's *Mizora*, Gilman's *Herland*, James Tiptree, Jr.'s "Houston, Houston, Do You Read?" and Suzy McKee Charnas's *Motherlines* give fictional illustration to precisely this revisionist psychoanalytic Irigarayan scenario, grounded on nonpossessive relations. Indeed, family structures in these societies do not follow any patriarchal model. In *Herland*, where there is also, meaningfully, mention of "the dawn of a new era" (119), the whole country is like an all-encompassing community, with no private homes, ruled by a "limitless feeling of sisterhood" (69). With its emphasis on mothers and daughters, the society in *Herland* provides an apt dramatization of Irigaray's incisive argument for the liberation of women "from the authority of fathers" ("Women-Mothers, the Silent Substratum," 50). As she goes on to claim, "In our societies, the mother/daughter, daughter/mother relationship constitutes a highly explosive nucleus. Thinking it, and changing it, is equivalent to shaking the foundations of the patriarchal order" (50).[41] In *Mizora*, each girl has one mother, and they live in single-family groupings with the latter, while in *Herland* the whole country is like an all-encompassing community, with no private homes, ruled by a "limitless feeling of sisterhood" (69), a scenario that similarly applies to "Houston, Houston, Do You Read?"
In words reminiscent of Irigaray's reflections on the traditional prepotence of fathers in family dynamics, Ellador, discussing with Van the role of the father, although acknowledging with reverence what the "co-mothers" (*With Her in Ourland*, 132) in Herland described as the "holy mystery of fatherhood" (132), considers that "much mischief has followed from too much father. He did put himself forward so! He thought he was the whole thing, and motherhood—Motherhood!—was quite a subordinate process" (132). Ellador goes on to lament what she sees as the "dominance" and "egoism" (132) of father figures and the authority they yielded over the mother. As Ellador puts it, from the point of view of the father it was "My house—my line—my family" (132), and if the mother "had to be mentioned it was on 'the spindle side'" (132). Ellador then goes on to evoke Posthumous in Shakespeare's *Cymbeline*, who wished "there was some way to have children without these women" (132), to which Van contritely replies that "it is only too

plain that 'Mr. Father' has grossly overestimated his importance in the part" (132), a role that is thoroughly revised in *Herland* and *With Her in Ourland*.[42]

> What would the world have looked like "if God had been a woman"?
>
> Hélène Cixous, *With, Ou L'Art de L'Innocence*, 89

James Tiptree Jr.'s "Houston, Houston, Do You Read?" for its part, portrays another all-female world, three centuries into the future. The epidemic that affected the world while the three male characters were on a mission aboard a spaceship caused almost universal sterility, having selectively attacked only men, bringing about a shortage of male births. As the narrator comments, when humanity found itself sterile "riots and fighting . . . swept the world. Cities bombed and burned, massacres, panics, mass rapes and kidnapping of women, marauding armies of biologically desperate men" (74) were some of the evils that beset humankind after the epidemic. The result was "a planet of girls and dying men. The few odd viables died off" (84).

The women who rescue the three male characters from their lost ship and bring them into their own had never seen men in their lives, like the inhabitants of Mizora and Herland, but, unlike Herlanders, who welcome the three American men (an interesting symmetry between the two tales), the women in Tiptree's story realize that they cannot run the risk of going back to the old, violent ways.[43] Unwilling to take that chance, the end of the story implicitly suggests that the men were given a lethal pill.

The stress in "Houston, Houston, Do You Read?"—as in *Mizora* and *Herland*—falls on an organic connection with the world. In spite of being on board a spaceship, the women have a greenhouse with a vine, which they tend with great care, as well as other plants and animals—an iguana, grasshoppers, and chickens. Like the women in *Mizora* and *Herland*, those in "Houston, Houston, Do You Read?" look magnificently healthy and robust until a very old age, a longevity they attribute to "some rules of healthy living" (80).

According to James Tiptree Jr., "Houston, Houston, Do You Read?" "shows, in glimpses only, seen through the mind of the male narrator, what an all-female society might be really like—in contrast to the usual 'Queen of the Amazons'-type masculine fantasy" ("Note on 'Houston, Houston, Do You Read?'" 373). As James Tiptree Jr. further asserts,

> The main purpose in constructing this all-women world was not specifically sexual, but rather to contrast its relaxed, cheery, practical *mood* with the tense,

macho-constricted, sex-and-dominance-obsessed atmosphere of the little all-male "world" of male-dominated culture in the *Sunbird* spacecraft. These men are meeting for the first time a world in which men qua males simply *do not matter*. (373)

These words can be applied to the all-female societies of *Mizora* and *Herland*. Indeed, like Terry, one of the three men to arrive in Herland, arrogant and sexist Dave in "Houston, Houston, Do You Read?" voices his misogynistic ideas. Referring deprecatingly to the women who saved them he remarks in a supercilious tone, "They are lost children. They have forgotten He who made them. For generations they have lived in darkness. . . . Women are not capable of running anything. . . . Nobody has given them any guidance for three hundred years" (92). Terry, in *Herland*, articulates identical feelings. Referring to the women, he says vituperatively, "Of course they can't understand a Man's World! They aren't human—they're just a pack of Fe-Fe-Females!" (81). Bud, like Dave, is totally insensitive to the women's feelings and almost rapes one of them. He is another equivalent of Terry in *Herland*, who rapes his wife Alima, leading to their separation. As one of the women explains to the men, "As I understand it, what you protected people from was largely other males, wasn't it? We've just had an extraordinary demonstration. You have brought history to life for us" ("Houston, Houston, Do You Read?" 96).

Like the women in *Mizora* and *Herland*, those in "Houston, Houston, Do You Read?" reproduce via a version of parthenogenesis, or virgin birth, in their case cloning. Women become pregnant and carry their babies to full term, like traditional mothers. As one of them states,

Not bottled embryos . . . Human mothers like everybody else, young mothers, the best kind. A somatic cell nucleus is inserted in an enucleated ovum and re-implanted in the womb. They have each borne two "sister" babies in their late teens and nursed them for a while before moving on. The creches always have plenty of mothers. (80)

"Houston, Houston, Do You Read?" is unusual in presenting a positive view of cloning. All the women in the story, which takes place on an Earth three hundred years from the present time, are clones and, like the cloned sisters in Kate Wilhelm's highly cautionary novel about cloning, *Where Late the Sweet Birds Sang* (1976), they experience a feeling of security in the closeness and support they give each other. Unlike Wilhelm's sets of cloned sisters, however, who are described as interchangeable and without individual personality, the women in "Houston, Houston, Do You Read?" are strong, distinct characters, with lives of their own. Indeed, one of them feels sorry for the three men they rescue and take on board their spaceship, for they are "all alone, no sisters to share with!" (80).[44] In "Note on 'Houston, Houston, Do You Read?'" James Tiptree Jr. comments, in relation to this story, that a particular

> "author's interest"—which I didn't have time to explore as fully as it deserves—is in the unique culture of a world of clones, where each person has perhaps two thousand living versions and extensions of herself. I saw this as permitting great relaxation, almost a playful response to life—since what "I" don't accomplish may be—or has been—accomplished by another Me. Each clone keeps a special place, and a record—e.g., "The Book of Judy Shapiro"—where they go periodically and learn about all the different potentials and experiments her "self" has explored. It was my feeling that such an institution would be quite congenial to women, but by definition rather horrifying, or meaningless, to traditional males. (Self-examination is "unmanly"—but is in fact a source of great interest, and incidentally a preventer of loneliness.) (374)[45]

This privileged connection and almost telepathic access to someone else's mind explored in "Houston, Houston, Do You Read?" is also an important thematic concern in many other clone narratives dealt with throughout this book, such as Pamela Sargent's *Cloned Lives*, Kate Wilhelm's *Where Late the Sweet Birds Sang*, Ursula K. Le Guin's "Nine Lives," and Damon Knight's "Mary."

> Where are the men?
>
> *Mizora*, 22

> This is no savage country, my friend. But no men?
>
> *Herland*, 20

Is it, however, necessary to kill the father in order to make way for the mother and her genealogy, a maternal genealogy? *Mizora* unequivocally indicates that if men were allowed back into their land they would almost inevitably bring with them the threat of war, while in *Herland* men are welcomed by the women thirsty for information from the outside world. These women, however, are as ferocious in the defense of their hard-won prosperity and stability as Mizorans are and do not feel in the slightest threatened by the three American travelers.

Irigaray, for her part, as if she were engaged in a critical dialogue with these texts, announces that "what is now becoming apparent in the most everyday things and in the whole of our society and our culture is that, at a primal level, they function on the basis of a matricide" ("The Bodily Encounter with the Mother," 36). As she goes on to suggest, "When Freud describes and theorizes, notably in *Totem* and *Taboo*, the murder of the father as founding the primal horde, he forgets a more archaic murder, that of the mother, necessitated by the establishment of a certain order in the polis" (36). We might say that both *Mizora* and *Herland* ressurrect the mother with a vengeance, implicitly rethinking the whole Oedipal theoretical edifice constructed by Freud, placing the mother–daughter connection as the central societal relationship. As Irigaray emphasizes, it is necessary, "if we are not to be accomplices in the murder of the mother, for us to assert that there is a genealogy of women" (44), that "the bond of female ancestries must be renewed" (*Thinking the Difference: For a Peaceful Revolution*, 108). Irigaray rewrites Freud's psychoanalytic scenario, arguing in an incisive way:

> Many people think or believe that we know nothing about mother–daughter relationships. That is Freud's position. He asserts that on this point, we must look beyond Greek civilization to examine another erased civilization. Historically, this is true, but this truth does not prevent Freud from theorizing and imposing, in psychoanalytical practice, the need for the daughter to turn away from the mother, the need for hatred between them, without sublimation of female identity being an issue, so that the daughter can enter into the realm of desire and law of the father. (*Thinking the Difference: For a Peaceful Revolution*, 109)

As Irigaray goes on to observe, "This is unnaceptable. Here Freud is acting like a prince of darkness with respect to all women, leading them into the shadows and separating them from their mothers and from themselves in order to found a culture of men-amongst-themselves" (110). After all, Irigaray remarks that "if the rationale of History is ultimately to remind us of everything that has happened and to take it into account, we must make the interpretation of the forgetting of female ancestries part of History and reestablish its economy" (110), a strategic move enacted by the women-only societies in *Mizora, Herland,* and "Houston, Houston, Do You Read?" Indeed, it can be said that the women in these texts have achieved what Irigaray so vehemently called for, a genealogy of women, namely, in her book *Sexes and Genealogies* (1993). In *Herland*, as one of the women elucidates, by keeping "the most careful records . . . each one of us has our exact line of descent all the way back to our dear First Mother" (76), a scenario similar to the books that record the lineage of each group of cloned women in "Houston, Houston, Do You Read?" In Tiptree's story, the women explain to Lorimer, one of the three men, that each of the clone strains has "a book, it's really a library. All the recorded messages. The Book of Judy Shapiro, that's us. Dakar and Paris are our personal names" (80). They also make clear how "each Judy adds her individual memoir, her adventures and discoveries in the genotype they all share" (80–81), in a pertinent illustration of the importance these women attach to their female ancestry.

Another novel that emphasizes the importance of genealogies of women is Suzy McKee Charnas's *Motherlines* (1978), whose protagonist, Alldera, escapes from a male-dominated territory, Holdfast, where women were called "unmen," arriving after a dangerous journey at the land of the Riding Women, a self-sufficient female community where the women reproduce by mating with horses.[46] Each daughter is a copy of her mother, a clone in today's terms, while each motherline can be identified by means of its distinct physical and psychological characteristics, as is the case in "Houston, Houston, Do You Read?" While at first Alldera felt confused about "the way these people appeared sometimes in identical pairs, trios, or even more" (263), she was getting used to their similar appearance. As Nenisi explains to her, "There were whole strings of blood relations called 'Motherlines,' women who looked like older and younger versions of each other. They were mothers and daughters, sisters and the daughters of sisters" (263).

The women in *Motherlines* descend from the "first daughters" (273), women who managed to endure and survive the "Wasting" (the ecological cataclysm that decimated the people and destroyed the land) and who before the catastrophe succeeded in learning the genetic technology that they went on to use in the Grasslands, where they settled down. This technology involves trait-doubling, which, by means of the addition of a "certain fluid" (274), provided by their mating with horses, initiated the process of reproduction. According to Nenisi's account, around the time of the Wasting, "The lab men didn't want to have to work with all the traits of both a male and a female parent, so they fixed the women to make seed with a double set of traits. That way their offspring were daughters just like their mothers, and fertile" (273). As Nenisi further explains, the women's "seed, when ripe, will start growing without merging with male seed because it already has its full load of traits from the mother" (274).

In Marleen S. Barr's words, "Each Motherline is an immortal feminist community whose immortality results from an unending succession of replicas of one individual" (*Alien to Femininity: Speculative Fiction and Feminist Theory*, 7). Alldera, who is pregnant, described as a "fem" in the Riding Women's terms, is accepted by the latter, who consider that her child may grow up to found a new Motherline, providing "new seed, new traits, the beginning of the first new Motherline since our ancestors came out of the lab" (275). In this society, in analogy with *Mizora*, *Herland*, and Joanna Russ's Whileaway in *The Female Man*, there are bloodmothers and four "sharemothers" (*Motherlines*, 240), among whom the "Heart-mother" (263) occupies a special place, so that children are brought up in a cooperative way, as in Piercy's *Woman on the Edge of Time*, where three comothers share the upbringing of the child, thus doing away with the traditional nuclear family. For Frances Bartkowski, both Charnas and Piercy "present models of a utopian family where a child raised collectively learns to love nonpossessively. Without an exclusive parental tie, there is no expectation that one person can or need satisfy all the longings of the heart, thus rewriting parenting along with the myth of romance" (*Feminist Utopias*, 101).

However, as Jenny Wolmark points out, the women-only communities in *Motherlines* "suggest that women can be free only in the absence of men, a proposition that ironically leaves existing gender relations intact and posits an unproblematic relation between women and the category Woman" (*Aliens and Others: Science Fiction, Feminism and Postmodernism*, 84). The paradox at work in these women-only societies, then, is that in order to survive they have to get rid of men, often in a brutal way, thereby seemingly having recourse to violence, which the women purportedly wished to avoid. Charnas herself explains that *Motherlines*

> turned out to be about separatism as a solution to sexism—the heart of the book is the all-woman culture of the "Riding Women." Some readers will call the Riding Women monsters, since many people find monstrous the idea of women living long, full lives without men. I do not, though separatism is not my blueprint for Paradise and not the only answer to sexism that I hope to explore in fiction. ("A Woman Appeared," 106)

It can be argued, then, that Lane, Gilman, Charnas, and Tiptree provide fictional dramatization to the kind of revisionist psychoanalytic contours Irigaray visualizes in *Sexes and Genealogies*, stressing the importance of motherlines.

The dynamics of some of these women-only societies can thus be read as suggesting alternative ways of sidestepping the conventional models governing traditional nuclear families, the Freudian family drama including the Oedipus and Electra complexes, which would cease to exist, as well as rehearsing complex practices of circumventing the appparently fixed binary oppositions that rule Western patriarchal thinking and modes of government and social organization. Irigaray wonders in this connection,

> It would be interesting to know what might become of psychoanalytic notions in a culture that did not repress the feminine. Since the recognition of a "specific" female sexuality would challenge the monopoly on value held by the masculine sex alone, in the final analysis by the father, what meaning could the Oedipus complex have in a symbolic system other than patriarchy? (*This Sex Which Is Not One*, 73)

Mizora, *Herland*, "Houston, Houston, Do You Read?" and *Motherlines* are texts that supply imaginative scenarios, tentatively providing answers to Irigaray's questions and formulations.

"THE OTHER OF THE OTHER," LUCE IRIGARAY

Profoundly implicated in Western philosophical thought is the category of woman as the other of man, implicitly considered as the norm against which all other categories have to be judged and subsumed. As Simone de Beauvoir put it, "Humanity is male and man defines woman not in herself but as relative to him; she is not regarded as an autonomous being" (*The Second Sex*, 16). These women-only societies fictionalize the subversive gesture of starting from the assumption that woman is the norm (there is no other sex), thus by extension transforming man into woman's other. This deeply transgressive strategy allows for a dramatization of a profoundly dissident scenario that constitutes a possible blueprint for potential models of future societies.

In her examination of alternative scenarios and permutations for women in society, Luce Irigaray provides useful conceptions with which to examine the whole phenomenon of human cloning, and in particular the role of female clones. For Irigaray, patriarchy is the realm of the same, or, even more tellingly, the "empire of the same" (*This Sex Which Is Not One*, 141), a formulation that echoes Baudrillard's "reign of the same" ("The Hell of the Same," 120). A clone can arguably be described as the same of the same in a system of potentially endless duplications. The whole edifice of theoretical discourse on otherness and difference, then, will have to be rethought in light of the future existence of clones.

Human cloning effectively brings into specific relief the vexed politics of "othering" and "saming." Where does the "dream of sameness" (*Speculum of the Other Woman*, 248) that Irigaray invokes and deconstructs fit in a discussion on human cloning? This "dream of sameness" analyzed by Irigaray is effectively a dream of reproduction of the same male characteristics, likeness, and dominance. Masculinity testifies to "the desire for the same, for the self-identical, the self (as) same, and again of the similar, the alter ego" (26). It follows that as a consequence sexual difference, from a male point of view, is "determined within the project, the projection, the sphere of representation, of the same" (26–27). Femininity, similarly, seems to me to attest to as strong a desire for exactly the same "sameness," the same of the same. Pursuing this argumentative line I would like to suggest that a female cloned person might very well correspond to Irigaray's formulation of the "other of the other," that is, the feminine as defined by and for herself, without recourse to a supposedly universal masculine norm, which underscores a masculine "sameness-unto-itself." With a sole female genetic parent, the cloned female would circumvent being posited as "the other of the same" male model, following instead a female one, a strategically empowering move.

Human cloning would effectively destabilize what Irigaray called the power of the philosophic logos "to *reduce all others to the economy of the Same*" (*This Sex Which Is Not One*, 74), hence doing away with the (male) domination over the (female) other and the "reduction of the other in the Same" (74). With families of cloned brothers and sisters there would arguably develop an economy of what we might call, in tune with Irigaray's terminology, sameness of the same, male or female, and not just a female "economy of sameness, oneness, to the same of the One" (139), that is, following an exclusively masculine model. While this feminine economy of sameness of the same might arguably be regarded as another, parallel version of the masculine exclusion of the feminine as wholly other, as another dream of the (same) symmetry, the very possibility of its existence gives an added impetus to the potentialities pertaining to human cloning that would gradually lead to women's empowerment and equality with men in terms of reproductive functions.

If for man "the other of the other" is, in Margaret Whitford's words, "that which is unthinkable" (*Luce Irigaray: Philosophy in the Feminine*, 115), a female other incarnates the kind of subjecthood denied by Lacan's pronouncement, "There is no Other of the Other" (*Écrits: A Selection*, 311). The question, however, remains: Who is the "other of the other," that is to say, who is woman's "other"? Does she need or want an other, or can she define herself without recourse to the masculine? Pursuing this line of reasoning, a related question could be formulated: If, according to semiotics, meaning is established in terms of a paradigm of related signs that are differentiated, how would woman define herself without man, and vice versa? In tune with Irigaray's efforts to predicate a specifically feminine identity, Christine Battersby theorizes the need for a new feminist metaphysics, a "metaphysics of becoming" (*The Phenomenal Woman: Feminist Metaphysics and the Patterns of Identity*, 124) that places woman as the norm in terms of defining her individuality and selfhood. For Battersby, "an identity politics need not entail homogenizing 'woman' or 'women' into Irigaray's 'other of the Other'" (124). The families of women in *Mizora, Herland*, "Houston, Houston, Do You Read?" and *Motherlines* seem, in polemical fashion, to suggest ways in which the other sex may no longer be necessary.

In *Herland* we can see dramatized and simultaneously contested Simone de Beauvoir's contention that woman is always the other of man, an assertion disputed by Luce Irigaray, who notes how her work "on female subjectivity develops in the opposite direction to that of Simone de Beauvoir's work as regards the question of the other" (*Democracy Begins between Two*, 123). As Irigaray further explains, "Instead of saying, as she does: I do not wish to be the other of the masculine subject and, in order to escape this secondary position, I claim to be equal to him, I argue: the question of who the other is has not been well formulated in the Western tradition, in which the other is always the other of a singular subject and not another subject, irreducible to the masculine subject and of equal dignity" (123–124).

This is precisely what Lane, Gilman, Tiptree, and Charnas, I would argue, are doing, that is, creating a feminine subject autonomous from man. Indeed, their project can be said to be Irigaray's as well, that is, to envisage and reflect on, in Irigaray's words, "the characteristics of a world in the feminine, a world different from that of man as regards relationships with language, with the body (age, health, beauty and, obviously, maternity), and relationships with work, nature, culture" (*Democracy Begins between Two*, 131).

A world for women.

Luce Irigaray, "Love of Same, Love of Other," 109

According to Irigaray, there is a pressing need for a culture of women among themselves, given a system ruled by the male norm and a culture of "men-amongst-themselves." To achieve this "woman-to-woman sociality," Irigaray argues that "women should create a social order in which their subjectivity can unfold along with its symbols, its images, its realities and dreams" ("How to Define Sexuate Rights?" 210), that is, a female symbolic not based on the masculine one. Donna Haraway also addresses this issue, stressing the need on the part of women "to re-narrate, to produce woman's writing, to produce a female symbolic where the practice of making meanings is in relationship to each other, where you're not simply inheriting the name of the father again and again and again" (*Women Writing Culture*, 56).

Throughout her work, Irigaray theorizes the need for a community of women, a "world for women" (*An Ethics of Sexual Difference*, 109). Prioritizing women was clearly also one of Gilman's central efforts in *Herland*. Gilman can thus be said to have already provided vivid expression to Irigaray's utopian scenario of precisely such a world, described as "something that at the same time has never existed and which is already present, although repressed, latent, potential. Eternal mediators for the incarnation of the body and the world of man, women seem never to have produced the singularity of their own body and world" (*An Ethics of Sexual Difference*, 109). This world would have, according to Irigaray, "two vertical and horizontal dimensions: daughter-to-mother, mother-to-daughter; among women, or among 'sisters'" ("Love of Same, Love of Other," 108). For Irigaray, the vertical dimension is

> always being taken away from female becoming. The bond between mother and daughter, daughter and mother, has to be broken for the daughter to become a woman. Female genealogy has to be suppressed, on behalf of the son–Father relationship, and the idealization of the father and husband as patriarchs. But without a vertical dimension . . . a loving ethical order cannot take place among women. Within themselves, among themselves, women need both of these dimensions . . . if they are to act ethically, either to achieve an *in-itself for-itself*, a move out of the plant life into the animal, or to organize their "animal" territoriality into a "state" of people with its own symbols, laws and gods. ("Love of Same, Love of Other," 108–109)

The female-only utopias we are concerned with here incarnate precisely these ideals put forward by Irigaray, foremost among them the fundamental significance of the mother–daughter bond. Also like Irigaray, who is intent on configuring a new horizon for women's trancendence, unlike Simone de Beauvoir who, according to Irigaray, believes that "woman remains always within the dimension of immanence and that she's incapable of transcendence" (*Women Writing Culture*, 142), Gilman firmly pursues and dramatizes a vision of women's transcendence in *Herland*. For Irigaray, as for Gilman, woman "needs her own linguistic, religious and political values. She needs to be situated and valued, to be *she* in relation to her self" (*Elemental Passions*, 3), a subject position that the women-only worlds considered here go a long way toward providing.

In *Of Woman Born: Motherhood as Experience and Institution*, Adrienne Rich, in Irigarayan mode, exhorts women to repossess their bodies, to start thinking "through their bodies" (284), the conditions for which practice she believes exist: "There is for the first time today a possibility of converting our physicality into both knowledge and power" (284). Rich goes on to argue that "we need to imagine a world in which every woman is the presiding genius of her own body. In such a world women will truly create new life, bringing forth not only children (if and as we choose) but the visions, and the thinking, necessary to sustain, console and alter human existence—a new relationship to the universe" (285–286), words that fit the women-only worlds in *Mizora*, *Herland*, "Houston, Houston, Do You Read?" and *Motherlines*.

Donna Haraway similarly intervenes in this argument by emphasizing that there is "a screaming need for a much richer array of narrative toolkits to work these eruptions of the unexpected and irreducible, these eruptions of what you've got to call an 'unconscious' into our lives through other stories besides Oedipal stories" (Interview with Gary A. Olson, "Writing, Literacy and Technology: Toward a Cyborg Writing," 73), a need that amounts to rewriting many psychoanalytic scenarios from women's point of view and reinscribing these narratives in the political and collective unconscious, a goal that Irigaray argues for and that I feel Gilman gave fictional expression in *Herland* best among the texts under consideration. These texts can thus profitably be read as dramatizing some of Luce Irigaray's most persuasive and influential pronouncements that attempt to reconceptualize women's role and positioning in our contemporary world, as well as putting forward bold outlines of a future world where women would no longer be the other of man but subjects in their own right, for-themselves (*pour-soi*), to borrow Irigaray's terminology.

A SOCIETY OF THEIR OWN

Hélène Cixous and Catherine Clément submit that "somewhere every culture has an imaginary zone for what it excludes, and it is that zone that we must remember *today*" (*The Newly Born Woman*, 6). I wish to suggest that the works that I am concerned with here can be said to derive their imaginative impulse from and take place in precisely the kind of imaginary zone invoked by Cixous and Clément, where women can contemplate the utopian creation of a society of their own. Simone de Beauvoir, however, points out the pitfalls that stand in the way of the creation of such a society:

> Sometimes the "feminine world" is contrasted with the masculine universe, but we must insist again that women have never constituted a closed and independent society; they form an integral part of the group, which is governed by males and in which they have a subordinate place. They are united only in a mechanical solidarity from the mere fact of their similarity, but they lack that organic solidarity on which every unified community is based; they are always compelled . . . to band together in order to establish a counter-universe, but they always set it up within the frame of the masculine universe. Hence the paradox of their situation: they belong at one and the same time to the male world and to a sphere in which that world is challenged; shut up in their world, surrounded by the other, they can settle down nowhere in peace. (*The Second Sex*, 562)

The women-only worlds of Mizora, Herland, the Earth (in Tiptree's story), as well as the Riding Women's territory, thus constitute utopian exceptions to Beauvoir's reductionist scenario, counteruniverses or wild zones where new societal organizations are tried and tested. As Nina Auerbach notes, "A community of women is a rebuke to the conventional ideal of a solitary woman living for and through men, attaining citizenship in the community of adulthood through masculine approval alone" (*Communities of Women: An Idea in Fiction*, 5), a point of view at least partially shared by Luce Irigaray. Having served throughout the ages as mirrors to man's narcissism and aggrandizement, woman, on the contrary, "has no mirror wherewith to become woman" ("Divine Women," 67), as Irigaray insists. As the male narrator in *Herland* rightly observes,

> These women, whose essential distinction of motherhood was the dominant note of their whole culture, were strikingly deficient in what we call "femininity." This led me very promptly to the conviction that those "feminine charms" we are so fond of are not feminine at all, but merely reflected masculinity—developed to please us because they had to please us, and in no way essential to the real fulfillment of their great process. (60)

For Irigaray, "Community means only dependence as long as each man, each woman, is not free and sovereign. Love of the other without love of self, without love of God, implies the submission of the female one, the other, and of the whole of the social body" ("Divine Women," 68). A very strong sense of community, indeed, is one of the predominant features of these women-only societies, far removed from patricentric rules. This polemical vision, then, illustrated in these women-only societies constitutes a utopian dramatization of an antithetical world to the one dominated by a patriarchal dream of sameness to the masculine subject. *Mizora, Herland,* and "Houston, Houston, Do You Read?" for their part, deal with and fictionally concretize, in polemical fashion, a matriarchal fantasy of sameness to the feminine norm.

A "SEX WHICH IS ONE"

In *This Sex Which Is Not One,* Irigaray's ideas appear to describe some of the women-only societies that constitute the object of this study:

> When women want to escape from exploitation, they do not merely destroy a few "prejudices," they disrupt the entire order of dominant values, economic, social, moral and sexual. They call into question all existing theory, all thought, all language, inasmuch as these are monopolized by men and men alone. They challenge *the very foundation of our social and cultural order,* whose organization has been prescribed by the patriarchal system. (165; italics in the original)

Indeed, this seems to me a fitting definition of the kind of project Mary E. Bradley Lane, Charlotte Perkins Gilman, James Tiptree Jr. and Suzy McKee Charnas have illustrated in their fictional work, creating the conditions for the existence of a "sex which is one" (in analogy with Irigaray's phraseology), whose "other lies in the one [*l'un(e)*]" (*This Sex Which is Not One,* 154), giving rise to "an economy of exchange in all of its modalities that has yet to be put into play," in Irigaray's words (154). This strategy also implicitly stresses a valorization of difference, thus avoiding reducing it to the (patriarchal) logic of the same. Saming, however, can work both ways. Do these societies fall prey to that very logic in reverse? Indeed, the question that needs to be asked is the following: What if the "dream of sameness" (Irigaray, *Speculum of the Other Woman*, 248), which Irigaray diagnoses as grounding patriarchal prejudices in an androcentric society, is duplicated in women-only worlds, such as those envisaged by Lane or Gilman? Are those same preconceptions reproduced? Are myths of womanhood perpetuated or deconstructed?

> It is time to theorize an "unfamiliar" unconscious, a different primal scene, where everything does not stem from the dramas of identity and reproduction.
>
> Donna J. Haraway, *How Like a Leaf!*, 123

The striking similarity between such utopian novels as *Mizora* and *Herland*, as well as "Houston, Houston, Do You Read?" and *Motherlines*, written many decades apart, testifies to the repeated patterns and concerns that straddle different periods and emerge when social circumstances, namely the increasing visibility and activism of feminist movements, propitiate an environment conducive to utopian dreaming and speculative cogitations about new ways of organizing life in society.

The utopian texts we have looked at leave, of course, many questions unanswered and many disturbing problems unsolved. How can the values developed by a vision of a women-only society be integrated into an organizational model where men and women exist without one of them imposing its power structures on the other? How can Irigaray's visions be implemented in our contemporary societies? In *This Sex Which Is Not One*, Irigaray reflects on the potential benefits and further implications of the separatist impulse for women. Remarking on the "disconnection from power that is traditionally women's" (32–33), she convincingly puts forward a series of strategic steps to help women achieve a society of their own:

> For women to undertake tactical strikes, to keep themselves apart from men long enough to learn to defend their desire, especially through speech, to discover the love of other women while sheltered from men's imperious choices that put them in the position of rival commodities, to forge for themselves a social status that compels recognition, to earn their living in order to escape from the condition of prostitute. . . . These are certainly indispensable stages in the escape from their proletarization on the exchange market. But if their aim were simply to reverse the order of things, even supposing this to be possible, history would repeat itself in the long run, would revert to sameness: to phallocratism. It would leave room neither for women's sexuality, nor for

women's imaginary, nor for women's language to take (their) place. (*This Sex Which Is Not One*, 33)[47]

These are indeed forceful words, which propound a specific political agenda for women. Adrienne Rich is, on the other hand, critical of "utopian" matriarchal worlds, presumably like Lane's or Gilman's female realms. She suggests that "the commune, in and of itself, has no special magic for women, any more than has the extended family. . . . Above all, such measures fail to recognize the full complexity and political significance of the woman's body, the full spectrum of power and powerlessness it represents, of which motherhood is simply one—though crucial—part" (*Of Woman Born: Motherhood as Experience and Institution*, 282–283). Furthermore, as Rich notes, "It can be dangerously simplistic to fix upon 'nurturance' as a special strength of women" (283), a social construct that has been one of the most important myths used to subordinate women to certain functions while excluding them from others, allegedly more adapted to "masculine" characteristics.

Nina Auerbach, for her part, observes that communities of women "are emblems of female self-sufficiency which create their own corporate reality, evoking both wishes and fears" (*Communities of Women: An Idea in Fiction*, 5). Indeed, reactions to these women-only societies have been rather extreme. While Carroll Smith-Rosenberg, in "The Female World of Love and Ritual" suggests that the intense relationships between nineteenth-century American women can be seen as a source of power and not as some kind of aberrant behavior, Simone de Beauvoir views the establishment of a female-only society as a virtual impossibility, since women cannot escape the "immanence" they are submerged in.

In "Women's Time," in turn, Julia Kristeva reflects on the idea of a "countersociety" (361), which she describes as "a sort of alter ego of official society that harbors hopes for pleasure" (361). As Kristeva further explains, this female society "can be opposed to the sacrificial and frustrating sociosymbolic contract: a countersociety imagined to be harmonious, permissive, free, and blissful. In our modern societies, which do not acknowledge an afterlife, the countersociety is the only refuge for jouissance, for it is precisely an anti-utopia, a place outside the law, yet a path to utopia" (361). This concept of countersociety is precisely what the texts I looked at in this chapter envisage and illustrate, imaginary alternative societies in which new modes of relationality are rehearsed.

Rosi Braidotti is one who argues the case for a separatist zone for women. As she points out, "Women's movements, a separatist space, are essential if women are to speak their desires and to shatter the silence about the exploitation they have undergone" (*Patterns of Disso-*

nance: A Study of Women in Contemporary Philosophy, 251), while Irigaray's position appears more ambiguous, more nuanced. Indeed, it is not Irigaray's goal to "reverse the order of things" (*This Sex Which Is Not One*, 33). As she persuasively argues, "If women are to preserve and expand their autoeroticism, their homo-sexuality, might not the renunciation of heterosexual pleasure correspond once again to that disconnection from power that is traditionally theirs? Would it not involve a new prison, a new cloister, built of their own accord?" (32–33). As Margaret Whitford notes in relation to Irigaray's blueprint for action, separatism, although "not a long-term goal . . . can be an effective short-term strategy, imperative even, for some women" (*Luce Irigaray: Philosophy in the Feminine*, 12). Indeed, again like Gilman's, Irigaray's vision most definitely includes men, provided that men and women can be two, following a "new model of possible relations between man and woman, without the submission of one to the other" ("'Je-Luce Irigaray': A Meeting with Luce Irigaray," 145).

In tune with Irigaray, I consider the fantasy of a woman-only world a strategically empowering dream, albeit a temporary one, which serves the very important function of helping us reflect on how we would like our world to change and be reformed in very concrete ways. Could the advent of human cloning, then, in light of this discussion, be the condition that brings about the possibility for the existence of a "sex which is one," by analogy with Irigaray's terminology, and no longer "this sex which is not one"? Irigaray's utopian and strategic dream of an autonomous woman, of women bonding together, given prophetic illustration in *Mizora*, *Herland*, and "Houston, Houston, Do You Read?" might eventually come to pass, even if on an egalitarian basis, with male members in a society where each gender would be a valuable part, with its own specificities, and sexual discrimination would be a thing of the past.

In "Post-Phallic Culture: Reality Now Resembles Utopian Feminist Science Fiction,"[48] Marleen S. Barr argues that there is a current general "penchant for phallic eradication" (69) and that we live in a "post-phallic culture" (68), a view that may be said to have already been retrospectively dramatized in the utopian texts that depict women-only worlds. Barr's argument is that "new cloning cognition yields old misogyny" (195), since "childbirth, whether natural or artificial, is something patriarchy seeks to control" (203). While I partially agree with Barr, on the other hand, she does not consider the alternative scenario, according to which cloning would allow women to be independent from men in reproductive terms. In *Alien to Femininity: Speculative Fiction and Feminist Theory* (1987), however, Barr discusses several novels that put forward precisely that kind of vision, such as

Jody Scott's *I, Vampire* (1984), where women can use parthenogenesis to have children, thus eliminating the need for men and abolishing the father's role in the nuclear family. Other examples are Joan D. Vinge's *Snow Queen* (1980), where the queen, Arienrhod, a genetic engineer, has a cloned daughter, Moon, who will in turn rule the kingdom (queendom), and Sydney J. Van Scyoc's *Star Mother* (1976), in which the protagonist, Jahna Swiss, acts as a "mold-mother" (128), passing her characteristics on to her "moldings" (128) who, exposed to her, will acquire her psychological constitution, so that in them Swiss can see herself "reborn" (128). As Barr maintains, Vinge's feminist version of cloning "shows that women as well as men can benefit from this particular reproductive method" (*Alien to Femininity: Speculative Fiction and Feminist Theory*, 147). According to Barr, Vinge's and Van Scyoc's vision of cloning "immortalizes rather than erases women's experiences" (143), a point of view with which I agree and which can be applied to speculations about women's use of cloning technology, should it ever become available.

However, in "'We're at the Start of a New Ball Game and That's Why We're All Real Nervous': Or, Cloning—Technological Cognition Reflects Estrangement from Women," Marleen S. Barr stresses how "cloning's advent reflects the desire to replicate men and eradicate women" (195). If this is the case, how can this view be squared with her argument that there is a pronounced tendency in our contemporary world for phallic extermination and that in effect we live in a postphallic society? Indeed, Barr diagnoses in Western culture what she describes as the "recent prevalent de-emphasis of phallic power" ("Post-Phallic Culture: Reality Now Resembles Utopian Feminist Science Fiction," 68), which for Barr corresponds to a "separatist feminist science fiction scenario" (68). In words to which I completely subscribe, Barr asserts that "cloning is exceedingly threatening—to patriarchy" (76), "the means to achieve women's ultimate reproductive freedom" (77), a scenario envisioned in the texts analyzed in this chapter.

As I have suggested, cloning technology can be seen as articulating the possibility of gradually making patriarchy obsolete. Although Marleen S. Barr's description of our Western culture as "post-phallic" seems to me overly and unduly optimistic, her statement that cloning "gives women unprecedented reproductive power," transforming "female reality into separatist science fiction" ("Post-Phallic Culture: Reality Now Resembles Utopian Feminist Science Fiction," 77) is a persuasive and strategic vision of what may become reality in the near future. The next chapter, however, will address a complementary but opposing set of visions that focus on the patriarchal fantasy of making women redundant in terms of reproductive agency.

NOTES

1. It should, however, be noted that there are two different narratives describing the creation of the first human beings in Genesis. See, for example, Gregory Clayes and Lyman Tower Sargent, eds., *The Utopia Reader*, 9–11.

2. See Chapter 3.

3. One of the first that would come to mind for a Westerner would be the Virgin Mary giving birth to Jesus, "fathered" by the Holy Spirit. Other famous virgin mothers who gave birth to notable sons are Isis, Cybele, Innana, the Sumerian goddess of fertility, and Buddah's mother, Mamaya. See Julia Stonehouse, *From Idols to Incubators: Reproduction Theory through the Ages*.

4. In *The Second Sex*, Simone de Beauvoir refers to the process of parthenogenesis in words that fittingly apply to the societies in *Mizora* and *Herland*: "In cases of parthenogenesis the egg of the virgin female develops into an embryo without fertilization by the male, which thus may play no role at all. . . . More and more numerous and daring experiments in parthenogenesis are being performed, and in many species the male appears to be fundamentally unnecessary" (36).

5. For a discussion of the theme of the war of the sexes from the mid-1920s until the present, see Justine Larbalestier, *The Battle of the Sexes in Science Fiction*.

6. For a spirited discussion of the topic of the war of the sexes and further fictional examples, see Joanna Russ, "*Amor Vincit Foeminam*: The Battle of the Sexes in Science Fiction."

7. Examples of fictional societies where men are not needed for reproduction abound: Mary E. Bradley Lane, *Mizora* (1890), Florence Ethell Mills Young, *The War of the Sexes* (1905), Irene Clyde, *Beatrice the Sixteenth* (1909), Charlotte Perkins Gilman, *Herland* (1915), Poul Anderson, *Virgin Planet* (1959), Gwyneth Jones, *Divine Endurance* (1984), James Tiptree Jr., "Houston, Houston, Do You Read?" (1976), Edmund Cooper, *Who Needs Men?* (1972), Suzy McKee Charnas, *Motherlines* (1978), Joan Slonczewski, *A Door Into Ocean* (1986), Philip Wylie, *The Disappearance* (1951), Virgilio Martini, *Il Mondo Senza Donne* (1935), Nicola Griffith, *Ammonite* (1992), Cecilia Holland, *Floating Worlds* (1971), Joanna Russ, *The Female Man* (1975), Marion Zimmer Bradley, *The Ruins of Isis* (1978), Sally Miller Gearhart, *The Wanderground* (1979), Donna J. Young, *Retreat: As It Was* (1979), Josephine Saxton, *Travels of Jane Saint and Other Stories* (1986), Sheri Tepper, *The Gate to Women's Country* (1988), Rochelle Singer, *The Demeter Flower* (1980), Pamela Sargent, *The Shore of Women* (1987), and Anna Wilson, *Hatching Stones* (1991). The authors of these visions are predominantly female, suggestive of a deep-seated desire for autonomy from males in terms of procreation. On the other hand, the examples of such women-only societies portrayed by men invariably suggest that those women-only communities unconsciously yearned for the arrival of men who, without fail, set things right, as they "should be," seen from a male point of view, including joint male and female procreation in the case of Anderson's *Virgin Planet* and Cooper's *Who Needs Men?* Indeed, the mere suggestion that women might possibly be better off without men is thoroughly scoffed at and scorned as heretical.

8. See Brian Attebery, "Women Alone, Men Alone: Single-Sex Utopias"; Lynn Williams, "Separatist Fantasies, 1690–1997: An Annotated Bibliography."

9. See Brian Aldiss, Introduction to Mary Shelley, *The Last Man*.

10. See also Thomas S. Gardner, "The Last Woman" (1932).

11. For a further problematization of this topic, see also Peter Fitting, "So We All Became Mothers: New Roles for Men in Recent Utopian Fiction"; idem, "For Men Only: A Guide to Reading Single-Sex Worlds"; idem, "Reconsiderations of the Separatist Paradigm in Recent Feminist Science Fiction."

12. The boundaries between the genres of science fiction and utopias have become increasingly blurred and indeed often intersect.

13. Russ is referring specifically to a cluster of utopias written in the 1970s, including Tiptree's "Houston, Houston, Do You Read?" but *Mizora* and *Herland* can be included as precursor texts.

14. See also Joanna Russ's "For Women Only: Or, What Is That Man Doing under My Seat?"

15. Florence Ethell Mills Young is here drawing on the work of scientist Jacques Loeb, who in 1901 developed, in Susan Merrill Squier's words, "a technique for parthenogenesis through the mechanical stimulation of sea-urchin eggs" (*Babies in Bottles: Twentieth-Century Visions of Reproductive Technology*, 183), which demonstrated that "unfertilized sea-urchin ova would develop to the larval stage when mechanically stimulated" (52).

16. For a discussion of the theme of the battle of the sexes in utopian and science fiction texts from the mid-1920s to the present, see Jane L. Donawerth, "Science Fiction by Women in the Early Pulps, 1926–1930"; Justine Larbalestier, *The Battle of the Sexes in Science Fiction*.

17. See Angela Ingram and Daphne Patai, "Fantasy and Identity: The Double Life of a Victorian Sexual Radical."

18. Quoted in John Carey, ed., *The Faber Book of Utopias*, 294.

19. I quote from the University of Nebraska Press edition of *Mizora*.

20. Quoted in Debra Benita Shaw's *Women, Science and Fiction: The "Frankenstein" Inheritance*, 186.

21. However, it seems to me that this idea runs counter to Gilman's strong emphasis on the bond between mother and daughter. Whose mother and whose daughter will bond with whom?

22. Here again we might pause a little. Van is the narratorial voice throughout the book, a strategy that, as Frances Bartkowski notes, by giving him "narrative control reasserts his dominance—he reconstructs by memory the events" (*Feminist Utopias*, 28).

23. In her introduction to Charlotte Perkins Gilman, *With Her in Ourland*, Mary Jo Deegan traces Lester Ward's influence on Gilman, noting how "Ward's discussions of 'gynaecocracy' and 'parthenogenesis,' his systems model approach, and his five-stage theory of societal revolution" (32) were applied by Gilman in *Herland* and *With Her in Ourland*.

24. As Jean Pfaelzer explains, "Through utopian eugenics, Lane implements the racist positivism of Herbert Spencer and the disciples of Jean Lamarck who promised that controlled breeding could swiftly transmit genetic changes from generation to generation and deliver Caucasian dominance through rapid evolution" ("Utopians Prefer Blondes—Mary Lane's *Mizora* and the Nineteenth-Century Utopian Imagination," xxxi). In "'Fecundate! Discriminate!' Charlotte Perkins Gilman and the Theologizing of Maternity," Sandra M. Gil-

bert and Susan Gubar discuss Gilman's ambivalent maternal politics, pointing out that "the way in which Gilman conceives woman as the First Sex attributes a unique eugenic centrality to mothers that not only undermines the feminist movement's ideal of sex equality but even degenerates into precisely the racism that marks much Social Darwinist thinking about racial betterment" (201).

25. As Joanna Russ observes in relation to several recent female utopias, they "not only ask the same questions and point to the same abuses; they provide similar answers and remedies" ("Recent Feminist Utopias," 136).

26. For examples of other tales that similarly dramatize the female wish to exert power over men and eventually lead to their eradication, see Justine Larbalestier's *The Battle of the Sexes in Science Fiction*, 51.

27. See Joanna Russ's "*Amor Vincit Foeminam*: The Battle of the Sexes in Science Fiction," 48–49.

28. However, as Creed pertinently notes, "What is most interesting about the mythological figure of woman as the source of all life (a role taken over by the male god of monotheistic religions) is that, within patriarchal signifying practices . . . she is reconstructed and represented as a *negative* figure" (*The Monstrous-Feminine: Film, Feminism, Psychoanalysis*, 129).

29. Quoted in Sandra M. Gilbert and Susan Gubar, "'Fecundate! Discriminate!' Charlotte Perkins Gilman and the Theologizing of Maternity," 200. As Gilbert and Gubar note, "Why would an emancipated woman who had spent a lifetime struggling for women's rights and sex equality abjure her own goals in favor of a quasi-Victorian celebration of maternity?" (201).

30. For a discussion of some of the reasons that might have led Charlotte Perkins Gilman to adumbrate this line of argumentation, see Sandra M. Gilbert and Susan Gubar, "'Fecundate! Discriminate!' Charlotte Perkins Gilman and the Theologizing of Maternity."

31. See also Patrice DiQuinzio, *The Impossibility of Motherhood: Feminism, Individualism, and the Problem of Mothering*; Grace Jantzen, *Becoming Divine: Towards a Feminist Philosophy of Religion*, for a discussion and problematization of the concepts of a symbolic of natality, of birth, and the notion of motherhood.

32. Quoted in Patricia Bowen-Moore, *Hanna Arendt's Philosophy of Natality*, 22.

33. An assertion whose validity may change in the not very distant future, with the possible introduction of artificial wombs or even of male pregnancy.

34. Indeed, as Mary Jacobus remarks, sounding a warning about some of the dangers inherent in the sexual politics of mothering "even as women refuse the traditional equation of femininity and motherhood, they may be compelled to make some claims under the sign of maternity in order to protect themselves in the workplace, under the law, and . . . in the face of the subtle but powerful workings of paternalist sexual ideologies and the symbolic systems that sustain them" (*First Things: The Maternal Imaginary in Literature, Art, and Psychoanalysis*, 39).

35. However, as Debra Benita Shaw argues, in *Herland* and its sequel Gilman presents "an alternative to what she called 'the Adventures of Him in Pursuit of Her'" (*Women, Science and Fiction: The "Frankenstein" Inheritance*, 13). Indeed, in *The Man Made World: Or, Our Androcentric Culture* (1911), Gilman de-

scribes the conventional outcome of those "Adventures of Him in Pursuit of Her" as "stopping when he gets her. Story after story, age after age, over and over, this ceaseless repetition of the Preliminaries" (99), a scenario that Gilman rewrites in *Herland*, where Ellador dictates the rules of her relationship with Van Jennings.

36. Naomi Jacobs notes how Mizorans' "distaste for physicality is evident as well in the Mizoran attitude to bodily processes" ("The Frozen Landscape in Women's Utopian and Science Fiction," 196).

37. The question of sexuality in *Herland* is a vexed one. Although it is elided from the text, it does not seem to evoke abject connotations, were it to surface, as it does in *Mizora*. Val Gough, however, describes Herlanders as "clearly lesbians as conceived by the utopian separatist lesbianism of the late 1970s, which stressed the collectivity of lesbian identity and perceived women's needs as nurturance and interrelatedness, articulated in terms of the desire for the pre-oedipal mother" ("Lesbians and Virgins: The New Motherhood in *Herland*," 197). Sandra M. Gilbert and Susan Gubar, on the other hand, referring to Herlanders, write about an "erotics of maternity," which they see as complementing "the same reverential replacement of God the father with 'Maternal Pantheism' that marks *His Religion and Hers*. Passion is not a prelude to motherhood but equal to motherhood" ("'Fecundate! Discriminate!' Charlotte Perkins Gilman and the Theologizing of Maternity," 205).

38. I find this vision depressing, for it suggests an utter lack of individuality as far as the women belonging to the same gene pool are concerned, as if they were interchangeable, functioning almost like automatic replacements.

39. This vision does not preclude, obviously, women having cloned sons from a male partner, for example, as in Lisa Tuttle's "World of Strangers," but it would certainly enable women who so wished to circumvent the agency of the father and fulfill the dream of virgin-mother fantasies that seem, after all, as widespread as the corresponding male ones.

40. Why "Bride," though? Why not "Mother," to provide a counterpart to "Father," or "Woman"?

41. As Braidotti stresses, there is no sentimentality involved in Irigaray's reassessment of the maternal feminine. According to Braidotti, Irigaray "acknowledges that motherhood is also the site of women's capture into the specular logic of the Same, which makes her subservient to the Masculine. Maternity, however, is also a resource for women to explore carnal modes and perception, of empathy and interconnectedness that go beyond the economy of phallogocentrism" (*Metamorphoses: Towards a Materialist Theory of Becoming*, 23) as Irigaray herself has so eloquently illustrated in, for instance, "The Bodily Encounter with the Mother," as well as Julia Kristeva in her powerful "Stabat Mater." Braidotti considers this "'other' maternal feminine in Irigaray as linked to the political project of providing symbolic representation for the female feminist subject as a virtual subject-position that needs to be created and activated" (23–24).

42. Posthumous's wish, which is after all shared by many men, will be examined in Chapter 3.

43. As Veronica Hollinger notes, "Houston, Houston, Do You Read?" "rewrites Russ's 'When It Changed' to the extent that it manages a 'happy end-

ing'—or at least an imaginatively satisfying one" ("(Re)reading Queerly: Science Fiction, Feminism, and the Defamiliarization of Gender," 201), while nevertheless suggesting "the high price of freedom for women" (201).

44. This scenario also prompts references to Fay Weldon's *The Cloning of Joanna May* (1989) where, after their first coming together, the four clones of Joanna May start to realize that their similarity constitutes a powerful bond, a source of strength and not a drawback, suggesting the enabling connection developed by the four clones and their genetic mother.

45. Here I suspect Tiptree is, even if unwittingly, falling into the trap of attributing some "essentialist" characteristics and stereotypes to women in general.

46. Like Marge Piercy in relation to *Woman on the Edge of Time*, a book directly inspired by and written as a response to Shulamith Firestone's *The Dialectic of Sex*, Suzy McKee Charnas explains that during the winter of 1972–1973 (*Walk to the End of the World* [1974], is the first in the series of novels to which *Motherlines* belongs) she "was doing what so many other women were doing and are still doing: reading books like Shulamith Firestone's *The Dialectic of Sex* (Bantam, 1971) and *Sisterhood Is Powerful* edited by Robin Morgan (Vintage, 1970) and participating in consciousness-raising sessions with other women" ("A Woman Appeared," 104).

47. Irigaray's words convey a vision that is strikingly similar to the one put forward in *Mizora*.

48. As far as postphallic culture is concerned, Barr's vision seems to me highly exaggerated. Even conceding that great changes are occurring, which are transforming gender definitions and making our contemporary world more egalitarian and increasingly more conscious of women's needs and desires, I would prefer a different way of phrasing this new "postphallic" condition that would not call attention to the phallus as a signifier of power (even if that is no longer the case in such a pronounced manner).

3

The Zeus Syndrome: Womb Envy and Male Pregnancy

> That father without mother may beget, we have Present, as proof, the daughter of Olympian Zeus: One never nursed in the dark cradle of the womb; Yet such a being no god will beget again.
>
> Aeschylus, *Eumenides*, 662–665

> Oh, Cassandra, for Heaven's sake let us devise a method by which men may bear children! It is our only chance.
>
> Virginia Woolf, "A Society," 20–21

In this chapter I wish to examine the fantasy of male pregnancy and womb envy, translated into jealousy of woman's reproductive capabilities and the desire to bypass woman's body in order to create other human beings. I would also like to reflect on some of the scenarios to which the potential implementation of this ambition could give rise in the not-so-distant future.

Fantasies of masculine autoreproduction go back a long time, from classical mythology and the Genesis story of the creation of Adam and Eve[1] to medieval alchemists' search to create a homunculus, finding powerful symbolic expression in such texts as Mary Shelley's *Frankenstein* (1818)[2] and H. G. Wells's *The Island of Dr Moreau* (1896), to cite only some of the most salient instances. Mary Shelley's *Frankenstein* constitutes a narrative about, among many other things, a man's wish to create life, a yearning that can be read as hiding the more profound desire of giving birth and also of becoming woman.[3]

Human cloning could fulfill the primarily masculine fantasy of creating life, procreating without the help of the woman, a world not "of woman born," while also concretizing the often manifested feminine desire of giving birth without male intervention. Indeed, fantasies of male power and dominance will of necessity engage with and encompass, even if only at a mythical level, buried or deeply suppressed wishes that may surface in dreams and fiction, the fantasy of reproductive autonomy. These dreams of male self-procreation can be seen as symptomatic of fantasies of patriarchal domination and self-perpetuation, eliminating the need to have women as essential parts of the reproductive process.[4] Here I will concentrate on some more recent examples that I want to connect with the powerful fantasy of human cloning,[5] intimately correlated with the aspiration of self-generation: Sven Delblanc's *Homunculus: A Magic Tale* (1969), Peter Ackroyd's *The House of Doctor Dee* (1993), and Anna Wilson's *Hatching Stones* (1991).

THE MYTHOLOGY OF THE FATHER AS THE SOLE PARENT

The recurring masculine desire for self-engendering has found expression in many cultural manifestations, among which can be cited the numerous narratives of male autoreproduction, as well as artistic images of gods bearing children from their own bodies.[6] Zeus bore Athena, who sprang fully formed and armed from her father's head,[7] as well as Dionysus,[8] whom he carried in his thigh, a process that can be seen as anticipating the creation of the paternal womb.[9] Zeus thus becomes the quintessential mythological father who bypasses the mother in the concretization of the male dream of a motherless birth.[10] In *Marine Lover of Friedrich Nietzsche*, Irigaray, reflecting on the sexual politics implicit in the generation and birth of many Greek goddesses and gods, mentions Zeus, who "acts like a mother, by giving birth to daughter and son" (161) as well as Dionysos, the "man-god . . . who no longer falls back into the chaos of woman's womb" (162) and who is born twice. As Irigaray notes,

> Conceived in the womb of a mortal by the god of gods, he did not mature in this original nurse. Is he saved from being doubly swallowed up? Saved first from a primordial nature that might have held him back from differentiation, second from being devoured–assimilated by the father who immediately repossesses his product and takes back from the mother the property deposited with her. (130)

Marina Warner, in turn, recapitulates Dionysus's origins as follows:

> Semele was . . . expecting Zeus's child, and this baby, six months in the womb, was saved by Hermes, who took him from his mother's body and sewed him into the thigh of his father to gestate there for the further three months needed. Then, as Euripides describes in the *Bacchae*, Zeus undid the golden clips that held the child in the *paternal womb*, and the infant Dionysus was safely born. (*Monuments and Maidens: The Allegory of the Female Form*, 121; emphasis mine)

In this context, Adriana Cavarero considers the birth of Dionysus as "particularly interesting, because its symbolic underpinnings are pervaded with envy/appropriation of maternal power, and overshadowed by the preventive matricide of Semele" (*In Spite of Plato: A Feminist Rewriting of Ancient Philosophy*, 108).[11]

In a related vein, and with reference to the mythical origins of the Athenians, who believe they descend from Erichthonius, son of Hephaestus, whose semen fell on Athenian earth, which nurtured and gave birth to Erichthonius without female intervention, Vigdis Songe-Moller concludes that what this myth suggests is "some ideal of one-sex humanity, where all children are boys, and each child originates from the father alone" (*Philosophy without Women: The Birth of Sexism in Western Thought*, 5).[12] According to the myth, then, Erichthonius has a father but not a mother, a state of affairs that could come to be fulfilled with human cloning, which would enable men to produce their own offspring provided they could have access to artificial wombs and eggs (even in this scenario they would still need women for the eggs, until the time artificial or surrogate eggs can be developed in laboratory).[13]

In Hesiod's long poem about the origins of the Olympian deities, *Theogony*, Kronos,[14] who is the leading god in the second generation, upon hearing that one of his children will replace him, devours them all. The deeper implications of this myth, as Marina Warner notes in *No Go the Bogeyman: Scaring, Lulling, and Making Mock*, inscribe Kronos as a "surrogate mother, giving birth from his mouth to the gods and goddesses as the issue of his body" (54). Kronos's "act of devouring his brood forms a prelude to birth; being eaten equals incorporation, and this in turn stands for a surrogate though unwitting pregnancy—of the male. In this way, his progeny re-enter the world, twice born of their father, begotten and brought forth" (54).

> If only children could be got some other way, Without the female sex!
>
> Euripides, *Medea*, 573–574

The wish voiced by Jason in Euripides' *Medea* is also the often repressed and unarticulated ambition experienced by many men, the aspiration for reproductive self-sufficiency, without having to depend on women for self-procreation.[15] As Teresa de Lauretis remarks, "Men will have to imagine other ways to deal with the fact that they, men, are born of women" (*Alice Doesn't: Feminism, Semiotics, Cinema*, 156),[16] which they do, as writers, by fictionalizing this wish, and as scientists by trying to develop the necessary technology to fulfill the dream of self-reproduction. In this context, as Rosi Braidotti remarks,

> Classical mythology represents no founding hero, no main divine creature or demigods as being of woman born. In fact, one of the constant themes in the making of a god is his "unnatural" birth: his ability, through subterfuges such as immaculate conceptions and other tricks, to short-circuit the orifice through which most human beings pop into the spatio-temporal realm of existence. (*Nomadic Subjects: Embodiment and Sexual Difference in Contemporary Feminist Theory*, 84)

The male impulse to self-reproduction can be seen, then, on the one hand, as symptomatic of male fears about woman and her reproductive capacities and organs, which can be perceived symbolically as a threat to virility, and, on the other, as indicative of womb envy, jealousy of those very reproductive capabilities that are denied to men. These two threads can be seen as intricately intertwined.[17]

The dream of male autoreproduction can take two main forms: the creation of another human being and the dream of a world without women, a fantasy that also has a venerable history. The vision that Euripides' Hippolytos conveys of such a state of affairs is particularly telling:

> Zeus, let me set you straight about women.
> Men chase their glitter, but it's all fake.
> You were mistaken to flood our lives with them—
> because if your purpose was to ensure
> perpetuation of the human race
> you could have by-passed women completely.
> A better idea would be this—
> To let prospective fathers
> Come to your temples and pay you
> In bronze, iron, or solid gold
> For seeds which will flourish into men,
> Each father paying for his sons[18]
> In proportion to his wealth and status. (940–952)[19]

Hyppolytos's appeal for a world without women can be seen as mirroring, as Nancy Tuana observes, referring to the *Theogony*, "Hesiod's account of the creation of humans in which the creation of woman is represented as secondary, both temporally and metaphysically, to the creation of man" ("Aristotle and the Politics of Reproduction," 195). In the *Theogony*, "Hesiod depicts mankind as existing prior to the creation of women. Putting the two accounts together, one can conclude that the original race, the golden race, was a race of men only . . . for it is only with the creation of women that man lived a life of suffering, illness, and toilsome hardship" (196).

Following the accent on male rationality and female subordination, based on woman's supposed association with "nature," emotions, and intuition, Aristotle and Plato refer to intellectual activity, to the conception of ideas, in markedly male terms, which suggest the appropriation of female parturition powers by men.[20] In Plato's *Theaetetus* this is particularly conspicuous in Socrates' rhetoric: "My art of midwifery resembles the usual sort except in one way: the person giving birth is a man, not a woman, and souls, not bodies, are delivered" (150b). As Socrates further emphasizes, "Those who join in with me undergo the same experience as women giving birth: they have labour pains, and are troubled day and night with anxieties, even more so than pregnant women. My art has the power to arouse and to soothe these pains" (151 a–b). Socrates's assimilation of his art with the role of a midwife can be read as, on the one hand, betraying a certain amount of jealousy of women's reproductive organs and functions while, on the other, excluding women from the pursuit of knowledge, since he only envisages men as "giving birth" to wisdom and rational thought, a male prerogative according to the Greek philosopher.[21]

Looking at different practices of gendering, Anne Cranny-Francis, in *The Body in the Text* (1995), notes how many male artists use body metaphors of reproduction to figuratively refer to their work. According to Cranny-Francis, "The feminine and woman are identified not with production but with reproduction, which in turn is considered a second-order act of creativity. As embodied subjects with the potential for reproduction, women find themselves positioned in this society as second-order producers—despite the fact that only they have the power to bodily (re)produce. Hence it is doubly ironic that male artists favour the use of this metaphor" (36), a discourse, however, that takes us back to similar Platonic and Aristotelian reflections on this subject.[22]

Andrew Marvell similarly dreams of a womanless world in the poem "The Garden": "Such was that happy garden-state, / While man there walked without a mate: / After a place so pure, and sweet, / What other help could yet be meet! / But 'twas beyond a mortal's share / To wander solitary there: / Two paradises 'twere in one / To live in paradise alone" (l. 64–71). John Milton, in *Paradise Lost*, in turn, clearly manifests a desire for masculine reproduction disassociated from women, who are seen as inferior and defective beings: "O why did God, / Creator wise, that people'd highest Heav'n / With Spirits Masculine, create at last / This noveltie on Earth, this fair defect / Of Nature, and not fill the World at once / With Men as Angels *without Feminine,* / Or find some other way to generate / Mankind?" (10: 888–895; emphasis mine).

Sir Philip Sidney also gave voice to this notion of the figuratively pregnant male body in the throes of parturition, while producing a poem, in "Astrophel and Stella":

"Thus, great with childe to speak, and helplesse in my throwes,
Biting my trewand pen, beating myselfe for spite,
Fool, said my Muse to me, looke in thy heart and write." (1: 12–14)

The male fantasies of self-procreation, in terms of both literary and physical progeny, thus often appear intimately interconnected.

ALCHEMICAL PRIMAL SCENES

In alchemy, that immensely rich repository of archetypes as well as of prescientific lore and experimentation, there are several examples of pregnant men giving birth. In Michael Maier's *Atalanta fugiens,* for instance, there is a figure of Boreas carrying the embryo of the Philosopher's Stone in his belly ("The wind carries it in his belly"),[23] while Lambspring in *Dyas chemica* (1625) depicts a king who swallows his son and subsequently lies in bed anxiously expecting his rebirth.[24]

Alchemy is centrally concerned with the search for origins, with beginnings, with the quest for the creation of life. The homunculus was clearly the embodiment of male wishes to create life, to conceive their own progeny from the very start, without recourse to woman. The male alchemist would thus become the sole progenitor of his child. The alchemical jar, for its part, can be seen as a representation of an artificial womb, as a prefiguration of ectogenesis. The homunculus, a tiny, fully formed man who would issue from the male alchemist's jar, can thus be seen as the alchemists' answer to Athena, born fully formed and fully armed from Zeus's head. As Rosi Braidotti remarks, "Alchemy is a *reductio ad absurdum* of the male fantasy of self-reproduction" (*Nomadic Subjects: Embodiment and Sexual Difference in Contemporary Feminist Theory,* 87).[25]

Paracelsus's recipe for creating a homunculus reads like a projection of the male impulse to procreate on his own, without the help of woman. Indeed, this intention is made abundantly clear in the philosophical preamble that introduces the recipe proper:

For there is some truth in this thing, although for a long time it was held in a most occult manner and with secrecy, while there was no little doubt and question among some of the old Philosophers, whether it was possible to Nature and Art, that *a Man should be begotten without the female body and the natural womb.* I answer hereto, that this is in no way opposed to Spagyric Art and to Nature, nay, that it is perfectly possible. (*Hermetic Chemistry*, vol. 1 of *The Hermetic and Alchemical Writings of Aureolus Philippus Theophrastus Bombast, of Hohenheim, called Paracelsus the Great*; emphasis mine).[26]

Paracelsus's formulation to produce the homunculus is as follows:

Let the semen of a man putrefy by itself in a sealed cucurbite with the highest putrefaction of the horse stomach for forty days, or until it begins at last to live, move, and be agitated, which can easily be seen. At this time it will be in some degree like a human being, but, nevertheless, transparent and without a body. If now, after this, it be every day nourished and fed cautiously with the arcanum of human blood, and kept for forty weeks in the perpetual and equal heat of venter equinus, it becomes thencefold a true living infant, having all the members of a child that is born from a woman, but much smaller. This we call a homunculus; and it should be afterwards educated with the greatest care and zeal, until it grows up and starts to display intelligence. (*Hermetic Chemistry*, vol. 1 of *The Hermetic and Alchemical Writings of Aureolus Philippus Theophrastus Bombast, of Hohenheim, called Paracelsus the Great*)

The principles of alchemical creation and metamorphosis have been appropriated by many writers in order to assist in the formation of *homunculi*, babies created without recourse to women and grown outside the womb. From Laurence Sterne's *Tristram Shandy* (1759–1768), Goethe's *Faust* (1790–1832), and Mary Shelley's *Frankenstein* (1818) to D. H. Lawrence's unfinished short story "The Undying Man" (1927),[27] including Sven Delblanc's *Homunculus: A Magic Tale* (1965) and Peter Ackroyd's *The House of Doctor Dee* (1993), male writers have often given dramatic form to this dream of male self-procreation drawing on alchemical symbolism.

WOMB ENVY

Jumping forward to the twentieth century, we see how these perceptions expressed by certain Greek philosophers appear to be confirmed and developed in terms of theoretical paradigms of more modern times. They are, as can be seen, recapitulating male anxieties that seem always to have been there.

Karen Horney, Melanie Klein, Bruno Bettelheim, Peter Lomas, and others have noted how culture is permeated with manifestations and images of what could be interpreted as signs of womb envy. In "The Dread of Woman" (1932), Karen Horney argues that men are profoundly jealous of woman's capacity for motherhood and are thus susceptible to womb envy. Melanie Klein, in turn, agreed broadly with Horney's ideas. In her "Early Stages in the Oedipal Conflict," she suggests that boys mask and overcompensate for their womb envy by a displacement onto the intellectual plane and an emphasis on the penis. Rosalind Minsky similarly addresses this question when she observes that "psychoanalytic theory generally suggests that academic, theoretical knowledge of all kinds, historically dominated by men, may be used as a flight from emotion, as a narcisssistic defense against feeling" (*Psychoanalysis and Culture: Contemporary States of Mind*, 123). Minsky then goes on to correlate this flight from emotion to the concept of "repressed womb-envy" (123), which suggests that

> consciousness and culture may stand for the loss of everything the small boy valued and desired at the beginning of his life and protect him from knowing anything about it. If, as Klein suggests, womb-envy is often concealed in men by displacement onto the intellectual plane, that is, onto reason and knowledge, then culture represents not only repressed desire *for* the mother but also a repression of the desire to be what the mother *is*. Culture then represents the male substitute for procreation when childbirth is not an option. (*Psychoanalysis and Culture: Contemporary States of Mind*, 123–124)

It is precisely this knotted question that Plato provides such vivid illustration of in the passages cited earlier.[28]

Freud himself wrote three studies that described the desire on the part of males to bear children or possess female reproductive organs: "Little Hans" (1909), "Dr Schreber" (1911),[29] and "The Wolf Man" (1918). In "Beyond the Pleasure Principle" (1920), as Freud explains, "The tie of affection, which binds the child as a rule to the parent of the opposite sex, succumbs to disappointment, to a vain expectation of satisfaction or to jealousy over the birth of a new baby—unmistakable proof of the infidelity of the object of the child's affections" (291). It sometimes follows, as Freud notes, that the boy's "own attempt to make a baby himself, carried out with tragic seriousness, fails shamefully" (291).

As Eva Feder Kittay remarks, Freud had already referred to children's theories of anal birth in 1908, when he wrote,[30]

> It was only logical that the child should refuse to grant women the painful prerogative of giving birth to children. If babies are born through the anus, then a man might give birth just as well as a woman . . . a boy [can therefore] imagine that he, too, has children of his own, without there being any need to accuse him . . . of feminine inclinations. He is merely giving evidence . . . of the anal eroticism which is still alive within him. ("Rereading Freud on 'Femininity' or Why Not Womb Envy?" 193)

However, as Feder Kittay notes,

> Freud himself never developed any of the theoretical considerations favoring the inclusion of womb envy into the heart of psychoanalytic theory—in spite of Freud's own case studies documenting fantasies on the part of men and boys for woman's organs and functions. In these studies, Freud denies womb envy explanatory power by reducing the apparent manifestations to symptoms of some other underlying psychological reality. ("Rereading Freud on 'Femininity' or Why Not Womb Envy?" 194)

Drawing on the Freudian theory of penis envy on the part of women, Kittay puts forward the hypothesis that men might experiment with a parallel envy, which she designates as "womb envy" ("Womb Envy: An Explanatory Concept," 95). As Kittay states, it is apt to wonder whether men "might not envy women's distinctive sexual apparatus and functions and whether such envy might not constitute a significant element in their personality development having explanatory force with regard to both individual actions and cultural institutions" (95).[31]

Jacques Lacan, in turn, elaborating on the concept of the body in bits and pieces, the serialized, compartmentalized impressions experienced by the infant, later reorganized into a unified body picture, includes in these sensations fantasies of male pregnancy:

> Such typical images appear in dreams, as well as in fantasies. They may show, for example, the body of the mother as having a mosaic structure like that of a stained glass window. More often, the resemblance is to a jig-saw puzzle, with the separate parts of a man or an animal in disorderly array. Even more significant for our purposes are the incongruous images of which strange trophies, trunks, are cut up in slices and stuffed with the most unlikely fillings, strange appendages in eccentric positions, reduplications of the penis, images of the cloaca represented as a surgical excision, often accompanied in male patients by *fantasies of pregnancy*. ("Some Reflections on the Ego," 13; emphasis mine)

Bruno Bettelheim, for his part, examines a welter of initiation, puberty, and other rites, namely circumcision and couvade, in order to show that "certain of these rites originate in the adolescent's attempts to master his envy of the other sex" (*Symbolic Wounds: Puberty Rites and the Envious Male*, 21), as expressions of male fascination with and jealousy of women's reproductive functions and organs. Bettelheim's examination of the behavior of a group of boys led him to conclude that there is a pervasive desire in boys "to be able to bear children and the feeling of being cheated because it cannot be done" (31).[32] This longing has found fictional expression in many literary works, among which are Sven Delblanc's *Homunculus: A Magic Tale* and Peter Ackroyd's *The House of Doctor Dee*.

The monster regarded his maker.

Sven Delblanc, *Homunculus: A Magic Tale*, 49

Swedish novelist and dramatist Sven Delblanc's *Homunculus: A Magic Tale* describes the protagonist's creation of a homunculus, following Paracelsus's recipe, in a glass jar.[33] The novel engages at many levels with Mary Shelley's *Frankenstein*.[34] Sebastian, the main character, is a high school chemistry teacher who, following an episode in which he caused the chemistry laboratory to explode, is dismissed from his duties. After this expulsion, Sebastian, who like Victor Frankenstein enjoys dabbling in mysterious potions, sets his mind on discovering the secrets of creation and on self-reproduction.

Like *Frankenstein*, *Homunculus* is traversed by alchemical references. Also like *Frankenstein*, often read as dramatizing the appropriation of women's reproductive capabilities by a male scientist,[35] Delblanc's book similarly illustrates the male drive to self-engender. As Alice E. Adams notes, "The threat of maternal power . . . becomes a major force in Sven Delblanc's story of a manmade man" (*Reproducing the Womb: Images of Childbirth in Science, Feminist Theory, and Literature*, 89). In Delblanc's tale, however, Sebastian goes a step further than Victor Frankenstein—toward the end of the book he himself becomes pregnant, dying while giving birth through a Cesarean section, which he inflicts on himself in the midst of labor pains.

At the beginning of the novel, we learn that "something was stirring heavily in his [Sebastian's] belly, perhaps a blind, thousand-year-old Leviathan, rolling over in ice-cold bilge water, eager to be born" (*Homunculus: A Magic Tale*, 3).[36] Sebastian, as a would-be Paracelsus, "grasped the little cut-glass bottle with its precious drop of red fluid and pressed it against the sex low inside him to conjure the monster to rest" (3). As if in the throes of an imaginary pregnancy, Sebastian feels "the fetus stir" (4) in his body, and is "afraid" (4).

Sebastian's (al)chemical experiments are conducted in the secrecy of his bathroom, where the bathtub contains a mixture clearly reminiscent of Paracelsus's life-yielding recipe:

> The room was lined with batteries of bottles, earthenware jugs, and glass jars with phosphorus, mercury, hydrogen peroxide, and fatty acid; cardboard boxes and brown sacks with iron, sulfur, calcium, and powdered carbon. The bathtub was half-filled with a sluggish, reddish-brown liquid. An electric immersion heater sunk down into the liquid. The black cord snaked its way over the edge and out across the floor in a coil. A saturated, almost nauseous smell of earth and compost met Sebastian's nostrils. (*Homunculus: A Magic Tale*, 47)

It is in this artificial womb with its umbilical chord that Sebastian is going to generate his "son" (*Homunculus: A Magic Tale*, 49) by adding his urine to the bathtub. With the help of a pipette, Sebastian "sucked up a precious red drop from the little glass bottle. (The Essence. Not only chemistry but living faith will grant you power.)" (48). Paracelsus is even directly invoked: "Now, Paracelsus, old master, do not fail me" (48).

The description that follows of Sebastian's creature is suggestive of *Frankenstein*:

> A body, the finished creation, rested on the dirtied bottom of the bathtub. (My son. Now he's raising his head.) It was a monster with a bald monkey head, hunched up like a half-finished fetus. A single eye shone above flat nostrils and a brutal lower jaw with yellow boar's tusks. Its arms and legs had birdclaws, and pointed buckhorns were growing on the knees. . . . The monster's single shining eye blinked with an eyelid of horn. *The monster regarded his maker*. (*Homunculus: A Magic Tale*, 49; emphasis mine)

This birth scene bears many resemblances to that of Frankenstein's monster. Both creatures are described in terms of their animalesque characteristics, their deformity, and the repulsion they will go on to inspire in others. The ambivalent reactions of their respective makers are also stressed. Indeed, this episode, which describes the scene of recognition between the newly created homunculus and his maker, Sebastian, is clearly reminiscent of that in *Frankenstein* when Victor Frankenstein first beholds the creature of his own creation. Both scenes are centred around exchanges of looks between the two, creator and creation, but there are telling differences. While Victor Frankenstein swiftly abandons the creature upon realizing, after a horrified glance, the monstrosity of his creation, leaving his "son" to fare on his own, Sebastian "carefully . . . dried him clean from the slime and wetness of birth. . . . Suddenly Sebastian was a young mother, deluged by practical worries. (What shall I give him to eat, how shall I dress him, how shall I go about showing him to the world?)" (*Homunculus: A Magic Tale*, 108).

Sebastian's concern about how to feed him is solved when the homunculus "suddenly put his face to Sebastian's chest, his lips opened, his head moved searchingly. . . . Then at once he understood, and a flutter of delight went through him, and he felt his flat, manly breasts swell and be filled with milk" (*Homunculus: A Magic Tale*, 109).[37] Besides being willing to "suckle him" (50), Sebastian also significantly gives him a name, Bechos, while by contrast Frankenstein's creature was never named. The feeding scene that follows is telling as an illustration of male envy of women's reproductive and nurturing function. Sebastian "ripped up his shirt at the breast and pressed his dripping nipple into the son's mouth, and Bechos laid his arm around his neck and with eyes shut drank the fatty milk of his maker, and Sebastian caressed his son over his auburn hair" (109). While Victor Frankenstein rejects his creature, Sebastian cares for his until the homunculus dissolves into liquid, transmuted back into the source from which it came: "Sebastian groped across the floor to find his son and got his hands full of red ooze. Bechos dissolved and disappeared into the cracks of the mud floor. All that remained were shirt and trousers, drenched with a red, foul-smelling ooze" (174).[38] The only trace left of his son was a "spherical object" (174) Sebastian found among the clothes: an "eye hard as enamel" (174). However, when Sebastian "looked into the dark pupil . . . he found no meaning, no answer, however hungrily he peered into the black well" (174). The vacant eye returns as a symbolic reminder that the male trespassing into woman's only prerogative, her maternal function, can only bring destruction.

Indefinition of gender roles comes in for severe criticism in Delblanc's novel. Sebastian is described throughout the novel as both man and woman, father and mother. Alice E. Adams comments in this context that

> at the moment Sebastian gives life to his son, he abandons masculine sovereignty and enters a no-man's-land of maternal chaos. His confusion of self and other transforms him into an androgyne. . . . Sebastian as androgyne occupies the mother's indefinable space. Sebastian longs for motherhood, but it is motherhood-as-self-annihilation that he seeks—the only tolerable form of motherhood within a masculinist logic of oppositions. (*Reproducing the Womb: Images of Childbirth in Science, Feminist Theory, and Literature*, 91–92)

Indeed, *Homunculus: A Magic Tale*, like *Frankenstein*, can be read as a cautionary tale about male ambition of intrusion in the realm of the maternal with the fatal consequences this aspiration may bring about, namely death of the male subject, as in Sebastian's case, who, at the very end of the novel, is described as he "arduously . . . began to give birth to his death" (188).

Another dimension of Delblanc's political satire needs some attention. The issue of personal freedom threatened by a totalitarian government is a recurrent theme in Delblanc's work. In *Homunculus: A Magic Tale*, Sebastian's drive to self-engender is further inscribed in a complex plot of spying and conspiracy carried out between the CIA and the government of the USSR, shadowed by the threat of destruction by an atomic bomb directed at Stockholm, where Sebastian is. The secret of life that Sebastian is searching for is also the secret of death.

In her essay on the politics of nuclear weapons and nuclear strategic doctrine and control, Carol Cohn explores some of the "gendered aspects of war and militarism" ("Sex and Death in the Rational World of Defense Intellectuals," 129) and develops a feminist critique of "dominant Western concepts of reason" (130), concepts that, in their close equation between male reason and woman's passivity, in describing male intellectual achievements as their brain "children," had already been explicitly expounded by Plato and Aristotle.[39]

According to Carol Cohn, "There is one set of domestic images that demand separate attention—images that suggest men's desire to appropriate from women the power of giving life and that conflate creation and destruction" ("Sex and Death in the Rational World of Defense Intellectuals," 140).[40] As Cohn goes on to observe, "The bomb project is rife with images of male birth" (140). The concept of male birth, indeed, and "its accompanying belittling of maternity—the denial of women's role in the process of creation and the reduction of 'motherhood' to the provision of nurturance . . . seems thoroughly incorporated into the nuclear mentality" (141). The strong political dimension of Delblanc's novel, with its underlying association between the nuclear bomb and atomic destruction of nature and Sebastian's ambition to create a son, suggests precisely the intricate connection between the male impulse to dominate the environment and to appropriate woman's birthing capacities.[41]

> None of woman born shall harm Macbeth.
>
> Shakespeare, *Macbeth*, IV.1. 79–80

> *Not of Woman born*
>
> Donna Haraway, "A Cyborg Manifesto: Science, Technology and Socialist-Feminism in the Late Twentieth Century," 177

Autonomous male (re)birth is also a crucial theme in Peter Ackroyd's *The House of Doctor Dee* (1993), which similarly engages with the vocabulary of alchemy to reflect on the question of origins, namely male creation without the agency of women. The novel draws on the life of the famous sixteenth-century mathematician and astrologer Dr John Dee (1527–1608)[42] to narrate the story of Matthew Palmer, who, having inherited his father's house in London, discovers that it had belonged to the celebrated magician Dr Dee. This finding prompts Matthew to try to unearth as much as he can about his house and Dr Dee's life, a process of learning that leads to the uncovering of many secrets behind the characters he was close to, namely his father and stepmother. We never hear about his mother, and the narrative drive seems to suggest, inconclusively, that Matthew himself might have had his origins as a homunculus, as a creature developed by his father on his own out of the latter's wish to procreate, as his father explains to their common friend Daniel (*The House of Doctor Dee*, 267).

Matthew Palmer is a professional researcher who, in the course of his investigations to insert his house into the London historical past, is also excavating other pasts, namely his own. He describes the pleasure he derives from his researches in terms of an archeological search for his own buried past, inscribed in the history of the city: "It is as if I were entering a place I had once known and then forgotten, and in the sudden light of recognition had remembered something of myself" (*The House of Doctor Dee*, 13). As the protagonist further explains, "If my work meant that I often viewed the past as my present, so in turn the present moment became part of the past" (13). These excavations into what could be called, in Freudian terms, the Minoan–Mycenaean layers of civilization[43] under a more superficial covering could be equated with the psychoanalytic process of disclosing increasingly deeper layers of the repressed contents of the psyche. It is significant in this context that Freud frequently used archaeological metaphors in his work. In addition, in *Civilization and Its Discontents* Freud compared the past of a city, Rome, with the past of the mind, proposing that we examine Rome, "by a flight of the imagination" not as a "human habitation but a physical entity with a similarly long and copious past—an entity, that is to say, in which nothing that has once come into existence will have passed away and all the earlier phases of development continue to exist alongside the latest one" (257). These insights could profitably be applied, in Peter Ackroyd's novel, to the historical past of the city of London, and simultaneously to his own past by Matthew Palmer, who ends up, through his encounter with the house in Clerkenwell and his father's papers, by discovering the contours of his own past.

In spite of the fact that, as Matthew ponders, "I remember very little about my childhood" (*The House of Doctor Dee*, 80), he fears what he may find there: "I can't bear to look at myself. Or look *into* myself. I really don't believe that there's anything there, just a space out of which a few words emerge from time to time" (81), echoing the Lacanian notion that the unconscious is structured like a language. Indeed, using language that suggests the concept of childhood trauma and the need to access it through psychoanalysis, Matthew confesses that "I began to repair my life when I became a researcher and entered the past" (129).

In his father's house, Matthew comes across several sheets of paper containing Doctor Dee's recipe for the creation of a homunculus, which is very similar to that of Paracelsus. Significantly, it starts with the words "so that it may grow *without the help of any womb*. This is the secret of all secrets, and must remain so until that time of the end when all secrets will be revealed" (*The House of Doctor Dee*, 123; emphasis mine). The glass Matthew finds in his house, "with the odd distortion or protuberance at the end" (122) stands imaginatively for the womb, the place of origins that was symbolically denied to him. Matthew gradually comes to feel like a homunculus, a "little man" (265) as a man calls him in a London street, his father's fanciful creation from knowledge borne, like Dr Dee's, out of "fear and ambition" (248). His father, in addition, in thus experimenting with the origins of life, had deprived him of a mother. Indeed, during his researches Matthew uncovers the unsettling information that he was "found" by his father, while he cannot locate his mother's identity. Going through his deceased father's papers, twenty-nine-year-old Matthew notes that there was "nothing of any significance amongst them, except for the fact that my father had bought the house from Mr Abraham Crowley on 27 September 1963—that date aroused fresh speculation in my mind, since it was the one we had always celebrated as my birthday" (219). In this house, which represents a return to his origins, he is forced to confront the place where his former self and alter ego, the homunculus, was created. Like the homunculus, who "remembers nothing about its past or future until it returns home at the end of its thirty years, but it always does return home" (125), Matthew remembers very little about his childhood.

The intricate connections between Matthew and the homunculus, on the one hand, and his father and Dr Dee, on the other, are interwoven throughout the novel. The very narrative form underscores these temporal conjunctions, for each chapter dealing with Matthew and his father is intercalated with one that engages with Dr Dee's life and times. The narrative drift seems to suggest that in the scene where the man calls Matthew "little man" and then "kneels down" (*The House of Doctor Dee*, 265) before him as a token of respect and awe, by some uncanny temporal twist Matthew has been transported to that same London street during the time in which Dr Dee lived.[44]

Dr Dee's lifelong search for the secret of life, for the creation of the homunculus, can be said to parallel that of the narrator for his origins, for the mysteries surrounding his father. As Matthew enigmatically notes, "I knew now that he was not my 'real' father; he was even closer to me than that, and had formed me in ways I could only begin to imagine" (*The House of Doctor Dee*, 219). Matthew's stepmother, who had until then been almost a stranger to him, throws more light on his father's life: "It was so difficult, Matthew. He was always there, you see. You were his child somehow, never mine" (178). Matthew's thoughts following this meeting with his stepmother are disturbingly enlightening for him, partially revealing the conundrums he had not been able to accede to. Matthew's search for origins also unveils his ambivalent feelings toward his father. He muses,

That was the secret, after all. I had grown up in a world without love—a world of magic, of money, of possession—and so I had none for myself or for others. That was why I had seen ghosts rather than real people. . . . That was why I had dreamed of being imprisoned in glass, cold and apart. The myth of the homunculus was just another aspect of my father's loveless existence—such an image of sterility and false innocence could have come from no other source. (*The House of Doctor Dee*, 178)

In related fashion, John Dee's wife Katherine chides him for his arrogance and presumption in trespassing over "forbidden" territories as far as the creation of the homunculus is concerned. As the dying Katherine Dee tells her husband, "This was a vision of the world without love, John Dee, but one you yourself have fashioned. You hoped to create life, but instead you have made images of death. Think of this, and repent in time" (218).

The House of Doctor Dee thus suggests that deficient parenting is at least partly responsible for the feelings of inadequacy, of blank spaces in his life, experienced by Matthew. It can, however, in a related vein, be read as a cautionary tale warning against male appropriation or sole seizure of "the secret of all secrets," as Dr Dee's recipe calls it, the secret of the creation of life but, meaningfully, without woman's intervention, a fantasy that alchemy was instrumental in recreating and spreading.[45]

> Fatherhood there may be, when mother there is none.
>
> Appolo's speech in Aeschylus, *Eumenides*

> There can be no other selves than selves of woman born.
>
> Grace M. Jantzen, *Becoming Divine: Towards a Feminist Philosophy of Religion*, 141

Michelle Boulous Walker argues that what she terms the "masculine imaginary" (*Philosophy and the Maternal Body: Reading Silence*, 67), a fantasy that may be more widespread than previously thought, which works "to silence women in quite specific ways" (1), namely, through their association with maternity, is structured around "the desire for masculine self-generation" (67). As Rosi Braidotti, in a related vein, notes, "Reproduction without men triggers a deep malaise in the patriarchal imaginary, resurrecting the centuries-old myth of gynocracy" (*Metamorphoses: Towards a Materialist Theory of Becoming*, 199), a myth similarly evoked by Marina Warner, who points out that "the spectre of gynocracy, of rule by women, stalks through the founding myths of our culture" (*Managing Monsters: Six Myths of Our Time*, 5).

Impending biotechnological and medical advances such as human cloning, for instance, could foreseeably help to concretize the fantasy of a motherless birth, thus conferring on the father the possibility to bypass the (m)other and fulfill the dream of self-reproduction.[46] As Patrick D. Hopkins remarks in relation to the prospect of men getting pregnant, "The possibility of this technology leads us to ask: Should women be the only ones *allowed* to bear children? Are women as a class *obligated* to bear children? Does the state have a compelling interest in preventing men from reproducing as they see fit? Should technology be judged wrong or made illegal *because* it upsets traditional biological sex roles?" ("[Re]Locating Fetuses," 172).[47]

Some of these scenarios have been given dramatic illustration in many fictional narratives, among which can be cited Marge Piercy's *Woman on the Edge of Time* and Lisa Tuttle's short story "World of Strangers," where some men give birth to their own cloned children while most men at least nurse them for up to six months, texts discussed in Chapter 3.[48]

Edmund Cooper's *Who Needs Men?* (1972), Fay Weldon's *The Cloning of Joanna May* (1989),[49] and Anna Wilson's *Hatching Stones* (1991) similarly provide fictional illustrations to some versions of such potential scenarios, where appropriation of cloning by one of the sexes eventually leads to a radical split and subsequent war between them, and even to the gradual diminution and elimination of one of the sexes, as is the case in *Hatching Stones*.

HATCHING STONES

> Bring forth men-children only;
> For thy undaunted mettle should compose
> Nothing but males.
>
> Shakespeare, *Macbeth* I.vii.72–74[50]

Anna Wilson's *Hatching Stones* (1991) provides a radical and suggestive take on a vision of what the consequences of a masculine appropriation of reproductive technologies might be, in this case cloning. Wilson portrays a society where men largely abandon women when the introduction of cloning allows them to clone only male babies, thus implicitly suggesting that males might conceivably forgo "normal" sexual intercourse if they had the technological ability to create sons who would be identical copies of themselves.[51]

The opening paragraph of *Hatching Stones* could almost serve as a journalistic report on the prospect of the introduction of human cloning:

> The technology came first, as is usually the case, and was threatening to become fully operational and change the shape of the world before anyone had had time to consider how it could be used, or how controlled. Genetic engi-

neering itself—on some pre-human level—had long been taken for granted as a useful means of producing a disease-resistant grain or a leaner breed of sheep. The sudden possibility of producing humans, however, seemed to take even those responsible for the successful experiments by surprise. Once the word was out there followed a period of intense speculation: airwaves and newsprint were choked with wild theories of the discovery's significance, and wilder fantasies about the future of the human race. (2)

What follows, however, is distinct from our contemporary experience, both in the speed with which it is applied and the wide use to which it is put. As the narrator explains, "This period of shock and adjustment ended abruptly; people shook off their paralysis and began to seek out and apply the new technique to themselves. Within three months there were clinics claiming to have access to the new technology in every major city, a network of back street operators, and the threat of a population explosion on a scale never envisaged in modern times" (2).

Hatching Stones is divided into roughly two parts: One details the male-only society that gradually arises as a result of the implementation of cloning and ectogenesis, while the second describes the women-only world that develops as a consequence of the banishment of the few women who were left to a distant, inhospitable island, Baba-I.

When population-control regulations, drafted by this male-dominated society where no woman had ever been the governor of a state, are first introduced, they decree that each person can have only one child, either in the "usual" way or under the new technology. As a result, "fewer 'natural' babies were born each year" (*Hatching Stones*, 12), since men wished to have their own cloned offspring, their own mirror-images who would supposedly carry on with their work in predictable ways. As the narrator further explains,

> The problem of women's increasing redundancy as child-bearers turned out, in fact, to have wider dimensions. . . . The ideal that the country had voted for had, after all, been a masculine one. Most women, perhaps from force of habit, had depicted their ideal citizen as a man. Perhaps they had not fully realized that they could not have a male offspring by the new technology . . . no blueprint for the woman of the future had . . . been drawn up. Women were, for the moment, excluded from participation in the new way. (13)

Women's lack of representation in the seats of power and decision making, together with a long-standing tradition of being subsumed under the heading "Man," forcefully conspire to keep women from visualizing a society where their outlook on life would be deemed positive and desirable.

Some people soon began predicting that "unless something was done or unless the novelty wore off and the rate of ordinary pregnancies increased, native-born women would successively become first rare, then ageing curiosities, and finally extinct" (*Hatching Stones*, 13). Men who wanted offspring of their own, a cloned son bypassing the woman, "began to look on women primarily as dangerous" (14). In spite of all the women's campaigning and demonstrations "for a return to the old methods of birth" (14), the future seemed to be irrevocably "gendered male" (16), for "in practice the law allowed only men to benefit from the new technology" (16).

The discourse of eugenics at work in *Hatching Stones* is veiled by melioristic arguments that highlight the improvements in the offspring's overall health, intelligence, and disposition, as well as by a strong narcissistic desire to have your own male self perpetuated. As the narrator puts it, "what finally the attraction" of the new reproductive technology "seemed to come down to was the predictability of this new species of offspring, its sameness, the familiarity that a parent felt at the sight of his small self" (*Hatching Stones*, 5),[52] a scenario reminiscent of Jean Baudrillard's numerous disparaging comments about human cloning, the results of which he variously describes as "the eternity of the same," ("Clone Story," 95), "the Hell of the Same" ("The Hell of the Same," 122), or "the reign of the same" (120), denoting precisely a predictable sameness that, in its most radical scenarios, would spell the end of individuality and alterity. Symptomatically, Baudrillard also implicitly envisages this eternity of the same as the male same, as does *Hatching Stones*, a concept that prompts reference to Luce Irigaray's critique of the reduction of women to the other of man, in turn a state of affairs which, as Simone de Beauvoir diagnoses, grounds patriarchal culture. Irigaray, opposed to Beauvoir's argument, however, wishes to see men and women as decidedly two different entities, not as others in a hierarchical sense.[53]

In *Hatching Stones*, when it became clear that there would be no room for them in this increasingly more male-only world, the president decided to offer the women a kind of exile on an island close to the polar icecap:

> Thousands, perhaps hundreds of thousands, committed suicide. . . . Some tried to take refuge in the small fundamentalist communities that had refused to accept the new technology. . . . Others simply took to the hills and the streets. The women who remained continued a source of disruption, violence and unease for some forty years. (*Hatching Stones*, 21)

Eventually, however, on the island of Baba-I to which they are sent, women manage to develop similar techniques of cloning to reproduce. As the narrator explains, the men "had not thought that any of the women that they sent away might be, even potentially, genetic engineers; they could see them, by then, only as women, as representatives of the old way, having no knowledge of, or desire for, the new" (23). After a short while, the women themselves start "to produce new offspring" and "the Baba-I began to be born" (23).[54] In order to escape what they perceived as the "degeneracy they saw in Lelaki" (76), the women scientists in Baba-I "proclaimed variety as a positive quality" (76), and their children were "accordingly much more various in appearance, despite being very many fewer in number, than their male equivalents on Lelaki" (76).

PLAYING WITH NARCISSUS

Narcissism can be seen as standing at the genesis of the transformation of the Lelaki society, as well as coming to dominate the all-male society, described by the women in Baba-I as having "succumbed to mass megalomania" (*Hatching Stones*, 77), a phenomenon of which narcissism can be seen as an integral part. Indeed, in relation to Lelaki and the behavior of the male clones, one could aptly quote here Narcissus's words from Ovid's *Metamorphoses*:

> "Am I the lover
> Or beloved? Then why make love? Since I
> Am what I long for, then my riches are
> So great they make me poor."(3, 464)

The male clones in *Hatching Stones* may be approached from a psychoanalytic perspective in terms of their extreme narcissistic behavior, which we could call pathological narcissism, for they can be seen as not having progressed beyond primary narcissism to love of others. Although this characterization applies more closely to the politicians and leaders of the Lelaki society who put through the population control and other policies, eventually eliminating women, it also comes to appropriately describe the dominant behavioral traits of the male society that is the result of those policies, marked by a strong drive to self-pleasure and lack of concern for others, a profoundly hedonistic culture.

Indeed, the male clones' psychological traits and behavior fit to perfection the definition given in the third edition of the *Diagnostic and Statistical Manual of Mental Disorders* (DSM-III, as it is popularly known) of the term "narcissistic personality disorder," which is characterized, according to DSM-III, by

> a grandiose sense of self-importance or uniqueness; preoccupation with fantasies of unlimited success; exhibitionist need for constant attention and admiration; characteristic responses to threats of self-esteem; and characteristic

> disturbances in interpersonal relationships, such as feelings of entitlement, interpersonal exploitativeness, relationships that alternate between the extremes of overidealization and devaluation, and lack of empathy. (315)

The drift toward an exacerbated narcissism in contemporary society has been much commented upon,[55] and *Hatching Stones* can aptly be read as in part a reflection on that state of affairs, where Narcissus seems to be back with a vengeance. Indeed, in the case of the male clones, remaining as they do mainly in the stage associated with primary narcissism, the drive to doubling and duplication, to producing replicas of themselves, can be seen as suggesting a form of protection against extinction, as suggested by Rank and Freud.[56]

On the other hand, as Kochhar-Lindgren suggests, "The threat of absolute solipsism marks the presence of Narcissus" (*Narcissus Transformed: The Textual Subject in Psychoanalysis and Literature*, 6), potentially leading to a deadly form of self-absorption. It is precisely the danger of this absolute solipsism that the presence of one's clone(s) can ward off, although simultaneously it can cause the risk of absorption into a group continuum, with the consequent loss of individuality, a recurrent thematic concern in clone narratives. As the cloned groups realize

> They disputed mainly for the sake of a familiar noise, for the pleasure of hearing themselves speak. They had not realized that one of the consequences of separation would be a new isolation of their own individual voice, that it would sound naked, altered somehow by solitude. And so they talked for the intermingling of familiar tones, the rising and falling and blending of a dozen similar voices; it was a way of remembering who they were, finding something they had not quite known they had lost. (*Narcissus Transformed: The Textual Subject in Psychoanalysis and Literature*, 36)[57]

The all-male society that eventually shapes Lelaki bears witness to Baudrillard's mostly negative comments on the gradual sameness and blandness that would come to dominate society if cloning came to be used with the aim of annihilating one of the sexes. Indeed, the works previously discussed can be said to betray a deep-seated fantasy that encompasses the total destruction of one of the sexes, while at the same time revealing a deeply rooted fear of the other, man or woman.[58]

"THE WORLD WITHOUT WOMEN" (JEAN BAUDRILLARD, *THE PERFECT CRIME*)

The omnipresent rule of male power is a fancy Baudrillard addresses in his essay "The World without Women." Drawing on Virgilio Martini's *Il Mondo senza Donne* (1935), where all childbearing women are decimated by a terrible disease, named "fallopitis," Baudrillard describes the central idea of the novel as the "extermination of femaleness—a terrifying allegory of the obliteration of otherness, for which the feminine is the metaphor and, perhaps, more than the metaphor" (111).[59] In his usual polemical vein, Baudrillard seems to be suggesting that the "other" is always and only the feminine in a world where the male is the norm.

Furthermore, Baudrillard addresses the controversial issue of the potential need for woman "to produce the other" (*The Perfect Crime*, 120), inscribed in a meditation on "the idea of a world given over entirely to the selfsame" (112), a subject that appears to endlessly fascinate Baudrillard. Almost as if engaged in a critical dialogue with Irigaray,[60] in "The World without Women" Baudrillard analyzes Martini's *The World without Women* (*Il Mondo senza Donne*), which seems to use some of the devices employed by Bradley Lane in *Mizora*. The perfect crime of Baudrillard's title can be described as the willful extinction or abolition of the other in contemporary times. In somewhat radical terms, Baudrillard argues that "to be deprived of the other is to be deprived of sex, and to be deprived of sex is to be deprived of symbolic belonging to any species whatsoever" (112). These controversial words appear tailor made to describe the disturbing absence of sexual desire, seemingly nonexistent in *Mizora* and *Herland*. Baudrillard further elaborates on this theme:

> In German, there are two apparently synonymous terms with a very significant distinction between them. "*Verfremdung*" means becoming other, becoming estranged from oneself—alienation in the literal sense. "*Entfremdung*," by contrast, means to be dispossessed of the other, to lose all otherness. Now, it is much more serious to be dispossessed of the other than of oneself. Being deprived of the other is worse than alienation: a lethal change, by liquidation of the dialectical opposition itself. An irrevocable destabilization, that of the subject without object, of the same without the other. ("The World without Women," 112)[61]

Wilson's *Hatching Stones* can also be read in terms of male unconscious disgust at having been born of woman. As Simone de Beauvoir observes, "Man feels horror at having been engendered; he would fain deny his animal ties; through the fact of his birth murderous Nature has a hold upon him" (*The Second Sex*, 178).

MALE CLONES AND BOTHERSOME MOTHERS

As previously noted, intimately connected with the fantasy of a single-sex world, fantasies of male control over reproduction and the consequent political power that domination would entail abound in literature and mythology. As Martha C. Nussbaum and Cass R. Sustein point out in their Introduction to *Clones and Clones: Facts and Fantasies about Human Cloning*, Greek mythology provides a rich array of instances of unusual conceptions and births, such as "masturbating into the fertile earth of Athens to create a nation of male clones without bothersome mothers" (12). As they further note, striking a very contemporary note, "The Athenians were thus, in their own civic mythology, a nation descended from male clones, very proud to have no jot of the female in their makeup" (12).[62] *Hatching Stones* may be said to illustrate a scenario similar to that dreamed up by the male Athenians. As Moira Gatens observes in this regard, "Our cultural unconscious is littered with examples that suggest that those *not* born of woman have awesome powers. Macbeth, who smiles with scorn at 'swords brandish'd by man that's of woman born,' can be slain only by the 'unbirthed' Macduff" (*Imaginary Bodies: Ethics, Power and Corporeality*, 53; emphasis in the original).[63]

Nancy Tuana similarly addresses the topic of the creation and position of women within the social hierarchy, drawing on Plato and Aristotle who, despite their distinct approaches to these themes, nevertheless come to a similar conclusion, namely, that the male is the true form of the human being. As Tuana concludes, "Plato thus envisioned a world without women, for in their primordial state, all humans were male. Women come into existence only in the later births" (*Woman and the History of Philosophy*, 14).[64] Indeed, as Gena Corea notes, "In the realm of cloning, as in most reproductive technology, the male is seen as the active principle in reproduction, the female the passive" (*The Mother Machine: Reproductive Technologies from Artificial Insemination to Artificial Wombs*, 261). She suggests that if it ever became possible cloning might be predominantly used to promote male urges to self-generate, circumventing woman's participation. As Corea points out, "This is the classic patriarchal myth of single parenthood by the male" (260), illustrated in, for instance, Wilson's *Hatching Stones*.[65]

Eva Feder Kittay's examination of representations of womb envy, which she considers as an explanatory concept that can be seen as having a similar importance to penis envy, as seen previously, effectively the male counterpart to it, is also relevant for our analysis of *Hatching Stones*. According to Kittay, "A consequence of and defense against womb envy is the appropriation on the part of men of the procreative powers for themselves" ("Rereading Freud on 'Femininity' or Why Not Womb Envy?" 200). Indeed, according to Kittay, in order to "defend against womb envy . . . the boy may devalue himself or devalue life-giving. He may maintain the high valuation on creation but displace the creative capacities from woman, appropriating them to man and to his exclusive domains. Such appropriation, in its many manifestations, is probably the most likely outcome of womb envy" ("Womb

Envy: An Explanatory Concept," 121). As Kittay further observes, "For the child, this means forming ambitions in which he can outdo women in their productivity, accepting ideologies in which men play the 'really' important part in procreation and the family and, in general, fiercely adopting the stance of male superiority" (121). These words help shed some light on Wilson's novel, where the men gradually appropriate women's procreative capacities, relegating them to increasingly subordinate roles until eventually they banish the ones who are left to a remote island.

Drawing on Bruno Bettelheim's book *Symbolic Wounds: Puberty Rites and the Envious Male* (1962), where Bettelheim examines various manifestations of what he interprets as male envy of women's reproductive organs and functions, Baudrillard similarly emphasizes the driving force of "man's jealousy of woman's power of fertilization" (*Seduction*, 16). Since, as Baudrillard goes on to argue, "this female advantage could not be atoned" (16), it followed that a "new social order had to be built at all costs, a masculine social, political and economic order, wherein this advantage could be reduced" (16), an impulse carried to its extreme consequences in both Cooper's *Who Needs Men?* and Wilson's *Hatching Stones*.

In "Is Pregnancy Necessary? Feminist Concerns about Ectogenesis," Julien S. Murphy points out that "clearly women researchers are underrepresented in the field of reproductive technology. What is to prevent men from making women extinct once our unique contribution to society—reproduction—can be supplied another way?" (198).[66] As Murphy goes on to argue, ectogenesis, "accompanied by sex selection techniques and methods for producing synthetic eggs, could guarantee the reproduction of an all-male population—the ultimate patriarchal culture" (192),[67] as is the case in *Hatching Stones*.

CONCLUSION

Could it be, then, as Barbara Creed argues, that "man's desire to create life—to give birth—suggests a more profound desire at work—to become woman" (*The Monstrous-Feminine: Film, Feminism, Psychoanalysis*, 57)?

In an interview conducted by Alice Jardine and Anne Menke, Julia Kristeva pertinently reflects on some of the issues I have been engaging with here. Kristeva believes that we might be seen as "moving toward a future that effaces sexual difference. . . . It would seem that humanity must prepare itself for the fact that men could be women and women men" (*Julia Kristeva's Interviews*, 126). Kristeva goes on to envisage the type of society in the twenty-first century where "there will be no difference between the sexes. There will be a kind of perpetual androgyny even to the extent that—as certain fictions say—*men will give birth*, and from that point on, the difference of reproduction, which up to that point had been women's realm will disappear" (126; emphasis mine).[68]

As Dick Teresi and Kathleen Mcauliffe muse, "But how do feminists feel? Do they see male pregnancy as their chance to escape biological destiny?" ("Male Pregnancy," 182). They go on to cite Gloria Steinem, cofounder and editor of *Ms.* magazine, who confesses, "I have a small, nagging fear, that if women lose our cartel on giving birth, we could be even more dispensable than we already are" (182). On the other hand, Steinem points out that experiencing pregnancy would lead men to be less violent (182), analogously with women, who, due to giving birth, "value life more" (182). An example of a fictional society where both women and men give birth and suckle the babies is Marge Piercy's Mattapoisett, in *Woman on the Edge of Time*, a striking example of an almost genderless society.[69] As Erin MacKenna states, the "Mattapoisians believe that the power to give birth is a power and women had to give it up to achieve equality and cooperation and so that men could become more human through the mothering experience" (*The Task of Utopia: A Pragmatist and Feminist Perspective*, 73).

It seems, then, that although Western societies are evolving toward less patriarchal configurations,[70] new reproductive technologies can renew male birthing fantasies,[71] as well as the perpetuation and extension of male power over procreation in the shape of fatherhood without women, for instance, made potentially feasible with the advent of human cloning and the development of artificial wombs.[72] The reactivation of these fantasies will also inevitably bring back fears of monstrous births, in this event male ones, reanimating grotesque visions of creatures not of woman born, seen as the result of trespassing into what are often perceived as forbidden and menacing female territories.

In *Quintessence: Realizing the Archaic Future: A Radical Elemental Feminist Manifesto* (1999), Mary Daly analyzes some manifestations of what she calls "mythic foreshadowings of cloning" (170), which she associates with the annihilation of women, as well as with what she terms "the widespread male motherhood syndrome" (170), both in Western and Eastern religions and mythologies, namely Christianity, Greek myth, and the Tibetan tulku system. Drawing on the story of Zeus and Dionysus, who was twice born and brought to term attached to his father's leg (the equivalent of an artificial womb), Daly argues that such "patriarchal mythic foreshadowing of the elimination of women is carried over into christianity" (171), in the biblical story of the birth of Jesus from the Virgin Mary, who was "taken over, made over, and diminished by christianity for that religion's own patriarchal purposes" (172), since the virgin functions exclusively as the vessel to carry and nurture the fetus until his birth. Daly then goes on to conclude, stressing the role played by myths as "self-fulfilling prophecies" (172),

> It is clear that patriarchal males are attempting to act out the myth of the pre-existing divine male giving birth to himself, while simultaneously working to eliminate Divine Female Creative and Procreative Power. The recent developments in cloning coupled with the new reproductive technologies, which are leaping to their catastrophic conclusion as the twentieth century comes to an end, are manifestations of this megadestructive pattern. (172–173)[73]

Although I agree with Daly's argument about the erasure of the Virgin Mary, her pregnancy and motherhood, from Christian ideology, this is far too drastic a conclusion to draw from the potential implementation of cloning, as I have repeatedly stressed. Daly totally forgets that cloning could be used to benefit women and not wholly to further a male agenda of gradual elimination of women and the creation of a predominantly male world.

Indeed, Mary Daly is dismissive and critical about the potentialities that human cloning could bring about for women. As she scathingly puts it, "Male-mothered genetic engineering is an attempt to 'create' without women. The projected manufacture by men of artificial wombs, of cyborgs which will be part flesh, part robot, of clones—all are manifestations of phallotecnic boundary violation" (*Gyn/Ecology: The Metaethics of Radical Feminism*, 71). Daly does not make room for women scientists, nor does she contemplate the possibility that those women scientists might see some advantages in applying cloning technique in selected cases that might benefit women and men.

The next chapter will address the consequences in terms of psychological traumas and group psychology of the creation of families of clones, created to satisfy particular needs and to save humanity, decimated by apocalyptic ecological disasters.

NOTES

1. As Robert S. McElvaine points out, "Another example of the emerging idea of male primacy is the order of creation in the second chapter of Genesis, which completely contradicts that given in the first chapter. In chapter 1, God creates man and woman simultaneously and presumably as equals, after all other forms of life, on the sixth day. But in Genesis 2, God creates man on the *first* day. . . . Then, after man, God creates plants and '*out of the ground* the Lord God formed every beast of the field and every bird of the air.' . . . The last of all living things to be made, according to the Genesis 2 account, was woman, widely separated chronologically from the creation of man" (*Eve's Seed: Biology, the Sexes, and the Course of History*, 144). As McElvaine states, the scene in the second chapter of Genesis in which God puts Adam into a deep sleep in order to remove from him a rib which will then become Eve, "perfectly symbolizes the new idea of male creation. It is a clear example of womb envy, because it asserts, in effect, that the first 'womb' was male" (131).

2. *Frankenstein*, although ostensibly the creation of a man by another man, circumventing the woman, has persuasively been read as a female birth myth

by, among others, Ellen Moers (*Literary Women*, 1976) and Anne K. Mellor (*Mary Shelly: Her Life, Her Fiction, Her Monsters*, 1988).

3. Interestingly for the theme of this book, George Annas notes how Shelley's *Frankenstein* "explores virtually all the noncommercial elements of today's cloning debate" ("Scientific Discoveries and Cloning: Challenges for Public Policy," 81).

4. Another manifestation of what could be described as male jealousy and hatred of women can be seen at work in one of Sade's characters, Bressac, the hero in *Les Infortunes de la vertu*, who kills his mother and justifies his act with the argument that the father is the sole agent in creating a child, the mother being only the vessel where the baby grows. With respect to this passage, Susan Rubin Suleiman describes this fantasy as a predominantly masculine one, "that of a world where man is not born of woman. . . . Woman, the irreducible Other, is eliminated to make way for the engenderment of the same by the same" (*Subversive Intent: Gender, Politics, and the Avant-Garde*, 68), as might come to happen with human cloning. The idea that woman is but the vessel that nourishes the fetus and carries it to term is, of course, an old one. In Aeschylus's tragedy, *Eumenides*, this concept is already present:

> The mother is not the true parent of the child
> Which is called hers. She is a nurse who tends the growth
> Of young seed planted by its true parent, the male. (658 ff.)

Vigdis Songe-Moller compares this passage to the autochthony myth that describes the origins of the Athenians. Here as well the man is seen "like a plant that sows its seed in the soil of the woman, which provides conditions suitable for the seed's growth. The man is the sole source of the child, who consequently has no mother in the biological sense" (*Philosophy without Women: The Birth of Sexism in Western Thought*, 6).

5. David Rorvik, the author of *In His Image: The Cloning of a Man* (1978), was already in 1971 suggesting that cloning "will make possible the birth of a child whose only parent is a male," going on to maintain that "before long, even men could, if they wished, get along by themselves. Dr James Watson . . . predicts that clonal propagation, by which we can make identical copies of ourselves by using single body cells rather than two sex cells, can be achieved by man within twenty years" (David Rorvik, "Present Shock," *Esquire*, August 1971, quoted in Jane Murphy, "From Mice to Men? Implications of Progress in Cloning Research," 84).

6. Examples of pregnant male bodies in art are, for instance, Wiligelmo's 1100–1106 carving, which represents Adam's protuberant belly pregnant with Eve in the Duomo at Modena. Marina Warner cites another example of a representation of a pregnant man in art: "In Bartolo di Fredi'd lunette in the Chiesa della Collegiata in San Gimignano, painted in 1356, the opening in Adam's side is ovoid and outlined and fleshy, unmistakably resembling the vulva stretched in childbirth" (*Monuments and Maidens: The Allegory of the Female Form*, 220). Similarly, George Platt Lynes, in his painting *The Birth of Dionysus*, offers a graphic depiction of a man giving birth to a baby issuing from his thigh.

7. In this context it is significant to note that both Zeus and his wife Hera brought forth children "on their own." However, while Zeus's children had harmonious bodies, Hera's offspring were monstrous and deformed, such as

the Titans. It seems, thus, that from the beginning androcentric Greek mythology wished to portray the male urge of self-engendering as acceptable and "natural," with no negative consequences, while the female one came under a distorting lens and was depicted as creating unnatural and abnormally colossal progeny.

8. As Jane Harrison notes, "The word Dionysus means not 'son of Zeus' but 'Zeus-Young Man,' i.e., Zeus in his young form" (*Mythology* 97). Mary Daly extrapolates from this, somewhat exaggeratedly—given that, Dionysus had a normal mother—that in this mythological realm Dionysus was "his own father" (*Gyn/Ecology: The Metaethics of Radical Feminism*, 64). While this doesn't make him exactly a clone of his father, it does move him in the direction of being some form of imprint of his father, leading Daly to go on to compare the elements shared by the myths of Apollo and Dionysus with the story of Christ: "Christ is believed by christians to be the incarnation of the 'Second Person of the Trinity,' and thus consubstantial with the father. Therefore, Christ, too, pre-existed himself and was simply a later manifestation of 'Zeus (Father)-Young Man'" (64).

9. In the animal realm, there are also examples of males that carry the eggs in sacs until the young are ready to be born. This is the case with midwife toads, so called because the male transports the eggs on its back, and sea horses. "The midwife is unique among frogs and toads of Europe because it is the male of the species that cares for eggs until they hatch. In the breeding season (late March to early August), females respond to calls of males . . . and are stimulated by males into laying eggs. Females lay large yolk filled eggs in strings, which are transferred to a male. The male wraps the strings around his legs, sometimes carrying two strings from different females. The male fertilizes the eggs immediately and carries them on his back. . . . A few weeks later, when the eggs are ready to hatch, the male deposits them in a pool or pond." (http://www.animalnation.com/Archive/reptiles/midwife.html, 8 August 2003).

10. Strictly speaking, Athena had a mother, Metis, who was also swallowed by Zeus, presumably in order to appropriate her reproductive capabilities. As Margaret Whitford explains, "Zeus had swallowed Metis in order to give birth to Athena (his thought or intelligence)" (Luce Irigaray, *Philosophy in the Feminine*, 384). I see a telling contradiction here, though, for if only men were associated with reason, as well as with intellectual tasks, how can Athena be described as Zeus's thought or intelligence?

11. Cavarero calls attention to the phallogocentrism of ancient philosophy, stressing that "the issue at stake is, quite literally, that of *male maternity*" (*In Spite of Plato: A Feiminist Rewriting of Ancient Philosophy*, 101; emphasis mine).

12. Songe-Moller relates this myth to the nature of the political institutions developed by the Athenians. By consecrating self-reproduction of the same male subject, women are effectively written out of this so-called democratic state. In this myth, "The male element symbolizes that which remains the same, identical and one, while the female is the other, the alien, that for which the story has no use simply because the story already allows the Athenians to explain their origins in terms of man alone. What the myth does is simply reaffirm political realities: there are only male citizens. But it goes further. Not

only does it deny women a political function, it deprives her of any function whatsoever and renders her utterly superfluous" (*Philosophy without Women: The Birth of Sexism in Western Thought*, 6).

13. In Thomas S. Gardner's story "The Last Woman" (1932), a method is devised to produce artificial ova as well as artificial wombs, or incubators. As in Anna Wilson's *Hatching Stones* (1991), this alters relations between the sexes radically, since men no longer need women to procreate. Similarly, once parthenogenesis or cloning is effectively implemented, women will not require sperm to fertilize their eggs, as is the case in numerous science fiction and utopian novels, some of which were examined in Chapter 2. See also Joanna Russ, "*Amor Vincit Foeminam*: The Battle of the Sexes in Science Fiction," 41–59.

14. Some of the most famous artistic renditions of Saturn, the Roman Kronos, devouring his children are Rubens's haunting painting *Saturn Devouring His Children* (c. 1620) and Goya's savage *Saturn Devouring His Children* (1797–1798).

15. Referring to this passage from *Medea*, Songe-Moller argues that "there are few places in Greek literature where the dream of women's superfluity is expressed so bluntly, yet there are good reasons to suppose that many aspects of ancient Greek culture were nourished by some such ideal of male self-sufficiency" (*Philosophy without Women: The Birth of Sexism in Western Thought*, 3).

16. According to Lévi-Strauss, the myth of Oedipus is centrally about the desire to assert the autochthonous origin of man. As Lévi-Strauss argues in *Structural Anthropology*, "The myth has to do with the inability, for a culture which holds the belief that mankind is autochthonous . . . to find a satisfactory transition between this theory and the knowledge that human beings are actually born from the union of man and woman. Although the problem obviously cannot be solved, the Oedipus myth provides a kind of logical tool which relates the original problem—born from one or born from two?—to the derivative problem: born from different or born from same?" (212). In the context of the topic of human cloning, Lévi-Strauss's questions "born from one or born from two?" and "born from different or born from same?" may be seen as acquiring added meanings and connotations: A cloned person would literally be born from "one" and from "same" (not taking into account, for the purposes of this line of argument, family predecessors).

17. There is another whole thematic area connected with the fantasy of male creation of life that I do not have the space to develop here, the creation or fabrication of women by men, from Haephaestus and Pandora, Pygmalion and Galatea, Prosper Mérimé's *La Vénus d'Ille* (1837), Edison's Hadaly in Villiers de l'Isle-Adam's *L'Eve Future* (1886), Olympia in E.T.A. Hoffmann's "The Sandman," Maria in Fritz Lang's *Metropolis* (1926), and the fabricated wives in Bryan Forbes's *The Stepford Wives* (1975), to name only some of the most salient examples.

18. The reference to sons exclusively should be noticed.

19. English translation lines.

20. In relation to Plato, Adriana Cavarero argues that Plato's position suggests "a sort of 'womb envy' which manifests itself in the masculine mimesis of maternity" (*In Spite of Plato: A Feminist Rewriting of Ancient Philosophy*, 103). Michelle Boulous Walker, in turn, reads Plato's work as suggestive that "his phantasy of self-engendering is a denial of the maternal body" (*Philosophy and the Maternal Body: Reading Silence*, 2). Connected with this theme and reflect-

ing on Plato's allegory of the cave, Irigaray, considering the male philosopher's desire for noble origins, away from the mother's, describes God the Father as "the Unbegotten Begetter" (*Speculum of the Other Woman*, 295), extrapolating then God's attribute to other men, who in analogous fashion fantasize about being themselves procreators, while being "*without origins* . . . or at least no beginning will be known for him and he knows no beginning . . . he procreates everything without being himself engendered and *thus puts an end to what has been staked in the game of generation*" (295; emphasis in the original). Ideally, then, man, "held captive by this excessively 'natural' conception and birth will be uprooted and referred to a more distant, lofty, and noble origin" (294), not in the mother's womb, which is invisible and unrepresentable.

21. Italian philosopher Adriana Cavarero calls attention to the phallogocentrism of ancient philosophy, stressing that "the issue at stake is, quite literally, that of *male maternity*" (*In Spite of Plato: A Feminist Rewriting of Ancient Philosophy*, 101; emphasis mine).

22. In "Creativity and the Childbirth Metaphor: Gender Difference in Literary Discourse," Susan Stanford Friedman remarks in related vein that "*Creation* is the act of mind that brings something new into existence. *Procreation* is the act of the body that reproduces the species. A man *conceives* an idea in his brain, while a woman *conceives* a baby in her womb. . . . The *pregnant* body is necessarily female; the *pregnant* mind is the mental province of genius, most frequently understood to be inherently masculine" (373).

23. See Gareth Roberts, *The Mirror of Alchemy: Alchemical Ideas and Images in Manuscripts and Books*, 23.

24. See Gareth Roberts, *The Mirror of Alchemy: Alchemical Ideas and Images in Manuscripts and Books*, 89.

25. As S. G. Allen and J. Hubbs observe, alchemical symbolism rests on the appropriation of the womb by male "art," that is to say, the artifact of male techniques. See "Outrunning Atalanta: Destiny in Alchemical Transmutation."

26. For a detailed account of the origins and ramifications of the concept of the homunculus see Clara Pinto Correia, *The Ovary of Eve: Egg and Sperm and Preformation*, 220–221.

27. See Maria Aline Ferreira's "'Glad Wombs' and 'Friendly Tombs': Reembodiments in D. H. Lawrence's Late Works," 166–170.

28. Adriana Cavarero, for her part, drawing on Hanna Arendt's category of birth, which Arendt developed in relation to and against Heidegger's notion of being-towards-death, emphasizes the need to shift philosophical perspective to an acknowledgment of the importance of birth, since "all persons, female and male, are invariably born from their mother's womb" (6). This statement seems very provisional nowadays, with the increasingly closer prospect of the development of an artificial womb where fetuses could be brought to term. The introduction of this technique would inevitably mean that many philosophical concepts would have to be revised and retheorized.

29. Michelle Boulous Walker reads Schreber's case, analyzed by Freud, in terms of a fantasy of "male procreation" (*Philosophy and the Maternal Body: Reading Silence*, 61), while Schreber's unmanning is seen by Walker as "a process designed to offer him a reproductive female body, one from which he can give birth" (61). As Walker notes, neither Freud nor Lacan referred to the im-

plications of Schreber's fantasies of self-generation, a meaningful absence according to Walker. For a stimulating analysis of Schreber's case, see Michelle Boulous Walker, *Philosophy and the Maternal Body: Reading Silence*, 58–67.

30. As far as fantasies of anal birth are concerned, a passage in Samuel Beckett's *Malone Dies* provides a dramatization of that fantasy, where Malone seems to be giving birth to himself "into death" (285) through a vexed fusion of vagina and anus: "the feet are clear already, of the great cunt of existence" (285). Ruth Parkin-Gounelas analyzes the theme of containment in Beckett's writing, arguing that it acts "as a fundamental locus of meaning" (*Literature and Psychoanalysis: Intertextual Readings*, 70), with "its preoccupation with uterine containment and expulsion" (70). With reference to this passage in *Malone Dies*, Parkin-Gounelas remarks that "giving birth to oneself, into death, is perhaps the closest the human phantasy can come to the figuration of turning oneself inside out, of attempting to expel something which one has incorporated and subsequently become" (77–78). Parkin-Gounelas further notes that the scene that follows this "near-end of self-parturition is dominated by a phantasmatic mingling of watery flows" (78).

31. Dick Teresi and Kathleen Mcauliffe cite Dr John Munder Ross, a psychiatrist, who argues that "if little girls want to have penises, boys also, at some level, want to have wombs and breasts" ("Male Pregnancy," 181). Dr Ross goes on to refer to the phenomenon of couvade syndrome, in which husbands experience symptoms similar to those of their pregnant wives. As Dr Ross asserts, "Most of the men I've analyzed during their wives' pregnancies have expressed wishes to have babies and have developed symptoms" (181).

32. Although these boys lived in a "residential treatment institution for emotionally disturbed children" (*Symbolic Wounds: Puberty Rites and the Envious Male*, 24), Bettelheim sustains that "this fact does not detract from the possible broader implications of their behaviour" (24).

33. Sven Delblanc was born in Manitoba, Canada, but went to Sweden at a young age. Delblanc wrote many plays and novels, combining his career as a writer with that of university professor in Uppsala and Berkeley.

34. For a discussion of Victor Frankenstein's narcissistic impulses and his drive to create new life, see Jeffrey Berman, *Narcissism and the Novel*, where Berman describes Victor as the "Modern Narcissus" (56).

35. See, for instance, Ellen Moers, *Literary Woman*; Anne K. Mellor, *Mary Shelley: Her Life, Her Fiction, Her Monsters*.

36. Like Frankenstein's creature, a man of colossal proportions, the underwater monster waiting to spring out of Sebastian's belly, a symbolic Leviathan, will presumably be a huge being.

37. Other characters who, like Sebastian, grow breasts with which to suckle their children can be found in Marge Piercy, *Woman on the Edge of Time* (Barbarrossa) and in Lisa Tuttle, "World of Strangers," where fathers can suckle their children if they so wish, thanks to hormone injections, a circumstance that promotes a state of equality between men and women as far as reproductive capabilities and policies are concerned.

38. This disturbing event is reminiscent of a similar one in Maurice Sendak's *Outside over There*, in which goblins take a baby and replace her with another made of ice, which eventually melts.

39. As a reaction against women's reproductive capabilities, there are many examples in our androcentric culture of "metaphors and images of procreation where the one who is 'birthing' is male, not female," as Eva Feder Kittay puts it ("Womb Envy: An Explanatory Concept," 96), referring to "the birth of Eve from Adam's rib or the male poets and scientists who speak of their 'brainchild,' as well as the pervasive use of the procreative female powers as metaphor for men's creativity" (96).

40. In this context, Charlene Spretnak analogously argues that "patriarchal culture alienates men from the life-giving processes, so their concern becomes the other half of the cycle: death" ("Naming the Cultural Forces That Push Us toward War," 58).

41. As Cohn further notes, in words that shed light on Delblanc's tale, "The entire history of the bomb project, in fact, seems permeated with imagery that confounds man's overwhelming technological power to destroy nature with the power to create—imagery that inverts men's destruction and asserts in its place the power to create new life and a new world. It converts men's destruction into their rebirth" ("Sex and Death in the Rational World of Defense Intellectuals," 142).

42. As Cherry Gilchrist explains, Dr John Dee, "a controversial Elizabethan figure, was suspected of being a sorcerer. . . . Dee was a remarkable man who took a keen interest in alchemy; in later life he had laboratories built at his house in Mortlake. He was a favourite of Queen Elizabeth's, who consulted him as to a suitable astrological date and time for her coronation. Dee was well versed in mechanics, optics, navigation, history and mathematics and was far-sighted enough to propose a national scheme for the preservation of ancient monuments and a national 'Library Royal'" (*The Elements of Alchemy*, 57).

43. See Freud, "Female Sexuality," 372.

44. I do not have the space here to elaborate on the intricate play of doubles between Matthew and the homunculus, as well as Matthew's father and Dr Dee, which constitutes a fascinating play of mirror images and reflections.

45. Another novel that finds in alchemy one of its main sources of inspiration for fantasies of male self-reproduction is James Morrow's *Only Begotten Daughter* (1990), where a celibate Jewish man oversees the development of his own fetus in an artificial womb after his semen has miraculously, through what is described as a process of "inverse parthenogenesis" (18), grown into an embryo. As Murray, her father, explains, "I'm her mother. Mother and father—both" (40). The baby who issues from this glass womb, a girl named Julie Katz, comes to be seen as God's own daughter and Jesus' half sister. As the new messiah, she will be the target of endless persecutions and have to negotiate her way through heaven and hell, as well as a very profane Atlantic City, New Jersey. Morrow's novel provides a satirical look at religious dogmas, religion and its fundamentalisms, as well as humanity's seven deadly sins.

46. A man who decided to have a cloned offspring, however, would still need a woman's womb to carry the child, at least until ectogenetic births are possible. That said, however, the feasibility of a man becoming pregnant exists, according to Prof Robert Winston, a reputed British specialist in reproductive technologies, who believes that technological and medical advances at least theoretically make a male pregnancy possible. As Prof Winston notes,

"IVF techniques could certainly be used to enable men to carry a baby" (*The IVF Revolution: The Definitive Guide to Assisted Reproductive Techniques*, 206). On the other hand, as he goes on to caution, although "there is no doubt that men could get pregnant . . . the risks and dangers in the techniques . . . of producing an ectopic pregnancy safely—seem insurmountable" (207). See also Lee M. Silver, *Remaking Eden: Cloning and Beyond in a Brave New World*, Chapter 16, "Could a Father Be a Mother?" as well as Susan Merrill Squier's discussion of William A. W. Walters, "Transsexualism and Abdominal Pregnancy," where Walters argues that a man might have a successful abdominal pregnancy. As Squier points out, Walters speculates that "abdominal pregnancy could be justified on the grounds of a right to reproductive autonomy" ("Reproducing the Posthuman Body: Ectogenetic Fetus, Surrogate Mother, Pregnant Man," 128).

47. See also Dick Teresi and Kathleen Mcauliffe, "Male Pregnancy."

48. Other fictional examples of tales where male pregnancy or male forms of mothering are contemplated can be found. The most bizarre is probably Guillaume Apollinaire's irreverent surrealist drama *Les Mamelles de Tirésias*, which starts with a wife, Thérèse, shouting, "Je suis féministe et je ne reconnais pas l'autorité de l'homme," while simultaneously taking away two balloons from her chest which she proceeds to burst. At the end, her husband is described as giving birth to thousands of babies, declaring, "Vous qui ne faites pas d'enfants / Vous mourrez dans la plus affreuse des débines." Gabriel Josipovici ponders on whether to read *Les Mamelles de Tirésias* as "a plea for women's rights or a patriotic and reactionary tract, one voice among the many on the nationalist Right that, amid the ravages of the war, tried to reassert that old and particularly French obsession about the need for large families" ("Scratch and Change," 6). Katharine Burdekin is another writer who engages critically with a masculine imaginary characterized by a dream of self-reproduction and the elimination of women. In her novel *Quiet Ways* (1930) there is a character, Carapace, whose hatred of women leads him to wish he had issued from a man's head, while in *Swastika Night* (1937) the official myth explaining Hitler's origins describes his being born from Zeus's head, disassociated from the corrupting presence of women. Naomi Mitchison, in turn, in "The Clone Mums," her unpublished draft of *Solution Three* (1975), tinkers with the idea of male pregnancy. As Susan M. Squier observes, quoting one of Mitchison's unpublished notes on the novel, "While planning the novel Mitchison toyed with the notion of writing about male abdominal pregnancy: . . . 'Council at work. Discussion. Sex almost always painful sooner or later—Why have it? Women councillor tells man she dislikes external sex organs. . . . Should there be implants rather than pregnancies? The men might then have them'" (*Babies in Bottles: Twentieth-Century Visions of Reproductive Technologies*, 248). Interestingly, Susan Squier reads Angela Carter's *The Passion of New Eve* (1977) as a novel about a "pregnant man" ("Reproducing the Posthuman Body: Ectogenetic Fetus, Surrogate Mother, Pregnant Man," 126). As Squier further notes, however, although the plot of the book "might be said to concern a man who gets pregnant, the novel ultimately represents pregnancy as occurring in, and linked to, the female body" (126). In *Up the Walls of the World* (1978) James Tiptree Jr. deals with an alien species in which the males are the ones who mother the young ones, a similar scenario to that we find in Octavia Butler's

"Bloodchild" (1984), which deals with interspecies pregnancy. In Butler's story only the men can carry the alien Tlic's eggs in their abdomen, then proceeding to deliver the alien babies through Cesarean section, which often kills the male hosts. See also Susan Merrill Squier, "Reproducing the Posthuman Body: Ectogenetic Fetus, Surrogate Mother, Pregnant Man." In South African writer Stephen Gray's *Born of Man* (1989) a gay man becomes pregnant for the first time in the whole world, with the fetus of a dying woman which is transplanted to his body and carried by him to term. The fetus, as the narrator emphasizes, is transplanted to the male body "*not* . . . by an incision through the stomach lining, but inserted through the rectum and implanted in the lower intestines. That turned out to be the correct method" (89). *Born of Man* can be read as expressing strong feelings of womb envy, as well as jealousy of women's reproductive organs and capacities in general. As the narrator observes, "It's my observation that the one thing every male gay I know wants most, and can never have, is a child" (142). Rosi Braidotti, for her part, mentions some films where the theme of self-birth occurs, including *The Invasion of the Body Snatchers, Alien, The Thing*, and *The Fly*, and goes on to describe Steven Spielberg as "the master of male birth fantasies" (*Metamorphoses: Towards a Materialist Theory of Becoming*, 194).

49. *The Cloning of Joanna May* (1984) constitutes a powerful denunciation of the consequences of the male domination and monopoly of control over reproductive power.

50. Dianne Hunter reads the play as exploiting "a male birth fantasy" ("Doubling, Mythic Difference, and the Scapegoating of Female Power in *Macbeth*," 147). For Hunter, one of the main underlying ideas in *Macbeth* is that "men must get rid of women in order to be reborn" (148). In *Macbeth*, "The biological function of women in sustaining the family tree is either denied mythically, as in the strange fantasy that there is a man not born of woman . . . as if men (in Coriolanus's words) were authors of themselves and knew no other kin" (150).

51. A similar thematic issue is also dramatized in Lisa Tuttle's "World of Strangers," where both men and women can have their own cloned son or daughter, thus eliminating the need for joint procreation, in what can be seen as a vision of endless narcissistic mirror images of themselves in miniature.

52. An analogous vision is put forward in a fictional form in Lisa Tuttle's "World of Strangers."

53. See Luce Irigaray, *This Sex Which Is Not One*; idem, *An Ethics of Sexual Difference*; idem, *Je, tu, nous: Toward a Culture of Difference*; idem, *To Be Two*.

54. Baba-I may be connoted with the Russian cannibalistic ogress Baba Yaga who, as Margaret Storch explains, is "one of the transformations of the devouring witch who lives in a little house in the forest" (115) and "is characterized by the chicken legs on which her house stands. The house itself is one of the forms of the imprisoning mother" (*Sons and Adversaries: Women in William Blake and D. H. Lawrence*, 115).

55. See, for instance, Christopher Lasch, *The Culture of Narcissism: American Life in an Age of Diminishing Expectations*; C. Fred Alford, *Narcissism: Socrates, the Frankfurt School, and Psychoanalytic Theory*; Jeffrey Berman, *Narcissism and the Novel*; Gray Kochhar-Lindgren, *Narcissus Transformed: The Textual Subject in Psychoanalysis and Literature*.

56. See Otto Rank, *The Double: A Psychoanalytic Study*; Sigmund Freud, "The Uncanny."

57. This scenario is reminiscent of similar descriptions of the clones' interactions in, for instance, Pamela Sargent, *Cloned Lives*; Kate Wilhelm, *Where Late the Sweet Birds Sang*; Ursula K. Le Guin, "Nine Lives"; Damon Knight, "Mary."

58. Anthony Ludovici, *Lysistrata, or Woman's Future and Future Woman*, with which Dora Russell engages in a critical dialogue in *Hypatia, or, Women and Knowledge*, presents a deeply misogynist view of a future society where it is feared that women might come to dominate, a dread that suggests a deep undercurrent of apprehension about female sexuality and simultaneously an unvoiced desire for a world without women, paradoxically a repressed male fantasy.

59. This dread of the putative power of a female-dominated society is fittingly expressed in, for instance, Poul Anderson, *Virgin Planet*; Edmund Cooper, *Five to Twelve*; idem, *Who Needs Men?*

60. Baudrillard engages indeed in a critical dialogue with Irigaray in his book *Seduction*, where he takes up and reflects on some themes developed by Irigaray in *Speculum of the Other Woman*. It seems thus perfectly feasible that this intellectual engagement might be carried on here.

61. The societies portrayed by Lane in *Mizora* and Gilman in *Herland* seem to suggest, on the other hand, that this is not necessarily the case. The other, in these societies, is the other woman, with all the specificities inherent in a different human being.

62. In this context see Nicole Loraux, *The Children of Athens: Athenian Ideas about Citizenship and the Division between the Sexes* (1993).

63. As Moira Gatens remarks, "The male child's primal wish, to take the place of the father, is expressed in political terms by the fantasy of the generation of a man-made social body: a body that is motherless and so immortal. . . . The motherless Athena can fearlessly confront the Furies, rebuking them for their vengeful pursuit of the matricide Orestes. Asserting her authority by sending them (literally) underground she establishes the priority of (male) citizenship over blood ties and thus institutes the classic patriarchal state, which even bears her name: Athens" (*Imaginary Bodies: Ethics, Power and Corporeality*, 53). As Gatens further explains, "Unmothered, such beings are autonomous, immortal and quintessentially masculine" (53), a fantasy powerfully illustrated in *Hatching Stones*.

64. Nancy Tuana elaborates on this question: "Plato's *Timaeus* contains a creation myth in which the creation of woman is depicted as both temporally and metaphysically secondary. In this dialogue, Plato recounts the creation of the universe and of the beings that populate it. The divine being formed the universe from a mixture of the four elements: earth, air, fire and water. To create man, the divine being added the soul of the universe to a mixture of these four elements and then 'divided the whole mixture into souls equal in number to the stars and assigned each soul to a star' (*Timaeus* 41d.). Each soul was implanted in a body and given the faculty of sensation and emotions, including those of love, fear, and anger. In the first birth, each man would be equal in perfection to all others, for god would allow 'no one [to] suffer a disadvantage at his hands' (*Timaeus* 41e.). In this way, the race of men came into being" (*Woman and the History of Philosophy*, 14).

65. Jane Murphy, in "From Mice to Men? Implications of Progress in Cloning Research," describes the male fantasy of "'creating' or 'giving birth' to *someone*—an offspring, themselves" (84) as the "Father as Sole Parent" mythology (84), which Murphy associates with patriarchy's struggle with such concepts as the meaning of life and mortality and the wish, on the part of men who indulge in these dreams, to "somehow escape mortality and live on, into future generations" (84), words that apply to the male clones in *Hatching Stones*.

66. I discuss this issue at greater length in Chapter 7.

67. See also Peter Fitting, "'So We All Become Mothers': New Roles for Men in Recent Utopian Fiction."

68. In *The Left Hand of Darkness* (1969) Ursula K. Le Guin has attempted to portray a genderless society, where sexual behavior does not conform to the patterns traditionally prescribed for men and women. Indeed, men could become pregnant, like the king. Referring to a sentence from the novel "The king is pregnant," W. A. Senior comments that this "is a line few of us will ever forget, a shocking moment of cognitive estrangement" ("Introduction," 204).

69. In Louise Erdrich's *The Antelope Wife* (1998) the protagonist, having found a baby in a little cart being pulled by a dog, in the desert, puts him to his breasts, having nothing else to offer him to eat or drink, and they miraculously fill with milk.

70. In her book *Pregnant Men: Practice, Theory, and the Law* (1994), Ruth Colker combines feminist theory and jurisprudence in order to analyze many concrete situations and legal cases which she examines from the perspective of an antiessentialism critique and equality theory. As Colker explains, "Because there are no 'pregnant men' to whom we can compare pregnant women, I decided to ask whether equality theory could be used to redress women's subordination in the reproductive health context" (xi).

71. In the film *Junior* (Ivan Reitman, 1994, US), Arnold Schwarzenegger, having used a new drug on himself, becomes pregnant. As a male mother, the character played by Schwarzenegger will go through the process of pregnancy in this comedy that in many ways can be seen as prefiguring a male fantasy that may come true one day.

72. Eva Feder Kittay makes a similar point to the one I am arguing in relation to the potential abusive appropriation by men of human cloning, but with respect to in vitro fertilization. Kittay suggests that "behind the controversy surrounding . . . in vitro fertilization, there is a strong element of man's attempt to control and hence appropriate those powers he envies in women and which he can never truly call his own" ("Womb Envy: An Explanatory Concept," 115).

73. As Daly in irreverent fashion argues, "The christian trinity can be seen as a paradigm for cloning" (*Quintessence: Realizing the Archaic Future: A Radical Elemental Feminist Manifesto*, 207).

4

The Eternity of the Same: Human Cloning and Its Discontents

> We're incubating monsters, you know. Dr. Frankenstein will be here on the next train.
>
> Pamela Sargent, *Cloned Lives*, 63

In this chapter I will look at a cluster of clone narratives that address some of the complex psychological dynamics operating within clone groups. In these texts the members of the groups of cloned people are described as practically interchangeable, as sharing one single personality among them. It is this uncanny feeling the clones give rise to—of the repetition of the same, as if one were observing the vivid concretization of the repressed fantasy of the return of one's feared (and at the same time desired) double—that is likely to create sentiments of anxiety and foreboding in those who witness it. Indeed, according to Baudrillard, cloning does away with "the possibility of alterity and of a dual relation" (*The Vital Illusion*, 13). Pursuing this line of enquiry, it is pertinent to ask, Will the cloned person feel threatened by the knowledge that he or she will have exactly the same genetic makeup as his or her parent and might thus grow up to be an almost exact copy of that person? Where, then, is room for individual growth, recreation, self-building?[1] The biological determinism versus social constructionism argument suggests that in spite of the same genetic material two or more people sharing that same biological makeup will nevertheless grow up with a different psychological makeup. Pamela Sargent's *Cloned Lives* actively participates in this debate, as does Fay Weldon's *The Cloning of Joanna May*. A related issue has to do with the fundamental problem of the genetic makeup of the cloned person, who will have "only" the genetic material of one progenitor, mother or father. Will that "lack" of DNA from the second progenitor be perceived as a disadvantage, as a serious "loss," potentially conducive to psychological dilemmas and imbalances?

Addressing mainly these issues, I will analyze Pamela Sargent's *Cloned Lives* (1976), Kate Wilhelm's *Where Late the Sweet Birds Sang* (1976), Ursula K. Le Guin's "Nine Lives" (1969),[2] and Damon Knight's "Mary" (1964).[3]

Both Pamela Sargent's *Cloned Lives* and Kate Wilhelm's *Where Late the Sweet Birds Sang* take place in a postapocalyptic world, which has been destroyed by ecological disasters and is in the process of being rebuilt. In *Cloned Lives* Pamela Sargent narrates the first attempt at creating human clones grown in ectogenetic chambers. The father, Paul Swenson, is a famous astrophysicist who, having lost his wife, accedes to the invitation by a geneticist friend, Dr Hidey, to have his genetic material used in this unique experiment. Out of six embryos, five come to term: four boys and a girl. The story revolves around the difficulties faced by the clones in coming to terms with their singular situation and the media attention, tracing their distinct biographies and following their search for individuation.

The Swenson clones, Edward, James, Michael, Kira, and Albert, are extremely sensitive to their different genesis and constantly interrogate their identity as separate individuals, fearing other people's reactions and often deliberately isolating themselves to avoid suffering. The novel raises many important questions. Who exactly are Paul's children in relation to him? He muses, "They were his children, yet closer to him than children. They were his twins, his brothers, and a sister too, separated from him only by age" (*Cloned Lives*, 77).[4]

The vexed question of the individual's uniqueness is a pivotal concern in the novel. As Jim reflects about himself and his brothers and sister, "Jim saw them as they must appear to others—identical, a closed group, undifferentiated, and inaccessible. *We're components, interchangeable parts,* he thought. Even their different pursuits were probably accidental" (*Cloned Lives*, 122). Jim experiences his body as a "prison, forcing him to live and struggle" (125). He imagines himself writing a report about the cloning experiment, his own feelings of bitterness, his excruciating doubts which lead him to contemplate suicide:

> The experiment with cloning has failed. One of the experimental subjects can no longer live with himself; the others are only four bitter people, denied even the small pleasure of feeling like unique individuals. He knew they had wanted

a team, a Paul Swenson multiplied by five, working together, synthesizing what they learned in different fields, minds so alike they could see connections where others might not. (*Cloned Lives*, 125)

Interacting with others is a major problem for most of the Swenson clones. When they drop their social misgivings against each other, due to their fear of regimentation and of depersonalization, the clones are depicted as enjoying a fulfilling mental and spiritual harmony and an outstanding feeling of empathy:

They had been sitting with Ed on the front porch, talking about one of Jim's poems, listening to Ed play the violin, discussing some of the work Kira had done with Dr. Takamura. They talked for a long time, their minds drawing together, communicating ideas, disagreements, and feelings with perfect understanding. (*Cloned Lives*, 140–141)

For society in general, they are looked upon as duplicates of Paul Swenson, as "peas in a pod" (*Cloned Lives*, 101). As Paul points out, "They don't view them as kids right now, just as something to fear. Some of the stories I've seen talk about mass minds, or mental telepathy among clones. One even said they might be condemned to doing the same things at the same time" (66). They represent what society perceives, in Baudrillard's words, as a disturbing and vaguely grotesque "renewal of the Same" ("Clone Story," 101).

Baudrillard's proposition of "the hell of the Same" ("The Hell of the Same," 122), is thus aptly dramatized in *Cloned Lives*, where the five clones have to come to terms with their similarity, which makes for perfect adjustment and consonance in their relationship among themselves but also simultaneously gives rise to complex inner frictions. Can they find such rapport and compatibility with anybody else? Can they (are they willing to) put up with second best when they embody to eacg other the perfect twin soul, the Platonic other whom humans can spend their whole lives trying to find and always be disillusioned with, when, in their case, the other that completes them is their brother or sister?

Cloning will inevitably reshape many of society's deepest underlying myths. For instance, how do incest taboos apply to a family of cloned brothers and sisters? The narrator in *Cloned Lives* meditates, "The old codes and ancient prohibitions could not apply to them, had not even allowed for their existence" (146). In turn Kira muses, "Some people would look at us and talk about incest taboos, and others would probably find it strange if we loved anyone else but the other clones" (149). Significantly, the sexual encounters between Kira and Jim, a cloned brother and sister, are described in the terms used by Aristophanes in Plato's *Symposium*:

She was his female self. . . . He saw himself as a woman, receiving a man . . . and knew that she was seeing herself as a man. . . .

This has never happened before. . . . Never before. He saw generation after generation evolve, becoming more differentiated, genetic structures changing and mutating. ***He saw millions of men and women seeking mates, trying to find those who would complete them, make them whole again,*** *yet always separated from them by the difference passed on to them by eons of change. He saw Kira and himself, reflections of each other, able to move along their individual paths and yet move in perfect communication. She was no longer his sister, but his other self, closer to him than a sister could have been, merging with him so completely and perfectly that they were one being.* (*Cloned Lives*, 145; the emphasis in bold is mine)

Freud's *The three Essays on the Theory of Sexuality* (1905) starts precisely with a description of the sexual instinct as it appears in the "poetic fable which tells how the original human beings were cut up into two halves—man and woman—and how these are always striving to unite again in love" (2). For Kari Weil,

Aristophanes' myth . . . identifies the androgyne with an ideal, even an Edenic, state of being, a state of wholeness in which nothing is lacking. Visually, his story represents androgyny as perfect symmetry between two united halves. Structurally and thematically, his story recalls the separation of Eve from Adam (described as androgynous in certain versions of Genesis) and the biblical fall into disunity. Both in Genesis and in Aristophanes' account, sexual division is regarded as the punishment for the fall. For Aristophanes it is also its remedy—the union of the two sexes through divinely inspired love is the route toward regaining salvation. (*Androgyny and the Denial of Difference*, 18)

In *The Four Fundamental Concepts of Psycho-Analysis*, Jacques Lacan puts forward an alternative interpretation of Aristophanes' myth of the androgyne in Plato's *Symposium*, suggesting that the search for one's complementary party, the other, is really at bottom a search for immortality. Thus, according to Lacan,

Aristophanes' myth pictures the pursuit of the complement for us in a moving, and misleading, way, by articulating that it is the other, one's sexual other half, that the living being seeks in love. To this mythical representation of the mystery of love, analytic experience substitutes the search by the subject, not of the sexual complement, but of the part of himself, lost forever, that is constituted by the fact that he is only a sexed living being, and that he is no longer immortal. (*The Four Fundamental Concepts of Psycho-Analysis*, 205)

Cloning can thus be seen as articulating the longing for immortality that the idea of the other half evokes, as well as the yearning for perfect harmony in love.

Indeed, *Cloned Lives* is also fundamentally a novel about the quest for immortality, as is arguably the case with most clone narratives, in which the characters' often narcissistic impulse to see themselves reborn can be equated with the search for at least a measure of immortality in having themselves perpetuated through their cloned other selves,[5] although, as Jane Murphy points out, "proponents of cloning as a reproductive technology in fact suggest that men may gain a sense of immortality by duplicating their genes in offspring" ("From Mice to Men? Implications of Progress in Cloning Technology," 85).[6]

For the general public, Paul is given a semblance of immortality through his cloned children. Another aspect linked with the search for immortality, indeed one of its sources, is the question of narcissism. As Jon, a clergyman and Paul's friend, muses, considering the ethical implications of the cloning experiment on which Paul is about to embark, "If it works . . . every narcissist alive will be trying to use it. You'll be interfering with the course of human evolution with no conception of what the results might be. What would happen in the long run if even a sizable minority decided to reproduce this way?" (*Cloned Lives*, 31). Paul's answer is telling: "*Abusus non tollit usum*. . . . We can't refuse to use something simply because it may be misused" (31).

Family ties are radically reconfigured in the Swenson family. In an interesting reversal of traditional roles, it is Kira, Paul's cloned daughter, who resurrects her father, bringing him back from death to life, acting as mother goddess, or as a mother giving life to a son. Kira's relation to Paul goes beyond that of a daughter: In genetic terms she is also his sister and his twin (and his mother?). Moreover, by falling in love with Hidey, the geneticist who created her and her cloned brothers, in effect her surrogate father, Kira is subverting accepted social roles. Pondering about her connection with Hidey, Kira muses, "He too was her father in some sense. And he had been one of Paul's closest friends" (*Cloned Lives*, 188). The shadow of incest will always loom over their relationship.

Kira also revises traditional notions of mothering. As a the result of being the clone of a male, Kira is sterile. There are, however, options availabe to her, and she and Hidey are granted permission to clone a child of their own, Rina. As the narrator explains, "Rina had been produced by the same process that had produced the Swensons. The difference was that both Kira's and Hidey's germ plasm had been used" (*Cloned Lives*, 238). Furthermore, in Kira's society, "People who wanted only clones of themselves were still frowned upon, unless the circumstances were unusual. But there were other children like Rina; children of homosexual couples, of sterile couples, or of groups who raised children communally" (238). In addition, Rina was brought to term in an ectogenetic chamber: "Although still not common, the procedure was becoming more usual as a convenience for mothers" (238).[7]

In the near future of Sargent's novel, a new psychological and demographic map for humanity is developing as a result of new reproductive technologies, namely, cloning and artificial wombs. As the narrator elucidates,

> The earth was becoming less crowded. As more of its six billion people had become reasonably well off, the birth rate had dropped. Almost one-sixth of the world was too busy with other pursuits to bother having children at all. Young women were not pressured into fulfilling themselves through childbirth; among some young men, sterilization had become popular, a way of asserting their masculinity, of showing people that they were secure enough in their manhood not to require the perpetuation of their genes. A growing industry catered to childless people; among many others, having only one child was increasingly common. (*Cloned Lives*, 238–239)

This vision seems to me to at least partially emblematize the kind of future scenario Shulamith Firestone envisaged in *The Dialectic of Sex: The Case for Feminist Revolution* and indeed to be potentially achievable in the not too distant future. Kira's vision is not a selfish one, though. She altruistically wants to make sure that everyone "had a chance at what might be immortality, not just a select or wealthy few" (*Cloned Lives*, 307), a desideratum that similarly applies to cloning, should it ever become available.

> Je est un autre.
>
> Rimbaud, letter to Paul Demeny, May 15, 1871

Clones' often-felt nostalgia for a supposedly lost authenticity constitutes the main driving force behind their actions. The genetic commodification of their bodies marginalizes them, turns them into curious rarities. They are often described as freaks, grotesque creatures who have to negotiate their acceptance into society or else remain among themselves, giving up the outside world. The groups of cloned brothers and sisters, who behave like one sole organism, can also be likened to automata, a figure that, incidentally, is one of Lacan's main metaphors for narcissism.[8] Jim Swenson, one of Paul Swenson's five cloned children, muses, "*I'm living Paul's life.* He felt paralyzed. He saw himself as a **puppet**, walking an ever-repeating cycle. *I'll go through it again,* his mind murmured, *I'll go on feeling the way I do, acting the way I do, and I won't have any choice. It's all happened before and I have no way of changing it*" (*Cloned Lives*, 139; italics in the original, bold is my addition). The seeming inevitability of this apparently deterministic fate imposed on him brings to mind the Nietzschean concept of eternal return, an aspect also noticed by Baudrillard. In *The Gay Science*, Nietzsche muses, "Do you desire this once more and innumerable times more?" (274).

Even when they are middle-aged, the old feelings of "being trapped" (*Cloned Lives*, 298) in a relationship with "these parts of himself" (297), as Mike, one of the Swenson clones, muses, are still as strong and disturbing: "Every meeting and conversation with them threatened his sense of identity, every family gathering erased years of effort and made him an awkward boy again, part of them yet alienated from them" (297–298). Indeed, Mike feared that they "would infect his life again" (298). Significantly, instead of growing closer as they grew older, the brothers and sister "grew more distant every year. Oddly enough, their similarities seemed to aid in driving them apart, as if each resented the part of himself he saw reflected in the others" (225). The stress here is on the fear of engulfment, of incorporation into the others, of loss of individuality, of being perceived as a part and not a whole, unique person, a fear that also pervades Kate Wilhelm's *Where Late the Sweet Birds Sang* (1976).[9]

Wilhelm's novel offers a useful comparison and contrast with Sargent's one. The dilemmas and contradictions faced by the clones in Wilhelm's story duplicate to a certain extent those experienced in Sargent's tale, while extending in other directions the problematics inherent in the politics of reproduction of the society of clones, as well as the power games necessarily inscribed in the propagation of the species.

Wilhelm borrowed the title of her novel from Shakespeare's Sonnet 73, in which images of a cold autumnal landscape and empty ruins, as well as deeply disturbing intimations of impending death, evoke the postapocalyptic scenery that dominates most of *Where Late the Sweet Birds Sang*. As a consequence of the alarming spread of pollution and all manner of abuses perpetrated on the natural world, the global cataclysm that ensues annihilates world civilization almost completely, destroying the global economy and ecological equilibrium, while the populations, culture, and agriculture of entire countries are totally obliterated. The patriarch of the Sumner clan decides to try to preserve his family with recourse to a program of human and animal cloning, perceived as the only answer to the alarmingly decreasing rate of fertility, in an isolated Virginia location, the secluded Shenandoah valley, where the global catastrophe is only barely avoided and the narrative follows the lives of three generations of the same family. The third generation, however, is so far removed from the first cloned individuals that they are explicitly described as "inhuman," while the older "humans among them will be pariahs" (53). Indeed, one of the main issues addressed in the novel is the crucial question of what it means to be human: Are the clones "inhuman" (42), as David, a character who belonged to the original clan, wonders, watching a group of cloned boys?

The clones, developing a society of their own, gradually shun the elders, who still maintain a certain control over the valley, and progressively create their own organizational rules that disregard the old values the original members of the Sumner clan are striving to keep. Classified according to letters and numbers, the clones are D-1, F-4, W-1, and so on, in uncanny and exact correspondence to Baudrillard's vision of cloning as bringing about "the hell of the Same."

Molly, one member of a group of identical sisters, undergoes a radical transformation during an exploratory trip to compile and retrieve information about other parts of the country which had been obliterated by pollution, radiation, and disease. She realizes that no matter how hard she tries, "something had died" (*Where Late the Sweet Birds Sang*, 86), that "that other self" (86) she had found during her journey, which she attempts to "submerge . . . and become whole again with her sisters" (86) will not go away and that in fact she wishes to cling to it, to recover it. This other self, however, "frightened her and isolated her in a way that distance and the river had not been able to do" (86). Her sisters, on the other hand, "would never permit her to be alone again. Together they made a whole; the absence of one of them left the others incomplete" (87). Banished to the old Sumner house in the Shenandoah valley, Molly feels happy in her solitude, with her paintings and books, and gives birth to Mark, Ben's son. Ben, who had also been a member in the exploratory expedition, had undergone similar transformations to hers, although not to the extent of wishing a complete severance from his brothers. When Mark is five, they are found out and taken back to the compound. Molly is kept imprisoned with the other breeders, women whose only function is to bear children and who are kept incarcerated in a special hospital,[10] while Mark grows up with the other boys, although he often tries to be alone. His difference makes him precious to the rulers, who see in him qualities that have been lost through generations of cloning. He held clues, for instance, "about how man lived alone" (129), how to survive in the woods and not be afraid of them, "without danger of mental breakdown through separation" (130) from the other brothers or sisters: "that was well known, the fear of the silent woods" (136). Apart from these characteristics, which made him useful for the community, there were "no common grounds for understanding. He was an alien in every way" (131).

His artistic bent was another feature that set him apart and made the gulf between him and the others even more unbridgeable. Mark's sculptures, for instance, were completely meaningless to the clones.

As he muses, "They learned everything they were taught, he realized, everything. They could duplicate what had gone before, but they originated nothing. And they couldn't even see the magnificent snow sculpture Mark had created" (*Where Late the Sweet Birds Sang*, 155). Psychoanalyst Marie-Louise von Franz addresses the role of art in words relevant for our discussion. According to her, a "civilization which has no creative people is doomed. . . . The person who is really in touch with the future is the creative personality."[11] This insight is consonant with the narrative drift of the novel, which suggests that the new clone society is condemned to gradual decline and annihilation. Mark is the only one with the creative capacity to survive, to go on producing culture and art.

LOST AURAS AND SIMULACRA

This persistent problematization of identity is one of the main leitmotifs in these clone narratives, crucially structured around concepts of wholeness and lack, of individuality and standardization, of a nostalgic yearning for a society before asexual reproduction and genetic engineering set in and became the norm. A character in *Where Late the Sweet Birds Sang* believes that "psychology is a dead end" for their society of families of clones because "it revives the cult of the individual" (100). Indeed, "singular" and "individual" are key terms in any discussion of cloning, as are such words as "original," "whole," "complete," "authenticity"—words that frequently recur in Baudrillard's comments on cloning.

What Baudrillard mainly objects to in cloning is the endless, mechanical reproduction of the same, a duplication that inevitably entails the disappearance—or at least the dilution to extreme, homeopathic doses—of a supposed original, unique, authentic first being or object. Walter Benjamin is, of course, a crucial source in any debate that deals with copies and reproductions, having analyzed the problematics of the art object and its value inscribed in a society that is ruled by mass production and consumption in his influential essay "The Work of Art in the Age of Mechanical Reproduction," a work Baudrillard often engages with. With reproduction, the "aura" of the object, to take up Benjamin's term, is lost, an insight that can be profitably applied to the analysis of the behavior of the societies of clones in Sargent's and Wilhelm's novels.

Indeed, the dread associated with the loss of an aura of authenticity in an age of simulacra is a driving force behind the horror often linked with the prospect of cloning. In "The Work of Art in the Age of Mechanical Reproduction," Benjamin contrasts the traditional concept of art and the original art work, defined as auratic and unique, with mass art, characterized by its lack of what he terms "aura" and its reproducibility.[12] Drawing on Benjamin's terminology, Baudrillard implicitly inscribes clones in the realm of mass art, as reproducible and unoriginal.

There is, however, an interesting paradox at work as far as these different conceptions of art, namely, high art and mass art, are concerned: While what Benjamin calls auratic art (characterized by the presence of the artwork in relation to the spectator and the latter's contemplative distance toward the original) places its emphasis on the uniqueness and authenticity of the original artwork cloning (while also stressing the genuineness and inimitability of the original person) simultaneously and inevitably partakes of the ethos of reproducibility, thus ultimately, if unwittingly, aligning itself with the concept of massification and mass art, to follow a Benjaminian frame of reference. Pursuing some of the implications of Benjamin's argument in "The Work of Art in the Age of Mechanical Reproduction," with its defense of mass art in view of its engagement with the revolution of the proletariat and the consequent releasing of society's productive forces, cloning, as far as it can be said to partake in the nexus of ideas governing mass art, might in effect subvert the very drive toward a perpetuation of the original that initially dictated it.

> Standard men and women; in uniform batches.
>
> Huxley, *Brave New World*, 7

The cloned brothers and sister in *Cloned Lives* are deeply fearful of having lost their "aura," what made them unique human beings, a distinct genetic patrimony. Similarly, Mark feels himself to be the only person in the society he moves in to have the combined genetic information of a mother and a father, a circumstance that in turn makes him a singular and solitary creature. As Baudrillard comments with reference to cloning, "The original is lost, and only nostalgia can restore its 'authenticity'" (*Seduction*, 171). In a related vein, Jean-François Lyotard reflects on what he perceives as "the nostalgia of the whole and the one."[13] It is precisely this nostalgia for a different past that did not contain groups of cloned people that drives Mark to search for his genealogy, to preserve the memory of his family and of his own past.

In light of a Baudrillardian terminology, the clones can arguably be described as "simulacra," that is, copies without an original, as was suggested in Chapter 1. Pushing further this line of argument, one might say, along with Baudrillard, that, "to rephrase Benjamin, there is an aura of the simulacrum" ("Objects, Images, and the Possibility of Aesthetic Illusion," 10). Clearly, for Baudrillard, cloning falls into the category of an inauthentic form of simulation. In his characteristic polemic style, Baudrillard argues that "there is an authentic form of simulation as well as an inauthentic form of simulation" (11), further remarking that "we must not add the same to the same, and then to the same again: that is poor simulation" (11), an insight that can be extrapolated to cloning, in the wake of his other pronouncements on this theme. This is the area where I have some problems with Baudrillard's views on cloning, for he consistently regards human clones as inferior copies of the "original."[14] When he derogatorily pronounces, with respect to a cloned person, that a "segment has no need of imaginary mediation in order to reproduce himself, any more than the earthworm needs earth" ("Clone Story," 97), not only is Baudrillard implicitly linking a cloned human being with an earthworm but he also seems able to think only in terms of hordes of robotically identical clones and unable to consider the human feelings of a single one. Again, when Baudrillard declares that "the clone is the materialization of the double by genetic means, that is to say the abolition of all alterity and of any imaginary" (97), surely he has in mind multiple clones in an apocalyptic vision such as those put forward in *Brave New World* or *Where Late the Sweet Birds Sang*.

In "Subjective Discourse or the Non-Functional System of Objects," Baudrillard, in a Benjaminian register, again addresses the question of origins, of authenticity, and, by extension, of the past. A sense of the past, not just a personal, genealogical past but also a historical past, is what is at stake for Mark in *Where Late the Sweet Birds Sang*, a family past that he wants not only to reconstitute and keep as a memory, a trace, but also to recover and reinvent, in the shape of a new nuclear family of his own and the implementation of a new agrarian, self-sustaining society away from the serial reproduction and automatization of the other valley. Mark's attachment to the old house where he grew up, to his books and paintings and to the statue representing his mother, constitutes a clear index of the importance of old objects, valuable tokens of a family past he wants to preserve. According to Baudrillard,

> The need that bygone objects fulfill is that of a definitive being, a *complete* being. The time of the mythological object is the perfect tense: it is that which occurs in the present as having previously occurred, and which by this very fact is based on an "authentic" self. The bygone object is always, in the full sense of the term, a "family portrait." ("Subjective Discourse or the Non-Functional System of Objects," 36; emphasis mine)

Thus, and pursuing this line of argument, Mark's objects, acquiring an "aura" of mythological objects in line with Baudrillard's analysis, become "complete" ("Subjective Discourse or the Non-Functional System of Objects," 37). According to Baudrillard, this "complete event that it [the object] signifies is birth" (37), for "I am not the one who is here right now, as this is anguishing—I am the one who has always been, according to an inverse link with my birth for which this object is a sign, a regression plunging me from present into time. Thus does the bygone object present itself as a myth of origin" (37). In search for his own myths of origin, Mark is described as desperately clinging to old objects that can stand for primal scenes and thus provide him with a past.

It is precisely this dimension of a sense of time in the past extending to the present that Mark does not want to forgo, a dimension the clones are totally unaware of and so do not even miss, although for Mark it represents a fundamental lack. Baudrillard goes on to emphasize that

> we need to distinguish two features in the mythology of bygone objects: a nostalgia for origins and an obsession with authenticity . . . the older are the objects, the closer do they bring us to an earlier time, to an original "divinity," nature, wisdom. . . . The demand for authenticity . . . expresses an obsession with certitude—about the origin of a work, its date, author or signature. ("Subjective Discourse or the Non-Functional System of Objects," 37)

This is precisely the kind of knowledge that is denied to the clones, who have no father or mother in the traditional sense. Mark, on the other hand, has a "conventional" father and mother, a situation that sets him apart, that makes of him, in the society of clones, a collector's item, in the sense that Baudrillard connects the "taste for the antique with the passion for collecting—since, in their narcissistic regression, in their systematic elision of time, and in their imaginary command over birth and death, there is a deep affinity between the two" ("Subjective Discourse or the Non-Functional System of Objects," 37). Indeed, as Baudrillard points out, the "search for *the creator's mark*, from actual impressions to the signature, is also the search for filiation and paternal transcendence. Authenticity always stems from the Father: he is the source of value. What the bygone object evokes in the imagination is this sublime filiation at the same time as the degeneration to the mother's womb" (37).[15]

At the end of the novel, Mark has managed to escape and take some women with him to settle down in a fertile valley, where he starts a new life as the patriarch of this emerging society, where there was no more cloning and no two of the children "were alike" (*Where Late the Sweet Birds Sang*, 206), in what constitutes a critical vision of cloning but also a return to a patriarchal view of dominance by the patriarchal figure.

Wilhelm's novel, then, is a powerful cautionary tale that forewarns about the potential dangers a misguided use of the biological sciences might lead to, roundly criticizing the emergence of a society of cloned people, in Baudrillardian vein. Sargent's narrative, on the other hand, attempts to assess the pitfalls that lie in wait for the cloned people themselves, in terms of patterns of socialization and ego development but seems to suggest that the emergence of cloned humans as well as their successful social integration are possible.

> It's only excess twins.
>
> Ursula K. Le Guin, "Nine Lives," 4

The uncanny empathic and almost telepathic connections that characterize the interrelations among groups of cloned people, which operate in Pamela Sargent's *Cloned Lives* and Kate Wilhelm's *Where Late the Sweet Birds Sang*, are also at work in Ursula K. Le Guin's "Nine Lives" and Damon Knight's "Mary."

In Le Guin's "Nine Lives,"[16] the arrival of a group of clones at planet Libra (described as a threatening, devouring female), makes the two technicians already there, Pugh and Martin, revise their notions of what is human and confront the "terrible thing: the strangeness of the stranger" (4), which became amplified in the clones. As Le Guin herself explains, after reading Gordon Rattray Taylor's *The Biological Time Bomb*, she found the possibility of "duplicating entire higher organisms, such as frogs, donkeys, or people . . . alarming" ("On Theme," 204). As she further observes, "I began to see that the duplication of anything complex enough to have personality would involve the whole issue of what personality is—the question of individuality, of identity, of selfhood" (204–205).

In "Nine Lives,"[17] the group of clones is described as interchangeable, as sharing one personality, one single individuality, characteristics that elicit several questions from Pugh and Martin: "Did repetition of the individual negate individuality?" (9); "What would it be like . . . to have someone as close to you as that? Always to be answered when you spoke; never to be in pain alone" (9). From a technical and economic vantage point, the clone group becomes an ideal work team. As one of the group explains to Pugh and Martin, "We think alike. We have exactly the same equipment. Given the same stimulus, the same problem, we're likely to be coming up with the same reactions and solutions at the same time. Explanations are easy—don't even have to make them, usually. We seldom misunderstand each other. It does facilitate our working as a team" (6). In addition, in the world envisaged by Le Guin, clones "are drawn from the best human material, individuals of IQ ninety-ninth percentile, Genetic Constitution alpha double A" (7).[18] As one of the clones further explains, "We have more to draw on than most individuals do" (7).

The drawbacks of being so close to the other clones are brought into sharp relief when nine of the ten clones are killed in an accident and only Kaph survives, remaining, however, in a comatose state of total indifference to his surroundings. As Pugh explains, "He never was alone before. He had himself to see, talk with, live with, nine other selves all his life. He doesn't know how you go it alone" ("Nine Lives," 22), since he has "never known anyone but himself" (24). Martin's baffled reaction sums up a generalized view: "Then by God this cloning business is all wrong. . . . What are a lot of duplicate geniuses going to do for us when they don't even know we exist?" (24–25), a feeling shared by David in Wilhelm's "Where Late the Sweet Birds Sang," the short story later expanded into the novel with the same name.[19]

In Pamela Sargent's "Clone Sister" (1973),[20] an early version of *Cloned Lives*, the five cloned children of Paul Swenson experience deep-rooted uncertainties about their status as individuals belonging to such a close-knit group of brothers and sister. Jim articulates that dilemma when he ponders on whether they are "all interchangeable" (181) and whether his girlfriend will think of him as an individual or as a "freak" (181). Looking at his brothers, Jim expresses the view of them others might have: "identical people, clones of the same man, undifferentiated and interchangeable" (184), like cogs in a machine, mass-produced items in a factory conveyor belt. Like Kaph's realization in "Nine Lives," Mary's in Damon Knight's "Mary," and Molly's in Kate Wilhelm's *Where Late the Sweet Birds Sang*, the perception of their homogeneity and assimilation with the others propels them eventually to sunder themselves from that threatening uniformity.

The group of clones in "Nine Lives" has many features in common with those in Kate Wilhelm's "Where Late the Sweet Birds Sang," awakening in external observers uncanny feelings of acute strangeness, of armies of doubles having come alive. In Wilhelm's novel, similarly, the clones are looked at with feelings of awe: "They worked interchangeably, incessantly—the first really classless society" (240). Indeed, they are perceived by those who created them as "interchangeable" to a degree where "it didn't matter which ones did what" (243), a scenario that prompts fears of their banding together and gradually taking over the compound where the noncloned people were by now a minority, as in Le Guin's "Nine Lives." In Le Guin's short story, Martin speculates, "What if some clone cloned itself? Illegally. Made a thousand duplicates—ten thousand. Whole army. They could make a tidy power grab, couldn't they?" (15), a widespread anxiety in clone narratives.

In "'We're at the Start of a New Ball Game and That's Why We're All Real Nervous': Or, Cloning—Technological Cognition Reflects Estrangement from Women," Marleen S. Barr argues that "Nine Lives" "addresses fear of women and patriarchy's efforts to control women's reproductive role" (195), a contention I have some problems with. The clone team that arrives at planet Libra has been cloned from John Chow's cells. Chow was a biomath, engineer, and undersea hunter and, as one of the clones explains, "We *are* John Chow" ("Nine Lives," 7). Another member of the group, in turn, clarifies the cloning process, elucidating that since Chow's brain could not be saved after his death in an air car crash "they took some intestinal cells and cultured them for cloning. Reproductive cells aren't used for cloning, since they have only half the chromosomes. Intestinal cells happen to be easy to despecialize and reprogram for total growth" (7). As for Martin's puzzled question to the effect of how some of them can be women, Beth, another member of the clone team explains that "it's easy to program half the clonal mass back to the female. Just delete the male gene from half the cells and they revert to the basic, that is, the female" (7–8).[21] According to Barr, the fact that the cells that were used came from a man's intestine instead of from a woman's reproductive cells "indicates that omitting women's reproductive role is a crock of defecation" ("'We're at the Start of a New Ball Game and That's Why We're All Real Nervous': Or, Cloning—Technological Cognition Reflects Estrangement from Women," 196). As Barr goes on to maintain, "Nine Lives" "announces that it is wrong to make new individuals from 'bits and pieces instead of whole people' . . . instead of from whole people who happen to be women" (196).

For Barr, "Nine Lives" is a "feminist revenge tragedy: the elderly female living planet acts against people who are not of women born" ("'We're at the Start of a New Ball Game and That's Why We're All Real Nervous': Or, Cloning—Technological Cognition Reflects Estrangement from Women," 196). Barr, then, reads Le Guin's story as a tale against cloning and, more specifically, as dramatizing the elimination of the female role in reproduction. Although I sympathize with this view, I believe Le Guin's story is indeed a cautionary tale mostly about the psychological consequences of cloning on groups of cloned people and not so much about the abolition of women's bodies in procreation. When Barr argues that the "dead female clones in 'Nine Lives' signal Le Guin's realization that cloning is not in women's best interest—that, for women . . . cloning yields eradication, not reproduction" (197) she does not seem to be taking into consideration the fate of the dead male clones or the fact that the two technicians, Pugh and Martin, were not affected by Libra's quakes.

While, on the one hand, Barr rightly contends that Le Guin's "view of cloning in 'Nine Lives' asserts that women are basic, powerful and dominant" ("'We're at the Start of a New Ball Game and That's Why

We're All Real Nervous': Or, Cloning—Technological Cognition Reflects Estrangement from Women," 197), since the narrator specifically asserts that the basic state of cells, from which they then specialize, is female, on the other, Barr states that "Nine Lives" suggests that for women cloning spells "eradication, not reproduction" (197), a reading that I find somewhat excessive. Even so, Barr rightly notes that Le Guin (in "Nine Lives") and Pamela Sargent (in *Cloned Lives*) "foresaw feminist implications of cloning" ("Post-Phallic Culture: Reality Now Resembles Utopian Feminist Science Fiction," 76), an argument that I extended to Mary E. Bradley Lane's *Mizora* (1890), Charlotte Perkins Gilman's *Herland* (1915), James Tiptree Jr.'s "Houston, Houston, Do You Read?" (1976), and Suzy McKee Charnas's *Motherlines* (1978).

"ALIKE LIKE PEAS"

Damon Knight's "Mary," in turn, thematizes Mary's deep-seated wish to be different from her sisters, to be a unique individual, and to exert the capacity to love. In this sense Mary, one of a group of thirty cloned sisters, "alike like peas" (32), is made to feel like a robot, almost forced to think and behave in unison with her sisters, in an indistinguishable continuum. Mary, however, feels different from the others. In a future society of standardized groups of clones produced for specific jobs such as "Bakers, Chemists, Mechanics" (38), as well as for promiscuity, Mary craves separation from her sisters and dreams of a type of exclusive romantic love for a man that would last forever. Unlike her sisters, she felt "whole and complete . . . she was uniquely herself" (42), "different" (38) as others would describe her disapprovingly.

Similarly, Molly, in Wilhelm's *Where Late the Sweet Birds Sang*, wishes to break away from her sisters, in whose company she feels submerged, steeped in an all-inclusive circle from where it was very hard to break free. Her sisters, indeed, like Mary's in Damon Knight's stories, although they "were strangers to her" (83), "would never permit her to be alone again. Together they made a whole; the absence of one of them left the others incomplete" (87). Like the group of clones in "Nine Lives" and in Damon Knight's "Mary," the sisters in *Where Late the Sweet Birds Sang* "made one organism" (98). Indeed, Molly, like Kaph and Mary, feels there were "no boundaries" (99) among the sisters; "it was all one" (99).

Molly, however, like Mary, is different. As Ben, also a member of a group of cloned brothers, tells her, "I'm going to find out what happened to you, what made you separate from your sisters, what made you decide to become an individual" (*Where Late the Sweet Birds Sang*, 102), since in "a perfectly functioning unit there are no secrets" (105). He feels a strong empathy with Molly, for she and he have broken away from oneness with their respective brothers and sisters in search of their own difference, pursuing the wish to be alone, apart from the sameness and invariability that characterized their relation with their brothers and sisters.[22]

AUTOMATA AND CLONES

Freud's theories in *Group Psychology and the Analysis of the Ego* (1921) help to shed light on the behavior of the groups of clones in these texts. Drawing on Le Bon's *Psychologie des Foules* (1895), Freud reflects on what is a group and on why a group can exert such a powerful influence on the mental and psychological life of the individuals that make it up. According to Le Bon, no matter who the individuals are who compose a group, the fact that they belong to one provides them with a sort of a collective mind, which leads them to think, feel, and react as members of that group in ways markedly different from their behavioral responses if they were conducting themselves as isolated individuals. Le Bon, as Freud notes, considers that "the particular acquirements of individuals become obliterated in a group, and that in this way their distinctiveness vanishes. . . . In this way individuals in a group would come to show an average character" (9).

In cloned groups living together, it is not just their identical genes but also their proximity that influences and conditions their behavior, as with any group. According to Le Bon, in a group "every sentiment and act is contagious to such a degree that an individual readily sacrifices his personal interest to the collective interest" (quoted in Freud, *Group Psychology and the Analysis of the Ego*, 10). Tellingly, Le Bon also refers to the individual in a group as an "automaton" (11), anticipating Lacan's connection between narcissism and automata in *The Four Fundamental Concepts of Psychoanalysis* and also prefiguring Jim Swenson's, one of Paul Swenson's clones, description of himself as a "puppet" (*Cloned Lives*, 139), unable to lead his own life, fated as it were to repeat his father's existence. Le Bon's characterization of the principal traits of an individual who is part of a group perfectly applies to the groups of clones in the texts analyzed in this chapter: "The disappearance of the conscious personality, the predominance of the unconscious personality, the turning by means of suggestion and contagion of feelings and ideas in an identical direction, the tendency to immediately transform the suggested ideas into acts" (quoted in Freud, *Group Psychology and the Analysis of the Ego*, 11).

Freud's own comments are meaningfully applicable to the cloned groups we have looked at. According to Freud, "A group is extraordinarily credulous and open to influence, it has no critical faculty" (*Group Psychology and the Analysis of the Ego*, 13). In addition, it is characterized by the "dwindling of the conscious individual personality, the focusing of thoughts and feelings into a common direction" (70), traits that for Freud correspond to "a state of regression to a primitive mental activity" (70), attributes that the Swenson clones are painfully aware of and strive to circumvent by going to great lengths to avoid being assimilated with what Mike, one of the clones, describes as "these reflections that made claims on him which acquaintances would never make, demanding love because of a shared genetic strain" (*Cloned Lives*, 298). This is also the case with Mark, in Wilhelm's *Where Late the Sweet Birds Sang*, who is horrified at the prospect of being absorbed into a clone group, engulfed in anonimity and standardization, devoid of critical or creative faculties, as is also the case with Mary in Damon Knight's eponymous story, where Mary desperately wishes to break away from the uniformity and the mindless functioning in unison that defines the group of cloned sisters to which she belongs.

Freud's considerations on the "herd instinct" can also be brought to bear on the behavioral patterns of clone groups.[23] Drawing heavily on W. Trotter's book *Instincts of the Herd in Peace and War* (1916), Freud points out how Trotter considers that the mental phenomena characteristic of a group originate in what he describes as a herd instinct. These phenomena include "the lack of independence and initiative in their members, the similarity in the reactions of all of them, their reduction, so to speak, to the level of group individuals" (*Group Psychology and the Analysis of the Ego*, 62), as well as "the lack of emotional restraint, the incapacity for moderation and delay" (62), characteristics that are clearly manifested in many fictional clone groups. Freud, however, objects to the fact that Trotter pays very little attention to the role of the group leader, going on to suggest that rather than a herd animal "man" [*sic*] is "rather a horde animal, an individual creature in a horde led by a chief" (68). Interestingly, the clone groups in these narratives we have been looking at do not display any salient chief or leader, giving the impression that their members are all equally significant, even if by dint of their being identical participants in a group their identity becomes substantially diluted.

Baudrillard has similarly reflected on the behavioral traits of groups and of the masses in general, connecting them with the patterns he proleptically associates with groups of clones. Some of Baudrillard's pronouncements about life in a society made up of clones are forcefully dramatized in the texts examined in this chapter. Thus, Baudrillard's claim that "Your 'fellow creature' will henceforth be the clone with its hallucinatory resemblance, such that you will never be alone, and will never have any secrets" (*Seduction*, 173) is aptly illustrated in both Le Guin's "Nine Lives" and Wilhelm's descriptions of clones, whose main terror is to be alone and without their respective brothers or sisters. It is precisely this massification that Baudrillard fears will occur with cloning. As he remarks,

> The masses themselves form a clone-like apparatus that functions without the mediation of the other. In the last analysis, the masses are simply the sum of all systems' terminals—a network travelled by digital impulses (this is what forms a mass). Oblivious to external injunctions, they constitute themselves into integrated circuits given over to manipulation (self-manipulation) and "seduction" (self-seduction). (*Seduction*, 173)

This description of the working of the masses conjures up scenarios graphically dramatized in *Brave New World* and the other texts under investigation.

What several of the texts I am analyzing thus seem to suggest is that if, on the one hand, confrontation with their mirror images in their cloned brothers and sisters constitutes a comforting if disturbing factor, in the sense that they inevitably represent the threat of loss of personhood, of engulfment in the same whole, on the other hand, when the members of the clone disappear, the one that is left alone is faced with an ontological void, as a part of a whole that is, by dint of that absence, left incomplete. The cloned brothers and sister in Sargent's *Cloned Lives*, for instance, forcefully strive to alter this disturbing feeling of homogeneity, for in the others they inevitably see, at least partially, the potential and feared death of their identity.

In many ways individual clones, and more so those who are part of a group of clones, could potentially suffer from what Kristeva describes as the "abolition of psychic space" ("Extraterrestrials Suffering for Want of Love," 171), unable to progress beyond primary narcissism, since in their brothers and sisters they find all the love they need. This cancelling of the psychic space, which simultaneously has also led in many cases to a greater degree of interiority and solipsism, of being "*alone-with-oneself*" (170) can be seen as a contemporary phenomenon occasioned among other cultural circumstances by mass production and consumption giving rise to what can be termed a "cloning" and increasing standardization of the social mind frame.[24]

Although many narratives of cloning and doubling seem to suggest that the narcissistic vertigo can only lead to an ontological abyss, "The hell of the same," to use Baudrillard's formulation, eventually some of the protagonists of the stories I have looked at manage to forge their own way in the world, distancing themselves from the cosy, all-embracing (although in the end strongly oppressive and confining) sphere of the other clones. This is the case with Kaph in Le Guin's "Nine Lives," Mary in Wilhelm's *Where Late the Sweet Birds Sang*, and Mary in Damon Knight's "Mary," as well as Jim and Kira in Sargent's *Cloned Lives*, thus hinting at the possibility that each clone can certainly develop his or her own personality and does not have to remain confined to an all-inclusive group, separate from the rest of society.

In "Clone Story," using words that can be seen as offering a useful commentary on the groups of cloned brothers and sisters in the works under examination, Baudrillard has announced the end of the subject and of the mirror stage that might come to pass with the advent of human cloning, as well as the industrial production of identical copies in a potentially endless series. Baudrillard's comments in this passage could very well fit future scenarios like that envisaged in Huxley's *Brave New World* and Roger Spottiswoode's film *The Sixth Day* (2000), where people die to be immediately replaced by copies of themselves, mechanical parodies of human beings, robots in a serial repetition that, by virtue of that very reproducibility, lose any last vestige of human-like characteristics they might aspire to. It is apt to say about these robots, with Baudrillard, that there is no more subject, only a succession of palimpsest after palimpsest, mind-texts erased and written over twice or as many times as they are rebuilt and resurrected. *The Sixth Day* can be described, from this vantage point, as a dark comedy that deliberately misconstrues many of the humanitarian goals of the human cloning projects, a vision that is almost a parody of the human condition and its future in the typically antiscientific discourse of popular film.

> Am I a man or just a potential clone?
>
> Jean Baudrillard, "Transsexuality," 24

Baudrillard also rejects this humanitarian impulse and considers clones as "subhuman" (*The Vital Illusion*, 21), as "not beyond but underneath the human" (21), inscribed in a process of "erasure of those symbolic marks that make up the species" (21). In Wilhelm's *Where Late the Sweet Birds Sang*, for instance, the clones are similarly described as "not quite human" (242), a suspicion fueled in the mind of one of the main characters by their apparent telepathic powers, as well as their seeming interchangeability, thus conforming to Baudrillard's pessimistic and apocalyptic view of cloning.

In contentious fashion, Baudrillard repeatedly hints at the potential and presumed inhumanity of clones. In "After the Orgy," for instance, he claims that "on all sides we witness a kind of fading away of sexuality, of sexual beings, in favour of a return to the earlier (?) stage of immortal and asexual beings reproducing, like protozoa, by simple division of the One into two and the transmission of a code" (7). He then goes on to place clones side by side with machines and replacement body parts, calling them "today's technological beings" (7) and to argue that they

> all tend towards this kind of reproduction, and little by little they are imparting the same process to *those beings that are supposedly human, and sexed*. The aim everywhere—not least at the leading edge of biological research—is to effect a genetic substitution of this kind, to achieve the linear and sequential reproduction, *cloning*, or parthenogenesis of little celibate machines. ("After the Orgy," 7; my emphasis)

Baudrillard seems to be implying, on the one hand, that clones are not exactly human while, on the other, casting doubt on the "humanity" of human beings as well in such a world. Cloned human beings are, of course, created with recourse to artificial means, to reproductive technologies, but so are those who are the result of in vitro fertilization and other techniques. They are not less "human" because of their more "technological" origin. This is precisely the area where I have more problems with Baudrillard's positions. According to Baudrillard's diagnosis, and thanks to modern technologies, "you are no longer . . . either one or the other: you are *the same*, and enraptured by the commutations of that sameness. We have left the *hell of other people* for the *ecstasy of the same*, the *purgatory of otherness* for the *artificial paradise of identity*" ("Xerox and Infinity," 58–59; emphasis mine). This apparently positive "sameness," which is mainly the result of the spread of the technologies of the virtual, is thus equated with the "sameness" introduced by cloning, which, however, is described as "the hell of the same" (people), whereas the virtual sameness of the contemporary users of virtual technology is referred to as "the ecstasy of the same." There is, then, a significant paradox at work here. Why does otherness, as far as the "Telecomputer Man" (59) is concerned, defined in terms of the "purgatory of otherness," or even more forcefully put, as "the hell of other people," carry such negative connotations, whereas in his discourse on cloning Baudrillard associates sameness with hell, and (the same) identity also has strong negative implications? For the telecomputer "men," on the other hand, it comes to mean, according to Baudrillard, the "artificial paradise of identity." To be sure, Baudrillard is being ironic, but he is also at the same time crucially critiquing these technologies that lead to depersonalization. It is what Baudrillard sees as the massification of the genetic code with the human being as just another commodity in the endless conveyor belt of mass production, the ceaseless assembly-line of objects without a Benjaminian "aura" of our consumer society that he calls, in *Le Paroxyste Indifférent*, "the malediction of the clones" ("la malédiction des clones") (173).[25]

An impassioned cult of similarity.
Walter Benjamin, "The Image of Proust," 206

From a conviction, held until relatively recently, that human beings could never be cloned, to the almost certainty and indeed imminence of that event, we are at the intersection of an age when our whole outlook on the human species stands in need of being rethought. If human beings are reproducible like so many other objects on an industrial production belt, then we inevitably become commodities, subject to multiplication through our genetic code. Baudrillard's arguments against cloning are grounded in similar horror-inspiring scenarios, failing, however, to take into account the potential application of cloning technology for specific and strictly limited procedures. This acerbic vision of a future dictated by eugenics, which was given such vivid illustration in Huxley's *Brave New World* and Ira Levin's *The Boys from Brazil*, for instance, may be fitting for a horror novel or movie, but in terms of technological feasibility it is a long way away in the future, if it is ever allowed to be implemented.[26]

In related fashion, and in a Benjaminian/Baudrillardian register, the apocalyptic vision of the groups of clones in Wilhelm's novel, their sameness and total interdependence, turns them into commodities for the society's rulers, constituting, in Terry Eagleton's words, an "illusion of infinite renewal in the mirror of infinite sameness" (*Walter Benjamin, or, Towards a Revolutionary Criticism*, 28). As he goes on to point out, "In the circulation of commodities, each presents to the other a mirror which reflects no more than its own mirroring" (28). This nightmarish vision of groups of identical people can also be inscribed in what Benjamin calls an "impassioned cult of similarity" ("The Image of Proust," 206). In this dialectical system of reproduction of the same, however, there is an interesting paradox at work. On the one hand, as Eagleton notes, "Since any one of them may be exchanged for another, all are indifferently levelled to relative insignificance" (*Walter Benjamin, or, Towards a Revolutionary Criticism*, 25). On the other hand, by dint of that very same repetitiveness, their devaluation in a system of endless reproductions of sameness turns them into fetishistic commodities with a vengeance, a circumstance that may be said to account not only for the very fascination of the clone but also (and simultaneously) for the uncanny sense of artifice and redundance attached to the concept of human cloning.

Associated with the fear of standardization of human beings is the dread that clones might come to take over and rule society. Such popular fiction titles as Evelyn Lief's *The Clone Rebellion* (1980),[27] Dawn Stoner's *The Killing Clone* (2000),[28] and George Lucas's film *Star Wars,*

Episode II, Attack of the Clones (2002),[29] forcefully convey the fears associated with the possibility that with the advent of human cloning hordes of serially produced people would take over the rest of humanity. This dread is similarly articulated by David in *Where Late the Sweet Birds Sang* when he suddenly realizes that the clones he helped create from his own genetic material and from that of his close family, who are indeed his own twins, brothers, children, "were gradually taking over" (241), associating only with the other clones and marginalizing those who had made it possible for them to exist.

The drive to singularity, which appears with renewed vigor in a society dominated by copies and simulacra, constitutes an effort to restore the kind of individual wholeness that can be attained only through difference from the other individuals in a given society, an aspect that does not exclude cloned human beings who, unlike some of the characters in the stories analyzed in this chapter, do not have to be just numbers in a series of endless duplications of the same, an impulse dramatized in such characters as Bernard in *Brave New World,* Mary in Damon Knight's "Mary," Mark in *Where Late the Sweet Birds Sang,* and the five clones in Pamela Sargent's *Cloned Lives.*

The interesting conflict and paradox at work here is that, while, on the one hand, as Barbara Maria Stafford points out, we live in an age of "otherness, of assertive identities . . . and have been doing so since the eruption of romantic individualism during the late eighteenth century" (*Visual Analogy: Consciousness as the Art of Connecting,* 10), on the other, the compulsion and constraints of massification and standardization are gradually becoming normalized, threatening to elide that very individualism and singularity. As consumer culture necessarily includes making your body the center of the universe, what more fitting correlative, therefore, can there be to the whole discourse of consumer culture than the act of cloning? The spirit of modern consumerism, which turns the illusion of "private person" versus the general public into an untenable deception, will inevitably subsume into its all-encompassing reach the practice of human cloning as another form of stereotyping, of imposing consumer patterns, and—what is more worrying—of achieving them in terms of the production cost–benefit ratio that business relentlessly pursues.

"Am I a man or just a potential clone?" (Baudrillard, "Transsexuality," 24). At bottom, this is the crucial question Baudrillard has been addressing in his many writings on cloning, where the implication that clones are not human beings is forcefully articulated, a question also dramatized in numerous clone narratives. In some important ways, these narratives constitute a fictional anticipation and critique of a scenario of reproductive technologies that can be seen as inscribed alongside so many other inventions that cater to mass production and consumption in our contemporary world, our Western way of life,[30] ruled by an exacerbated narcissism and a generalized feeling of anomie. In this economy of abundance where serial products and ubiquitously similar consumer goods contribute to a state of apathy,[31] individuals, in turn, become the "cloned" addressees of a mass "cloning" of tastes and ideologies,[32] tantamount to what could be described as a cloned mass psychology, so much in evidence in the texts looked at in this chapter.

Is standardization, then, inexorably extending from consumer culture to a kind of class-structured human uniformity and, paradoxically, to elitist designer humanity, where wealth and narcissism prescribe the continuation of one's genes in the form of cloned children? These are indeed dystopian visions that the highly cautionary tales analyzed in this chapter illustrate, well before the technology has been implemented, alerting us to novel societal configurations that need to be prefigured and theorized before unwittingly we are caught up in them and they are upon us.

The next chapter will also look at the issue of mass psychology in societies where cloned people predominate, as in Huxley's *Brave New World*, but more optimistic alternatives will be seen dramatized as correctives to this scenario in works such as Naomi Mitchison's *Solution Three* and J.B.S. Haldane's *The Man with Two Memories*.

NOTES

1. These are precisely some of the crucial thematic concerns addressed in Eva Hoffman's *The Secret: A Fable for Our Time* (2001).

2. "Nine Lives" was first published in *Playboy*, November 1969.

3. "Mary" was first published in *Galaxy*, June 1964, under the title "An Ancient Madness."

4. Thomas J. Morrissey argues that the clones, "although they have the same cellular donor and are therefore identical to one another genetically (except for the lone female, whose gender is the only difference), and although they are raised as siblings, they are not in fact siblings, nor is the genetic donor their father" ("Pamela Sargent's Science Fiction for Young Adults: Celebrations of Change," 184). I, on the other hand, believe they are very much brothers and sister, together with a father (Paul Swenson), who in this case is their sole progenitor.

5. But would it not also be fair to argue that every narrative can be seen as a manifestation of that very quest for immortality, as a wish, even if only unconscious, to leave something behind, an inscription to remind others that we existed in the memory and traces encoded in those narratives?

6. Murphy, however, does not consider a parallel effect as far as women are concerned, for cloning would potentially make women independent of men for reproductive purposes.

7. Because of lack of space, I cannot develop here the importance of ectogenesis in terms of the benefits it would bring for many women and also as far as enabling men who so wished to do without women for reproduction.

8. See Jacques Lacan's *The Four Fundamental Concepts of Psycho-Analysis*.

9. *Where Late the Sweet Birds Sang* was reissued in 1998 by Tom Doherty Associates, New York.

10. This state of affairs is reminiscent of the predicament of fertile women in Margaret Atwood's *The Handmaid's Tale* (1985).

11. Quoted in Suzi Gablik's *The Reenchantment of Art*, 24.

12. For a discussion of the nature of mass art, see Noël Carroll, *A Philosophy of Mass Art*.

13. Cited in Scott Bukatman's *Terminal Identity: The Virtual Subject in Postmodern Science Fiction*, 299.

14. Luce Irigaray's reading of the Platonic allegory of the cavern, in *Speculum of the Other Woman*, also provides a useful context here. Woman in patriarchy has traditionally been relegated to an inferior position, regarded as a copy of a supposedly original male model. As Margaret Whitford points out with specific reference to Irigaray's notion of the "other of the other," "She is not the 'other of the other,' the non-representable other of the cavern, but the 'other of the same,' the world of appearances which is supposed to be transcended in the final vision of truth. Femininity, then, qua appearance, is thus an integral if unacknowledged part of the economy of truth" (*Luce Irigaray, Philosophy in the Feminine*, 114). In this economy of (male) truth, then, according to Whitford's reading of Irigaray, "Women cannot be other than appearance . . . they are bound to be seen (in Platonic terms) as semblances, inferior copies of the truth" (117).

15. Here, of course, I have many reservations about Baudrillard's discourse, tinted with a patriarchal rhetoric he makes no effort to disguise. This is also often the case in *Seduction*, where many of his pronouncements betray an androcentric bias.

16. Jack Dann and Gardner Dozois consider "Nine Lives" as "perhaps science fiction's first true clone story" (*Clones*, 1).

17. In Ursula K. Le Guin's "Nine Lives" (1975), James Tiptree Jr.'s "Houston, Houston, Do You Read?" (1976), and Kate Wilhelm's *Where Late the Sweet Birds Sang* (1976), human cloning is introduced as a possible answer to the population problems after an ecological calamity turned most or all the people infertile and put at risk humanity's survival.

18. This eugenicist policy of production of the best offspring in order to reshape the world population is reminiscent of Huxley's *Brave New World*, where dozens of identical twins, compared to "machines" (7), are constantly being produced: "standard Gammas, unvarying Deltas, uniform Epsilons. . . . The principle of mass production at last applied to biology" (7). Pugh elaborates on the policy of eugenics in the process of being developed in "Nine Lives": "They're trying to raise the level of the human genetic pool, which is a mucky little puddle since the population crash" (23). This principle is reminiscent of that put forward by Herbert Marcuse in *One Dimensional Man*, where he argues that the freedom people believe they enjoy in contemporary societies is belied by the fact that they are indeed tethered to the mechanisms of mass production and so-called free institutions, which in reality limit that freedom, creating a new form of totalitarianism, taken to extremes in *Brave New World*.

19. The short story was first published in *Orbit* 15 (1974).

20. This story was first published in *Eros in Orbit*.

21. The idea that the female is the basic and original form of all human beings is also put forward in Lane's *Mizora*.

22. James Tiptree Jr. provides a different perspective from Wilhelm's as far as groups of cloned sisters are concerned in her short story "Houston, Houston, Do You Read?" where the groups of "sisters" (79), as they call themselves, experience a comforting feeling of recognition and strong empathy with each other that contributes to shaping their vision of who they are. As Judy Shapiro, one of a group of cloned sisters, notes, turning the tables on her interlocutor, Lorimer, for whom cloning is an abhorrent notion, "How do you know who you *are*? Or who anybody is? All alone, no sisters to share with! You don't know what you can do, or what would be interesting to try" (80). This is a new concept of consciousness of the self derived from identification with the group you belong to but not necessarily submersion of individuality in that group, no dilution of selfhood but in a way intensification of their identity in a reassuring context of similitude but not absorption into colorless sameness. For Lorimer cloning conveyed "the horror of manipulating human identity, creating abnormal life. The threat to individuality, the fearful power it would put in a dictator's hand" (80). For them it is very hard to imagine life differently. As they further explain, each one of them "adds her individual memoir, her adventures and problems and discoveries in the genotype they all share" (80–81). This story is discussed at greater length in Chapter 2.

23. Nietzsche's concept of the herd instinct would also be relevant to this study but it is beyond the scope of this book.

24. Indeed, it is possible to look at the society of cloned people in the texts under examination as a profoundly distressing embodiment of a disturbing trend toward homogeneity. Julian Pefanis mentions "Foucault's critique of the modern Western episteme as the advance of the same into the realm of the other, destroying it in the name of a quest for a universal conception of the human, [which] can be related to Clastres's critique of the state as possessed of a 'will to reduce difference and otherness'" (*Heterology and the Postmodern: Bataille, Baudrillard, and Lyotard*, 119).

25. The question of copyright applied to clones is precisely the main focus in Charles Sheffield's short story "Out of Copyright."

26. Although I agree with some of Baudrillard's general points about cloning, namely, the potential dehumanizing conformity and troubling lack of individuality pertaining to groups of cloned people, I find that Baudrillard is here speculating with respect to far-fetched scenarios that envisage crowds of identical cloned beings without giving any thought to specific cases in which cloning technology might prove to be highly useful and justified, as in the case of infertile couples, for instance, who are placing great hopes on the possibility of having a cloned child before resorting to adoption.

27. The first book of the Clone Chronicles, it describes the Clone Wars, that is, the fight the cloned people have to wage against their creators in order to be fully accepted.

28. In *The Killing Clone* the cloned person is bent on taking revenge on her cell donor, driven by jealousy, in a haunting illustration of the fantasy of the evil twin.

29. The latest film in the *Star Wars* space saga is called *Episode II, Attack of the Clones* (although the title that was first rumored was *The Clone Wars*). Both titles feed on popular conceptions of clones as entities to be feared.

30. I am painfully aware that I am writing from a privileged position and that I am overgeneralizing the modus vivendi of the Western world.

31. A state reminiscent of the Eloi in Wells's *The Time Machine*, who have reached a standard of living that does not require much effort and leads to monotony and apathy.

32. Our time is indeed characterized by what we could call cultural cloning. As Jacques Derrida observes, "on cherche à 'reproduire' des individus qui pensent la même chose, qui se conduisent de la même manière à l'égard du chef et dans la horde, selon des schémas fort bien connus. Il s'agit là aussi de clonage" (*De quoi demain . . . Dialogue*, 95; "one tries to 'reproduce' individuals who think alike, who behave in similar ways in relation to the leader and the horde, according to well-known schemes. This can also be described as cloning"; my translation).

5

"The Malediction of the Clones": Huxley, Mitchison, and Haldane

The purpose of this chapter is to examine Naomi Mitchison's *Solution Three* (1975) and J.B.S. Haldane's *The Man with Two Memories* (1976) as, on the one hand, related visions of future worlds coming out of a shared scientific and cultural background (Mitchison was Haldane's sister), and, on the other hand, as critical responses to Aldous Huxley's *Brave New World*, a novel with which they can be seen as engaged in an ongoing dialogue. Both *Solution Three* and *The Man with Two Memories* provide thoughtful alternative scenarios to the profoundly totalitarian and repressive societal patterns put forward in *Brave New World*, as well as its wholly negative view of human cloning in the service of a dictatorial and eugenicist social policy.[1]

While the three novels under examination can be seen as giving voice to contemporary widespread fears and misgivings at the prospect of human cloning, their treatment of that question is strikingly different, suggesting distinct configurations and implications as far as the cloning issue is concerned, as well as new angles for examining its many ramifications.

Indeed, *Solution Three* and *The Man with Two Memories* are among the few examples of clone narratives to portray cloning in future societies as not only acceptable but potentially positive.

All three novels can be said to draw directly from Haldane's *Daedalus, or Science and the Future* (1924), providing vivid illustrations and simultaneously challenging many of the assumptions presented in that groundbreaking defense of the future of science and of some of the contours that the science of the future would take.

J.B.S. Haldane's specific disquisitions about science and the future in *Daedalus* take the form of an essay that an imaginary undergraduate would read to his supervisor 150 years from the present time of writing (1923). Haldane's forecasts prove wide of the mark. Thus, he predicts that "it was in 1951 that Dupont and Schwarz produced the first ectogenetic child" (*Daedalus, or Science and the Future*, 63). By that stage, "The birthrate was already less than the deathrate in most civilised countries. France was the first country to adopt ectogenesis officially, and by 1968 was producing 60,000 children annually by this method" (64–65). Haldane's prognostications include the vision that 150 years from 1923, that is in 2073, "ectogenesis is now universal, and in this country less than 30 per cent of children are now born of woman" (65), a prediction taken to greater extremes by his friend Aldous Huxley in *Brave New World*, where only women in the Reservations still give birth, all other babies being produced in a laboratory environment, grown in artificial wombs and "decanted."

Haldane further anticipates that in his eugenically oriented society of the future most people will derive from a "small proportion of men and women who are selected as ancestors for the next generation" (*Daedalus, or Science and the Future*, 66), a scenario given dramatic illustration in Mitchison's *Solution Three*. These ancestors, in turn, according to Haldane's vision, "are so undoubtedly superior to the average that the advance in each generation in any single respect, from the increased output of first-class music to the decreased convictions for theft, is very startling" (*Daedalus, or Science and the Future*, 66). These visions are given fictional expression in *The Man with Two Memories*.

Gordon Rattray Taylor, in *The Biological Time Bomb* (1968) mentions J.B.S. Haldane as "one of the most brilliant and practical scientists of our time" and "one of those who took cloning seriously" (28). Taylor goes on to detail some of Haldane's views, which the latter expounded in "Biological Possibilities for the Human Species in the Next Ten Thousand Years," many of which found their way in a dramatic form into *The Man with Two Memories*: "Haldane declared that most clones will be made from people of at least 50,[2] except for athletes and dancers, who could be cloned younger. They will be made from people who have excelled in some socially acceptable accomplishment, though we shall have to be careful that their success wasn't due to mere accident. Equally useful, he said, might be the cloning of people with rare capacities, even if their value was problematic—for instance, people with permanent dark adaptation, people who lack the pain sense, those who can detect what is happening in their viscera and even control it, as some eastern yogis can" (28).[3]

These prognostications, however, amount to a program of eugenics, an area with which Haldane had a complex and vexed relation. Daniel J. Kevles explains that Haldane, "who as a young man joined the Oxford Eugenics Society, sympathized for a time with aspects of the creed, particularly its denigration of the lower classes and eagerness to reduce their rate of reproduction" (*In the Name of Eugenics: Genetics and the Uses of Human Heredity*, 123). However, "the rapidly advancing field of genetics" (123), together with "factors of background, temperament, and sociopolitical belief" (124) contributed to turn him against mainline eugenics. Kevles recounts that in 1932, while Haldane was attending the Third International Congress of Genetics, he observed that "a society composed of uniformly perfect men . . . would be highly imperfect. The essence of perfection among plants, animals, and most certainly man was variety. The ideal society had to have room for all sorts of people, each best at some one thing than another" (147).

It is precisely this kind of society that Haldane visualized in *The Man with Two Memories*, the result of a "rather mild eugenical programme" (121), as one of the protagonists, Ngok Thleg, puts it to James Murchison, his alter ego.

THE MAN WITH TWO MEMORIES

The protagonist of Haldane's novel, James Murchison, wakes up one morning to find himself a different person, Ngok Thleg, his double, who was born on the planet Ulro more than sixty thousand million years in the past.

The society depicted in *The Man with Two Memories* is a highly specialized one, consisting of groups of cloned people who were created according to their specific abilities. As Ngok Thleg explains,

> Most people in the society from which I came were not born. About a third were born as a result of the sexual intercourse of two parents. A tiny fraction was born of women artificially inseminated by seed of a long dead father which had been preserved at liquid helium temperatures. One or two in a generation were produced parthenogenetically. About two thirds of all children were produced clonally from cells of other persons which were induced to divide and form an embryo. (*The Man with Two Memories*, 19)

Ngok Thleg's mother, for instance, "belonged to a clone of about 5,000 members" (20). His father, in turn, "belonged to one of the clones of spaceship pilots" (20).

Ngok Thleg was suckled for eight months by his mother, but "from time to time [he] was suckled by three other women" (*The Man with Two Memories*, 20). This, he explains, "was the normal practice in our society both on social and biochemical grounds" (20). In Ngok Thleg's society, furthermore,

> Milk secretion is hormonally controlled, and we took this control completely for granted and turned on milk production as we wished. So while most women did not bear a child, almost all of them suckled at least one. More remarkably, some men did so. A good many men were able to produce milk without a serious upset of their physiology or psychology; and as a matter of course, given the outlook of our species, they took the opportunity. However a man who was producing milk very rarely took over the complete care of a child. (*The Man with Two Memories*, 20–21)

This scenario is reminiscent of the one put forward in Marge Piercy's *Woman on the Edge of Time* (1976), where a male character is described in great detail suckling a baby and anticipates such tales as Lisa Tuttle's "World of Strangers," where men can decide to suckle their children.

In Ngok Thleg's society, in addition, it was common practice to invent new genotypes with improved capacities and skills. As he explains, "When a man and a woman have children, about one in forty thousand is so useful that we want lots more like him, and start a new clone" (*The Man with Two Memories*, 50). Moreover, the feature of predictability is valued in Ngok Thleg's society. Thus, "If a boy or a girl belongs to a clone, we know pretty well what they can and can't do beforehand" (49).

As in *Solution Three*, *The Man with Two Memories* shows a system of "positive" eugenics that effectively regulates society. Ngok Thleg tells James Murchison that "methods of birth control considerably more efficient than those in vogue on your planet were devised, and were used for eugenic measures of a primitive kind" (*The Man with Two Memories*, 92). He further notes that "meantime our species, having achieved an extremely high standard of living . . . began to remake itself. The first clonal reproduction was achieved about the year 12,000 and about that same time we had the biological and psychological basis for the beginnings of scientific agencies" (100). However, as Murchison learns, they "have bred for a type of person whom you would call hopelessly unambitious and unpractical, and who would not rise far in your societies" (100), thus suggesting an implicit and underlying wish to attempt to curtail violence and aggression. Ngok Thleg himself is regarded as potentially "sufficiently interesting to make it worth while making another like you" (112). Indeed, "Quite a number of people were concerned with human genetics, and the gradual improvement of our innate capacities" (123), a recurrent theme in the novel.

There is then a great variety of modes of reproduction and types of individuals on Ngok Thleg's planet. As we further learn, some people are still "born of women" (*The Man with Two Memories*, 128). In addition, "Besides clonals, ordinaries, defectives, and archaics, there is a fifth class of human being. Every thousand years or so someone is born (and once someone mutated in a clone) who obviously has novel capacities, but who is not adapted to live in our society. . . . We record such people's words and acts very carefully. We sometimes breed from them sexually. More usually we take a bit of their tissues, preserve it at liquid helium temperatures, and breed a new individual from it after ten or twenty thousand years, when we think conditions have changed so that they are better adapted" (52).

Haldane's *The Man with Two Memories* presents a much more sympathetic and positive view of cloned people than Huxley's *Brave New World*, with which it can be said to engage in a critical debate in many places. As with the use of soma in *Brave New World*, in the society described in *The Man with Two Memories* widespread use of drugs to enhance specific moods or curtail others is made, so that as Ngok Thleg puts it to Murchison, they would be considered, "from your point of view, a race of drug addicts" (33). Another feature in common with *Brave New World* is the use of hypnosis to assist with the learning process, a method comparable to the system of hypnopaedia employed in Huxley's novel, which is also occasionally used in *Solution Three*.

In his essay "Biological Possibilities for the Human Species in the Next Ten Thousand Years" (1963), Haldane similarly reflects on the future of humankind and puts forward many scenarios that are given fictional expression in *The Man with Two Memories*. Haldane considers, for instance, the "unification of mankind" (340) as a desirable goal, an objective attained in *The Man with Two Memories*, where there was a "world community" (44) which had been the result of the "unification of our species into a single community" (19), as Ngok Thleg explains. In the future societies envisaged in Huxley's *Brave New World* and Mitchison's *Solution Three*, similarly, there had been a unification of the world community after protracted periods of war and aggression.

In "Biological Possibilities," as he had already argued in *Daedalus*, Haldane considers that "owing to the large number of harmful recessive genes carried by most people, eugenics, largely directed to preventing their coming together, would be an important branch of applied science" (340), a type of policy exemplified in *The Man with Two Memories*. Many other biological developments envisioned by Haldane, including the gradual transformation of specific human characteristics, fall into the category of "positive eugenics" (345). These would include such features as people who produce "inodorous faeces" (342), like Ngok Thleg, a much greater awareness of inner bodily functions, as described by "yogis" (348), as well as the "verbalization of kinaesthesia" (346). All these skills are illustrated in great detail in *The Man with Two Memories*, Haldane's utopian view of an alternative world in the future.

> Man has two memories an outer and an inner, or a natural and a spiritual memory.
>
> Emanuel Swedenborg, *Heaven and Its Wonders and Hell: Drawn from Things Heard and Seen*

William Blake's prophetic utopianism was a decisive influence on Aldous Huxley, particularly noteworthy in *Island*, and on J.B.S. Haldane, whose novel *The Man with Two Memories* is structured around Blake's visions and mythology.

In addition, through Blake's mediation, the visionary world described by Haldane in *The Man with Two Memories* is also symbolically framed by Swedenborg's ideas, since Blake was a great admirer of the former's theories. By drawing on Swedenborg's visionary configuration of the human mind, with its two memories, Haldane provides a structure for his protagonists to engage in their internal dialogue.

According to Swedenborg, in *Heaven and Its Wonders and Hell: Drawn from Things Heard and Seen*, "Man has two memories an outer and an inner, or a natural and a spiritual memory" (nn. 2469–2494), and although "man does not know that he has an inner memory" (nn. 2470, 2471), "it is from the inner memory that man is able to think and speak intellectually and rationally" (nn. 9394). This inner memory, for Swedenborg, is "the book of his life" (nn. 2474, 9386, 9841, 10505).

In this framework, then, Ngok Thleg can be seen as James Murchison's inner, spiritual memory, which manifests itself in the shape of the latter's *Doppelgänger*, shadow and twin over space and time, the impetus of the inner imagination that created a whole new fantasized life where Murchison could give vent to his innermost dreams.

In his essay "Biological Possibilities for the Human Species in the Next Ten Thousand Years," Haldane reflects about the results that "an objective investigation of the inner life, or as I should prefer to say, the study of life from inside, will reveal" (356). He then goes on to mention private worlds such as those described by Blake "in his prophetic books" (357) and also by Freud. Haldane considers that "the exploration of the interior of the human brain will be as dangerous as that of the antarctic continent or the depths of the oceans and far more rewarding" (358). It is a version of these private worlds, mediated by Blake's cosmogony, that Haldane puts forward in *The Man with Two Memories*. For Blake, the artist's goal is to reveal an inner vision, which in *The Man with Two Memories* takes the form of Ngok Thleg's sweeping vistas over millions of years of human evolution, intricately connected with the world of the imagination.

The implication behind Blake's knowledge of Ngok Thleg's world and of Blake's renderings of that world in his prophetic books, however distorted at times, is that it had come to Blake possibly through that inner memory, the book of life that Swedenborg talks about. Indeed, Ngok Thleg, who believes Blake's source was Los, who in the Blakean mythology stands for Imagination, the "guardian of the Divine Vision" (Bernard Nesfield-Cookson, *William Blake: Prophet of Universal Brotherhood*, 204), corrects several "misconceptions" transmitted by Blake in his works about the former's world. As he explains, "Clearly if Los was the source, his memories were much less perfectly merged with Blake's brain than mine with Murchison's. But they were vivid enough to force Blake to write and illustrate the Prophetic Books" (*The Man with Two Memories*, 91).

Indeed, Haldane draws on Blake's parody of Genesis, *The Book of Urizen*, in order to suggest, like Blake, an alternative vision of the creation of humankind and, beyond that, of humankind effectively remaking itself, in this case through biotechnological means.

Urizen can be described as the repressive God and father figure in Blake's countermythology, the mentor of a philosophy of materialism and tyranny, symbolized by his association with iron and brass, with no room for the imaginative drive, the creative impulse symbolized by Los. Like the Blakean Urizen, the one in Ulro was similarly connected with mines, industry, mechanization. However, as Bernard Nesfield-Cookson notes, "It is not only Urizen, fallen Reason, but also creative imagination, Los, who is associated with iron—and with brass" (*William Blake: Prophet of Universal Brotherhood*, 208). Indeed, Los built Golgonooza, the Blakean city of art and of imagination, with these two metals, thus suggesting the bringing together of art and industrialism. Golgonooza, literally "the city inside the brain," suggests yet again the realm of the imagination, the inner memory on which Murchison is drawing in order to dream up Ngok Thleg.[4] In *The Man with Two Memories*, however, Haldane is not using Blakean symbols uncritically. On the contrary, he is borrowing them in order to produce recognizable types but going on to point out the differences in his own countermythology, or alternative system. Thus, as the narrator observes, Blake's "source clearly gave a very distorted account of Urizen. Blake, we think, distorted it still further" (91). He goes on to comment on Los and the Blakean narrative according to which Los built Golgonooza: "To the best of my belief Los was a metal worker who preached an impracticable revolutionary and semi-religious doctrine about the time of Orc. . . . But Los no more built Golgonooza, as Blake claims, than I did" (91).

In Golgonooza, which "was rebuilt with a splendour which even our day finds fantastic" (*The Man with Two Memories*, 82), we are told that "the mining industry led the way" (80). Again, what is stressed by the narrator is the need to connect imaginative work with manual tasks, since both can and should, in a Blakean perspective, be informed by creative imagination. Indeed, one of Ulro's most important theorists, Latsuhi, "had insisted . . . on the dignity of manual labour . . . and the new order, once fighting was over, insisted on a minimum of four hours of manual work even for the highest officials" (89).[5]

In Blake's complex mythological order, Ulro stands for a nightmare realm characterized by solipsistic alienation. The fact of his having been born in Ulro, equated with chaos,[6] evocative of a reductive sphere of the self where human imagination is stagnant and petrified, suggests that Ngok Thleg is imprisoned within the shell of Murchison's brain and only through the expansion of the latter's imagination will he be able to express himself. Haldane's revision of Blake's symbolic system allows him, in turn, to build an alternative one that strongly emphasizes not only the importance of imagination in the shaping and subsequent building of a future world but also the relevance of science developed on a par with art and literature, so that they can feed on each other.[7]

SOLUTION THREE

In *Solution Three*, as already suggested, Naomi Mitchison engages in a critical dialogue with her brother's prognostications for the future, mainly those put forward in *Daedalus*. In addition, *Solution Three* can be read as a feminist response to and critique of Aldous Huxley's *Brave New World*, providing a nuanced corrective to many of the grim visions dramatized in Huxley's novel.[8] As a counterpoint to Huxley's profoundly dystopian view of cloning, Mitchison provides a vision of an alternative society where clones are seen as the means to rescue and help shape, in the ways deemed best by the Council, the depleted population of an after-catastrophe world. *Solution Three* puts forward alternative scenarios that are gradually subject to change when modifications in the almost exclusive reproductive practice of cloning are considered beneficial to society at large. Mitchison can indeed be said to provide a feminist revision of male-practiced science and a fruitful reflection on "this horrid idea,"[9] as she describes the tyranny of the almost exclusive practice of cloning and the consequent drastic reduction in the gene pool, both human and vegetal.

The cloned people in *Solution Three* and in *The Man with Two Memories*, then, are strikingly different from the society of clones envisaged by Huxley in *Brave New World*. This latter society constitutes a graphic instance of what Jean Baudrillard sees as the massification of the genetic code and the human being as just another commodity on the endless conveyor belt of mass production and the ceaseless assembly line of objects without a Benjaminian "aura" of our consumer society that he calls, in *Le Paroxyste Indifférent*, "the malediction of the clones" ("la malédiction des clones") (173), which was so vividly dramatized in Huxley's *Brave New World*.

Naomi Mitchison's *Solution Three* takes place in a postapocalyptic world that has been destroyed by ecological disasters and is in the process of being rebuilt. In *Solution Three* the only way to solve the population problems was to condition people into becoming homosexual and reproduce through cloning, the Solution Three of the title. All the clones were of Him and Her, the venerated originators of the new societal order: She had been a mission doctor in India while He had been deeply immersed in the racial wars. Since the clones were all copies of Him and Her, it was hoped that their genetic makeup would turn them into particularly useful members of the new society that was arising. "She" had created the first Council, Solution One, which had worked on the resolution of the food and population problems, putting through the kind of policy that would bring back sanity and order. Here Mitchison can be said to engage with and dramatize Haldane's vision of a "small proportion of men and women" who will be "selected as ancestors for the next generation" (*Daedalus, or Science and the Future*, 66).

In *Solution Three* women who appear to be particularly well suited to be mothers become the Clone Mums, being indoctrinated to believe that it is an honor to serve as mothers to the clones of Him and Her.[10] Although well cared for in beautiful surroundings, the Clone Mums are nonetheless deeply conditioned to unquestioningly accept separation from their children when the babies reach a certain stage in their development and start showing what are called Signs. The children, however, are not related to them genetically, they are all clones of Him and Her, classified by letters and numbers.[11] The women are then surrogate mothers, forced to relinquish their babies allegedly for the greater good of society: "The Clones didn't belong to the Mums who had been their nests and love givers, but whose own cell nucleus had been eliminated at an early stage. If anything it was the other way, the Mums belonging to the Clones" (35). When babyhood ended, the cloned babies were taken away from their mothers and a process of "strengthening" would begin, a "toughening process" (36) that replicated the stresses experienced by Him and Her in order to prepare the clones to be as nearly as possible copies of their genetic parents. One of the Clone Mums, Lilac, starts questioning this whole procedure and refuses to report to the watchers and carers the appearance of the Signs in her baby, F Ninety, which would mean a radical separation between her and her cloned baby. Lilac thus becomes the spokeswoman for the anxieties felt about the new rules in the evolving society, a society in which, although there was hardly any violence or aggression, was as regulated and, albeit in different ways, as full of prejudices as the older one. That all the cloned boys are brown, after Him, and all the cloned girls white and fair-haired, like Her, only stresses the sameness and regimentation of the new world order in *Solution Three*, pointing toward a class structure based on genetically related privileges, as in *Brave New World*. In *Solution Three*, however, the Clone Mums are given great respect, even if in a regulated environment, an aspect radically distinct from the scenario in *Brave New World*, where "mother" is an obscene word. Here, by contrast, it is heterosexuality that is seen as "rather an unpleasant word" (*Solution Three*, 14), while those who are heterosexual are considered to be "deviants" (41) and discriminated against. "Solution Three," then, is far from unproblematic and widely accepted in the new postapocalyptic society.[12] The Professorials and some of the Clone Mums are not the only ones to swerve from the norm and voice their doubts. What is implicitly hinted at, I believe, is that heterosexuality potentially breeds violence and aggression. The vision conveyed in the novel of a world where sameness reigns, where heterosexuality is wiped out and people are coerced through conditioning to be lesbian or homosexual, seems to me to amount to a version of eugenics. In a society that was supposedly developing a caring, egalitarian ethics, a classless society, there seem to be several unresolved issues calling for other solutions, Solution Four or Solution Five. Although, as Jenni Calder argues, in *Solution Three* (as well as in *Not by Bread Alone*), "The emphasis is on men and women working together rather than sleeping together" (*The Nine Lives of Naomi Mitchison*, 270), this particular version of the separation of sexual love from reproduction comes in for implicit criticism; and, by the end of the novel, the narrative drift suggests that social conditioning into mandatory homosexuality might be gradually discontinued, at the same time as the return of the fertility of the earth would encourage the consumption of real food and the enjoyment of real gardens and trees.

BITTER NURSERY RHYMES: "BOOKS AND LOUD NOISES, FLOWERS AND ELECTRIC SHOCKS"

I will here concentrate on two exemplary scenes taken from both novels, where Mitchison produces a critical and feminist revision of the methods for bringing children up dramatized in *Brave New World*.

While in Huxley's novel the nurseries are aseptic, impersonal rooms, in Mitchison's novel they are described as privileged oases apart from the real world, real gardens with real air. Although artificially enhanced, with the result that even "the seasons . . . had been genetically jumbled, so that the strangest combinations of flowers and fruit coincided" (*Solution Three*, 26), there was "genuine, living grass" (26) and Clone Mums could pick flowers, whereas there was "no picking of real flowers in the real world" (59). On the other hand, as soon as the children started to show the "Signs," they would be taken away from their mothers and undergo a process of "strengthening," consisting of a series of "stresses and strains" (63) like the ones He and She had to undergo, designed to make them just like Him or Her. However, it is precisely against this uniform and enforced sameness that Lilac, one of the Clone Mums, rebels, believing her own clone baby should have the chance to develop his own tastes and to follow his own destiny.

In *Brave New World*, by contrast, the neo-Pavlovian methods used to condition the children are unduly violent, including explosions, sirens, alarm bells, and electric shocks to deeply imprint the desired message: "Books and loud noises, flowers and electric shocks—already in the infant mind these couples were compromisingly linked; and after two hundred repetitions of the same or a similar lesson would be wedded indissolubly. . . . Reflexes unalterably conditioned" (21–22).

In both cases, however, there is enforced conditioning of the babies and children, even though in *Solution Three* the alleged goal of the process of "strengthening," as the conditioning of the children is called, involves suffering deemed necessary to turn the cloned children into versions of Him and Her, while in *Brave New World* the aim is to instil in the children certain likes and dislikes considered as desirable political and strategic choices by the governing bodies.[13]

Clones in *Solution Three*, as clones of Him and Her, are viewed very positively, and their good influence is expected to extend to large sectors of society. There is still, however, a disturbing kind of positive eugenics at work here, which in spite of its "positive" and beneficial connotations is still a system of selection of the supposedly best elements available, a very questionable policy. As a geneticist in *Solution Three* points out, "There is a danger of having too high a proportion of one kind of excellence in the population. Any population. It is even possible that another kind of excellence might be needed" (137–138). The necessity for homosexuality, on the grounds that intersexual love bred aggression, was first mooted by Her, subsequently homologated by Him and the Council. The results of that new strategy were a remarkable drop in violence and in the population curve, but they also meant a high degree of enforced conditioning and compulsory endorsement of the Code.

This is an area I find problematic in *Solution Three*, for in the name of safety, of eradicating violence,[14] cloning by the same stroke contributes to a gradual elimination of diversity, an enforced solution that will radically change the physical and psychic map of society. A system of eugenics is definitely operating here, even if supposedly defensible through the argument that all the clones are clones of Him and

Her, the paradigmatic examples of justice and equity.[15] Who better to instil order and a more equitable, juster society in the world? However, this system would also potentially and eventually spell the end of individuation, the end of art, since the genetic characteristics of Him and Her were not conducive to such activity, as is shown in the novel. As Ric, one of the Council members, explains to a bewildered Clone boy, Bonni, who does not demonstrate any special sensitivity as far as aesthetic and artistic experiences are concerned, unlike some clone girls, "It was in the DNA, you know, Bobbi. I would always think this might happen to Her Clones. We know, for instance, that she was good at anatomical drawing; there are one or two notebooks extant. You don't have that in your genes" (81).

Another instance of the authoritarian dimension of the Code in *Solution Three* is the fact that people have been conditioned into passive reception of the historical narratives deemed socially correct, so that the old books are not read and the past has been rewritten, so as to fit the ruling policies, again in what might constitute a glance toward Huxley's *Brave New World*. Ric, who considers himself a "geneticist of history" (*Solution Three*, 49) ponders that "history must be re-made in order to flower profitably" (49). Indeed, history would "be re-made again in another century, always yielding something new by a re-arrangement of evidence" (49). In addition, society in *Solution Three*, particularly as exemplified by the rules governing the Clone Mums, can be seen as amounting to a "police state" (95), even though without police: "instead watchers and carers" (95), prompting references yet again to *Brave New World*. The meaningful difference, however, is that while the ideology and political practice in Huxley's novel remains in place at the end of the book, in *Solution Three* the end hints at the possibility of change according to the kind of policy that seems to be of benefit to the greater number of people.

As far as the family is concerned, *Solution Three* can be seen as a critical reply to Haldane's *Daedalus*, dramatizing the problematical ramifications of some of his prophecies. Haldane's reflections on the consequences that the new reproductive scenarios he envisages are likely to have on societal configurations are far-reaching. Haldane muses that

> The effect on human psychology and social life of the separation of sexual love and reproduction which was begun in the 19th century and completed in the 20th is by no means wholly satisfactory. The old family life had certainly a good deal to commend it, and although nowadays we bring on lactation in women by injection of placentin as a routine, and thus conserve much of what was best in the former instinctive cycle, we must admit that in certain respects our great grandparents had the advantage of us. (*Daedalus, or Science and the Future*, 65–66)

On the other hand, as Haldane goes on to maintain, "It is generally admitted that the effects of selection have more than counterbalanced these evils" (66). *Solution Three* seems to end with a nod to Haldane's reflections on the putatively positive "effects of selection" (*Solution Three*, 66) when Jussie, in her conversation with Miryam, muses, in the last sentence of the novel, "There are so many kinds of happiness. According to the genes" (160). She appears to be condoning diversity rather than stratification through selection. On the other hand, this somewhat ambiguous phrasing can also invoke a return to precloning times when there was no selection of the gene pool; but that hypothesis, given the emphasis on control that informs all policies in the new society, seems less probable.

Interestingly, and still in dialogue with Haldane's *Daedalus, Solution Three* seems to suggest, toward the end, that heterosexuality will be accepted and that traditional nuclear families might come to be allowed to thrive in the open again. As Jussie, one of the Council members, ponders, "It's not impossible that there might be some planned alterations of policy" (160), which might mean that Miryam, who belonged to one of the "deviant" heterosexual couples, might be able to take her children to the garden "without upsetting people" (160), as she puts it. The relationship between the clones Anni and Kid also seems to point in the direction of a greater tendency toward heterosexuality and its gradual acceptance as a nondeviant practice, since they are considering having children through procreation and not cloning. Mutumba, the Convener of the Council, begrudgingly concedes that they might have to consider slowly proceeding toward Solution Four, in view of situations such as Anni and Kid's. As Mutumba muses, "In the days of Her and Him theirs was the necessary excellence, the sudden emergence of the exact gene combination for that moment in history. But might not a slightly other kind of excellence be needed now. . . . And move towards solution Four?" (123). The very fact that such radical changes as "the sin of meiosis" (122) are being contemplated suggests that the Council might be open to policy alterations, for the time might be ripe for kinds of genetic combinations other than Hers and His.

It might be apposite to introduce here some of Baudrillard's ideas, taking him as a representative of a type of apocalyptic contemporary thinking on these issues, so much at variance with the more positive perspectives of earlier writers on the subject. Large portions of Baudrillard's work can almost be seen as a running commentary on the societal configurations Haldane describes in essayistic and fictional ways which arose as a result of the introduction of cloning and the separation of sex and reproduction. Baudrillard, whose fascination with human cloning permeates virtually his whole *oeuvre*, considers that we are entering a "second phase" (*The Vital Illusion*, 10) now, after a first phase, which consisted in the "dissociation of sexual activity from procreation" (10), this second phase, involving the "dissociation of reproduction from sex . . . through asexual, biotechnological modes of reproduction such as artificial insemination or full body cloning" (10). As Baudrillard sees it, "Now we will find ourselves liberated from sex—that is, virtually relieved of the sexual function. Among the clones (and among human beings soon enough), sex, as a result of this automatic means of reproduction, becomes extraneous, a useless function" (10). Here, as in many other passages, I have some problems with Baudrillard's radical views on cloning. Why distinguish clones from "human beings"? Are clones not human beings? Furthermore, are clones necessarily infertile? Some of these questions have been dramatized in Haldane's *The Man with Two Memories* and Mitchison's *Solution Three*.

In *The Man with Two Memories*, as in *Brave New World*, sex and reproduction are totally divorced and a chemical substance is used to enhance sensations. Clones, just like other human beings with distinct origins, indulge in sex that is completely dissociated from the procreative function, the same occurring in *Solution Three*, with the exception of the Professorials and, tentatively, the clones Anni and Kid. It seems to me that implicit in Haldane's novel is the suggestion that there should continue to be as many different forms of reproduction as seem to be called for and sensibly applied after careful evaluation of the biotechnological advances as they were introduced. *Solution Three*, in turn, appears to hint that a return to heterosexual love and nuclear families like the Professorials' will be accepted, while cloning a certain number of individuals will also continue. As the narrator ponders, "Population could be allowed to stabilize. But from what parents? Or should it be always and only the Clones, the proved excellence? Or—or what?" (58). Clones in these texts are never portrayed as lesser beings or freaks, as in Baudrillard, a vision that would certainly apply to *Brave New World* but not to Haldane's or Mitchison's novels. However, even though surreptitiously, what Sarah Lefanu describes as "the ideology of Clone perfection" ("Difference and Sexual Politics in Naomi Mitchison's *Solution Three*," 159) in *Solution Three* is subtly disrupted, and insinuations about the Clones' difference are strewn throughout the book. In a conversation with Jussie, Miryam declares, "I could find myself so—so humanly fond of a Clone boy" (158), thus implicitly voicing a certain measure of ambiguity in relation to the Clone boy's "humanness." Indeed, similar turns of phrase often suggestively intrude, hinting at people's feelings of ambivalence toward the Clones, who are nonetheless described in terms of genetic excellence and as the saviors of the world through their vision and sensible policies, as had been seen in Him and Her. Here, as in so many other instances,

Solution Three is not so much suggesting a way forward as playing with various possibilities and scenarios, a function that is, after all, one of the most important roles of science fiction and utopian texts: the anticipation and critical examination of those scenarios. As Marleen S. Barr puts it, "Feminist fabulation enables readers to pioneer spaces beyond patriarchal boundaries" (*Feminist Fabulation: Space/Postmodern Fiction*, xiv), and that is precisely what Naomi Mitchison is doing here.

In *The Other Machine: Discourse and Reproductive Technologies*, Dion Farquahr suggests a third way of thinking about new reproductive technologies and their effect on society's workings, a mode reminiscent of Mitchison's *Solution Three* that could indeed be seen as an apt commentary on the novel:

> In attempting to make sense of the contradictions, paradoxes, and aporias raised by the new reproductive technologies, only a *third path* that avoids the simplifications and binary evasions of both liberal and fundamentalist discourses seems viable. Instead of uncritical endorsement or out-of-hand dismissal, such a *third way* struggles to appreciate their multiple workings with regard to their creativity and generativity—for the revision of old and the creation of new hybrid entities and social relations, no less than for their uncritical recuperation of old categories of domination. (*The Other Machine: Discourse and Reproductive Technologies*, 179)

This seems to me an apt description of the ways in which the Council in *Solution Three* works, trying to find alternative solutions to the community's problems, as well as those of the world at large.[16]

CONCLUSION

While Huxley provides a horrifying and deeply pessimistic view of the effects of cloning used as a privileged technique in a strategic program of social eugenics, Mitchison and Haldane offer alternative perspectives on the use to which cloning can be put. Whereas Mitchison sees cloning as "Solution Three" to help solve the population problems after ecological devastation destroys the earth ecosystems and J.B.S. Haldane envisages a future world where clonal reproduction appears to have contributed to a stable and productive society, Baudrillard, by contrast, sees the advent of cloning as "the final solution" (*The Vital Illusion*, 8). For him, this final solution, which he visualizes as bringing an end to society as we know it, entails the end of sex and death. Of sex because with sameness there is no place for seduction, and of death because the possibility of endless duplication of the human being amounts to a version of immortality in one's genes. As Baudrillard remarks about death in the era of the clones, "Entwined as it is with sex, it must eventually suffer the same fate. There is, in effect, a liberation from death that parallels liberation from sex. As we have dissociated reproduction from sex, so we try to dissociate life from death. To save and promote life and life only, and to render death an obsolete function one can do without, as, in the case of artificial reproduction, we can do without sex" (11). As Baudrillard further observes, "So death, as a fatal or symbolic event, must be erased. Death must be included only as virtual reality, as an option or changeable setting in the living being's operating system" (11). Baudrillard further concludes, in a polemical vein, that "all these useless functions—sex, thought, death—will be redesigned, redesigned as leisure activities. And human beings, henceforth useless, might themselves be preserved as a kind of ontological 'attraction'" (11–12). Here we are in the realm of "cybersex" and "virtual" death, different conceptions from those fantasized by Haldane but nonetheless related. What the latter announces in *Daedalus* is not the end of death but the end of the fear of death in his future society, when he envisages members of the same generation living and dying together, after what feels like a well-lived, fulfilled life, as in *Brave New World*, a scenario not far from that put forward by Baudrillard, after all.

Returning at the end of this chapter to the affinity Haldane felt with Blake and on which he drew to write *The Man with Two Memories*, it is apposite to stress here again Blake's conviction that believing in the inner reality of the imagination can be said to amount in the end to a belief in the transforming of the outer reality, in some ways like the "virtual" reality of a computer environment where we can let our "inner memory" loose, as it were, allow the mask we may put on in our computer dealings (are we not always already made up of layers of masks, inner and outer personae) to follow distinct courses from our outer, "real" life world. Haldane's overarching faith in the future of science and in the power of imagination to help shape the world (and in this respect he can be seen as working in a contemporary tradition that regards the fusion of science and philosophy in its *latu sensu* as indispensable for the progress of both areas) is synthesized in the words with which he ends his essay "Biological Possibilities for the Human Species in the Next Ten Thousand Years." There he expresses his profound conviction that "the description of utopias has influenced the course of history" (360), thus emphasizing the potential of human beings to use new technologies positively.

In *Solution Three* Mitchison similarly demonstrates a measure of confidence in the fundamental role science will play in the shaping of the future world.[17] As Sarah Lefanu suggests, and I thoroughly agree, *Solution Three* is a novel that "encourages questioning" ("Difference and Sexual Politics in Naomi Mitchison's *Solution Three*," 165), one that raises as many questions as it tentatively answers, an aspect I consider one of its greatest strengths. As a feminist critique of Aldous Huxley's *Brave New World* and a fictional answer to Haldane's *Deadalus*, Mitchison's book provides a feminine perspective in relation to her (male) friend's and her brother's androcentric points of view.

Both Mitchison and Haldane, then, have put forward visions of the future based on biotechnological developments that betray a certain optimism in humankind's capacity to make those advances serve imaginatively improving ends and to rise to the ethical challenges they represent. Amid the dismay and gloom from some quarters, this is a refreshing view of human ingenuity and goodwill.

The next chapter will deal, on the one hand, with some of the problematical repercussions of a patriarchal dominance over decision-making processes in the spheres of new reproductive technologies, along with the Earth's ecosystems, and, on the other hand, with a more optimistic vision of the strength groups of cloned women may find in their union and "sisterhood" as a means to challenge that androcentric supremacy.

NOTES

1. Susan Merrill Squier, in *Babies in Bottles: Twentieth-Century Visions of Reproductive Technology*, provides a thorough background picture of Huxley's, Haldane's, and Mitchison's families and friendship, as well as the influence they exerted on one another.

2. Taylor further notes that Haldane "held that after the age of 55 great geniuses would spend their time in educating their clonal offspring, which he thought might avoid some of the frustration the latter might have felt" (*The Biological Time Bomb*, 28).

3. This bodily skill is also mentioned in Aldous Huxley's *Island* as an ideal to strive for.

4. Kathleen Raine stresses Swedenborg's influence on Blake's conception of Golgonooza when she notes that "it was Swedenborg, certainly, who was Blake's most immediate and obvious source of his astonishing and rich conception of a city within" (*Golgonooza: City of Imagination*, 111).

5. This ideal is also defended and practiced in Aldous Huxley's *Island*.

6. Harold Bloom, in *Blake's Apocalypse* (1970), describes the space of Ulro as "genuinely diabolic" (199), as a "self-enclosing mental space that insists upon being taken as phenomenal finality, and so menaces man with true madness" (199).

7. When Haldane specifically quotes from Blake's *Milton* a passage about the "Mundane Shell" (*The Man with Two Memories*, 90), Urizen's creation, it might be read, as Bernard Nesfield-Cookson notes, and in tune with my argument, as signifying "the total of the errors created through relying solely on the five physical senses for the attainment of knowledge" (*William Blake: Prophet of Universal Brotherhood*, 192).

8. For a thorough commentary of Naomi Mitchison's relations with the Huxleys and her engagement with science, see Jill Benton, *Naomi Mitchison: A Biography*; Jenni Calder, *The Nine Lives of Naomi Mitchison*.

9. Mitchison dedicates her book to "Jim Watson, who first suggested this horrid idea."

10. According to Hermann Muller, in "Social Biology and Population Improvement" (The Geneticists' Manifesto) (1939), which Mitchison may have read, it was advisable to replace "the superstitious attitude toward sex and reproduction now prevalent [with] a scientific and social attitude." The Manifesto also recommended to render it "an honour and a privilege, if not a duty, for a mother, married or unmarried, or for a couple, to have the best children possible, both in respect of their upbringing and their genetic endowment" (521–522). These injunctions can be said to be illustrated in *Solution Three* in the Clone Mums.

11. Mitchison subverts readers' expectations, for according to anticipated conventions people who are referred to by letters or numbers have traditionally been seen as stantardized beings in a mass-produced human chain. Here, by contrast, we learn that "it was only the ones who were physiologically perfect who were given numbers" (*Solution Three*, 79).

12. In this context, when Sarah Lefanu writes that in *Solution Three* such matters as "how relationships between women, and between women and men" ("Difference and Sexual Politics in Naomi Mitchison's *Solution Three*," 164) might be different "are not the points at issue" (164), since the "sexual organization is symbolic of other matters" and "it is not, finally, of fundamental interest in itself" (164), I am a bit puzzled, for it is precisely through those sexual models and how they potentially relate to violence, aggression, and modes of reproduction that the novel engages with those crucial questions and how the future might be shaped by those social arrangements.

13. While I believe *Brave New World* and *Solution Three* provide very different solutions to a set of common problems, I do not agree with Susan Merrill Squier when she states that "to compare Mitchison's novel to Huxley's *Brave New World* is a misunderstanding on many levels" ("Afterword," 177). My point is that *Solution Three* can profitably be read as a feminist critique of Huxley's novel, engaging with it at several stages in a critical dialogue and suggesting different solutions to similar issues.

14. While the inhabitants of *Brave New World* use *soma* to induce a peaceful and euphoric state, so as to eliminate violence, in *Solution Three* Council members and others smoke "cannabis, the aggression dispeller" (88).

15. In this respect I sympathize with Sarah Lefanu's avowed "difficulty of interpretation" ("Difference and Sexual Politics in Naomi Mitchison's *Solution Three*," 155) and what she considers the "confusing" (163) signals scattered throughout *Solution Three*.

16. Meaningfully, these views sound very similar to Shulamith Firestone's project in *The Dialectic of Sex: The Case for Feminist Revolution*, thus suggesting and reinforcing the need to think about the position of women in relation to new reproductive technologies, namely, the dissociation between sex and procreation.

17. As Sarah Lefanu remarks, Mitchison's novel is particularly interested in "the hard science, the work on plant cells and genetic engineering, and it is this that gives the novel its structure" ("Difference and Sexual Politics in Naomi Mitchison's *Solution Three*," 164).

6

Cloning and Biopower: Joanna Russ and Fay Weldon

A recurring trope in science fiction narratives has been an intense concern for environmental issues, as well as for gender asymmetries, questions that are in many ways inextricably linked, generating anxieties that have been translated into a vast number of postapocalyptic narratives and fictional accounts of alternative worlds where the reigning preoccupations are ecological and gender balance.[1] The urgent need to introduce ecological principles into the running of organizations and industries in order to minimize planetary pollution, as well as the development of new conceptual models of organizational structures, have played a fundamental role in the management and containment of the ecological crisis that has set in. Here I will be focusing upon the intersections of ecofeminism and organizational theory and practice as they appear dramatized in Joanna Russ's *The Female Man* (1975) and Fay Weldon's *The Cloning of Joanna May* (1989), two texts that, as I will be arguing, share many feminist concerns and offer apposite instances of science fiction's engagement with gender. Indeed, they can profitably be seen as in dialogue with each other.

Especially since the 1970s, when a spate of feminist science fiction novels were published, this genre's involvement with issues of gender has demonstrated a growing concern and preoccupation. Many authors have found the freedom to speculate granted them by science fiction an ideal medium through which to outline the contours of future, alternative worlds where women would have greater independence, as well as the autonomy to shape them according to their own personal, as well as sociopolitical, agendas.

The ideals of competitiveness advocated by capitalist society have come under severe criticism on the part of individuals, communities, and organizations concerned with environmental welfare. As Carolyn Merchant remarks, "Theories about nature and theories about society have a history of interconnections . . . as the ecology and economy of farm, forest, and fen were altered by new forms of human interaction with nature, traditional models of organic society and modes of social organization were likewise being undermined and transformed" (*The Death of Nature: Women, Ecology, and the Scientific Revolution*, 69). Market-oriented values coupled with capitalist modes of production spelled the destruction of great portions of fertile land, while nuclear energy further threatened already-endangered, fragile ecosystems. The long history of human domination of the natural world, traditionally gendered as female, has as a close counterpart male exploitation of women and their reproductive capacities. These are precisely two of the most important thematic concerns in *The Female Man* and *The Cloning of Joanna May*, two novels that further engage with some of the intersections of the problems previously outlined. Indeed, male imperialism of individuals and organizations over the female domain of procreation and the "feminized" natural world are the central thematic issues of Russ's and Weldon's books. Both novels thus dramatize some of the potential evils perpetrated by societal organizations and human agency on both the environment and women's bodies and suggest possible solutions to deal with these problems.

The Cloning of Joanna May constitutes a powerful dramatization and indictment of the repercussions of the male supremacy and monopoly of control over the natural world, on the one hand, and on women's reproductive capacities, on the other. In *The Cloning of Joanna May* the threat to the natural environment and to people is centered around the Chernobyl nuclear disaster and the role of the protagonist, Carl May, in the supervision of that crisis, addressing the question of potential ecological calamities and their repercussions. The action of the novel takes place in London against the background of the Chernobyl accident, in 1986. Weldon explicitly links male power over women's bodies and their reproductive functions with domination over the natural world, suggesting that male exploitation is responsible for the subjugation of both women and nature. Carl May, the husband of the novel's protagonist, Joanna May, is the chairman of British Nuclear Agents, Britnuc for short. He embodies the dangers of a monopoly of power in organizational life, of decision making as a hazardous power game when concentrated in a single person.[2] Britnuc, with its two nuclear reactors, has a particular responsibility in keeping the environment uncontaminated. Carl May's nods to public pressure and the incorporation of ecocentric values into his business enterprises are basically reduced to cosmetic gardening around the nuclear plower stations to maintain the illusion of healthy, clean air and verdant pastures. Weldon's novel dramatizes a widely held assumption on the part of the public that, as Carolyn P. Egri and Lawrence T. Pinfield note, "Governmental and business organizations are judged not to take the interests, aspirations and needs of their citizenry into account in their pursuit of organizational goals and objectives. . . . The 'environmental problem' is a consequence of how society is structured, for as multiple organizations pursue their self-interests, the interstices of society become an increasingly degraded residual" ("Organizations and the Biosphere: Ecologies and Environments," 209).

If "organizations . . . are the fundamental building blocks of modern societies" as H. E. Aldrich and P. V. Marsden maintain ("Environments and Organizations," 367), measures geared toward a greater respect for the environment and less polluting emissions have to come from inside the organizations themselves, less from an orgocentric perspective and more with an outlook toward the greater good of all. Carl May is the epitome of the self-centered, selfish individual, bent on the pursuit of personal and financial success and the commercial and financial prosperity of his business, Britnuc, without any concern for the damaging consequences the practices condoned by him can have on people and the environment. This is significantly an area addressed by Egri and Pinfield in their reflections on organizations and the biosphere: The warring impulses of self-interest are contextualized and examined in different situations, the authors concluding that "a shared appreciation of environmental issues is critical as resolution of environmental threats invariably requires interdependent collective action" ("Organizations and the Biosphere: Ecologies and Environments," 225), a line of action Carl May consistently and irresponsibly veers away from.

In *The Culture of Narcissism: American Life in an Age of Diminishing Expectations*, Christopher Lasch comments that the narcissistic organizational culture produces the kind of leader who "sees the world as a mirror of himself and has no interest in external events except as they throw back a reflection of his own image" (96), words that fittingly apply to Carl May's prepotent behavior. As Albert J. Mills, in a related vein, remarks, "Such leaders are often more concerned with image than substance—advancing through the corporate ranks not by serving the organization but by convincing his associates that he possesses the attributes of a 'winner'" ("Organizational Discourse and the

Gendering of Identity," 140), a perception that can be said to correspond to May's professional trajectory. As Lasch further contends, "For all his inner suffering, the narcissist has many traits that make for success in bureaucratic institutions, which put a premium on the manipulation of interpersonal relations, discourage the formation of deep personal attachments, and at the same time provide the narcissist with the approval he needs to validate his self-esteem" (*The Culture of Narcissism: American Life in an Age of Diminishing Expectations*, 91–92), an insight pertinent to the analysis of Carl May's authoritarian conduct and its pernicious repercussions.

Intricately connected with the question of narcissism and the exercise of power within organizations is the issue of the gendering of authority in business enterprises. Organizational culture, traditionally dominated by men, is heavily gendered, as many commentators have noted.[3] Sylvia Gherardi, for instance, observes that "culture, gender and power are . . . intimately bound up with each other in organizations as well as in society" (*Gender, Symbolism and Organizational Cultures*, 17). However, as she remarks, "organizational theories . . . claim to be gender neutral" (17), even though, as she observes, "There is an implicit subtext to this literature which assumes that workers are male, that managers are men with virile characteristics, and that organizations are the symbolic locus of production just as the home is the locus of reproduction" (17). In related fashion, Rosemary Pringle, in her examination of the working relations between male bosses and female secretaries, came to the conclusion that "male managers use sexuality and family relations to establish their control over secretaries," treating them "'narcissistically' as an extension of themselves" ("Bureaucracy, Rationality and Sexuality: The Case of Secretaries," 173), just as May behaves toward his secretary, Bethany. It is precisely this kind of organizational culture, which to a great extent mirrors male-dominated societal patterns of organization, that comes under incisive scrutiny in *The Cloning of Joanna May*.

IN HIS IMAGE: FABRICATED WOMAN

The other fundamental theme in Weldon's novel is human cloning, science fiction becoming (almost) science fact. As she herself explains, "In *The Cloning of Joanna May* I take birth away from women, and hand it over to men: as they are of course busy doing for themselves in the real world" ("Of Birth and Fiction," 206). Together with Dr Holly, who works for Martins Pharmaceuticals, of which Carl May is meaningfully a director, Carl May manages to remove an egg from his wife's anesthetized body and have four cloned embryos of Joanna May implanted in four healthy wombs and subsequently born.[4] Carl May's action may be inscribed in his overarching dream of domination over both women and the natural world.[5] He fantasizes about creating the perfect woman but, like Frankenstein, neglects to take responsibility for his creations, in what can be regarded as a parallel neglecting of the implementation of strict security measures in the nuclear power stations belonging to Britnuc. In "Egg Farming and Women's Future," philosopher Julie Murphy addresses the complex issue of reproductive technologies and rights in words that fit Joanna May's situation. According to Murphy, women "are defined in patriarchy as 'reproductive bodies.' . . . We are constantly discouraged, forbidden to use our bodies for ourselves. Reproductive technology, in the service of patriarchy, assumes that women's bodies are fertile lands to be farmed. Women are regarded as commodities with vital products to harvest: eggs" (68–69), just as Joanna May's body is harvested for her ripe eggs, a body Carl May wishes to have full control over as he would of a fertile plot of land to be exploited for one of his business enterprises. In a related vein, Indian philosopher and feminist activist Vandana Shiva critically interrogates the alleged connections between women and nature, as well as their abusive handling, namely, in terms of biotechnology experiments by mostly male scientists:

> Women's bodies, the seed and the soil, the sites of creative regeneration, have been turned into "passive" objects which experts can manipulate for profit. The sources of renewal of life have thus been turned into dead, inert and fragmented matter, mere "raw material" waiting to be processed and manipulated into a finished product. ("The Seed and the Earth: Women, Ecology, and Biotechnology," 5)

These words fittingly apply to Carl May's narcissist drive to dominate both nature and women.

In an analogous fashion, science historian Donna Haraway's vision of nature is relevant here, providing a critical counterpart to Carl May's imperialist views of control and exploitation of the natural world. She considers that "nature is not a physical place to which one can go, nor a treasure to fence in or bank, nor an essence to be saved or violated. Nature is not hidden and so does not need to be unveiled. Nature is not a text to be read in the codes of mathematics and biomedecine. It is not the 'other' who offers origin, replenishment, and service. Neither mother, nurse, nor slave, nature is not matrix, resource, or tool for the reproduction of man" ("The Promises of Monsters: A Regenerative Politics for Inappropriate/d Others," 296). As Haraway goes on to assert, "We must find another relationship to nature besides reification and possession" (296).

In "From Mice to Men? Implications of Progress in Cloning Research," Jane Murphy presents a negative view of cloning as far as its implications for women are concerned. Murphy sees the image of the "enucleated egg," the egg from which the nucleus, with all its genetic information, is removed in the process of cloning, as a perfect metaphor for the way women's bodies are manipulated under patriarchy's rule, another step in the depletion carried out by an androcentric society of women's art, writing, and progeny, often usurped by men as their own. For Murphy, then, women are, on the one hand, exploited as the sources of material necessary to proceed with cloning, such as eggs and wombs, while, on the other, they will mostly serve as carriers of the cloned embryos men want for themselves. Murphy argues that most examples of the popular literature dealing with cloning addresses the cloning of men, mentioning women only in passing and even then to suggest the cloning of beauty queens and other sex symbols. For Murphy, then, cloning "is the most extreme development along a continuum in science, religion, and other aspects of society, that attempt to immortalize men by establishing 'The Father' as the sole parent in creation. . . . Father of His Country. Father of Modern Science. Father of Gynecology. Father of the Industrial Revolution. Father of the Nuclear Age. Father, Son and Holy Ghost" (84).[6] I find Murphy's view excessively pessimistic and reductive. There would be many more options open to women if clone technology were to be implemented than just as passive suppliers of eggs and wombs. In tandem with men as sole parents, cloning would also enable women to be sole mothers, a scenario that Murphy fails to contemplate. Although Murphy's feminist critique of cloning is relevant, the increasing presence of women as agents and a creative force in most professions, including scientists and medical doctors, should provide a balance and different perspectives as far as decisions in the field of reproductive technology are concerned. While Robin Rowland believes that "biological technology is in the hands of men" ("Reproductive Technologies: The Final Solution to the Woman Question?" 356), my own feminist take on the possibilities and benefits cloning might hold for women is much more positive and optimistic.

Donna Haraway also significantly stresses the links between the politics of reproduction in late capitalism and the effective exercise of power in society and organizations. As she notes,

> Reproductive politics provide the figure for the possibility and nature of a future in multinational capitalist and nuclear society. Production is conflated with reproduction. Reproduction has become the prime strategic question, a privileged trope for logics of investment and expansion in late capitalism, and the site of discourse about the limits and promises of the self as individual. Reproductive "strategy" has become the figure for reason itself—the logic of late capitalist survival and expansion, of how to stay in the game in postmodern conditions. Simultaneously, reproductive biotechnology is developed and con-

tested within the large symbolic web of the story of the final removal of making babies from women's bodies, the final appropriation of nature by culture, of woman by man. (*Primate Visions: Gender, Race, and Nature in the World of Modern Science*, 352)

Carl May's act of "playing God," of taking into his hands decisions about his wife's procreative rights and considering only his own best interests, namely, the possibility of having at his disposal in the near future younger copies of his wife Joanna May, who might take her place, are on a par with his exercise of power games in the organizations in which he is chairman. Being at the top of the organizational hierarchy, he has been able to delegate and concentrate his energies on higher "ministerial dealings" (*The Cloning of Joanna May*, 76). Carl May is considered a "monster" (58) by one of his colleagues, who acknowledges that "our organization is a little sketchy . . . and I am not getting much cooperation where I had hoped to find it" (58), since May's managerial skills tend to obliterate hard realities and situations potentially hazardous to the public, in this case, the threats that the Chernobyl disaster might entail to the British population, as well as the conceivable dangers posed by a similar malfunction in the nuclear power stations owned by Britnuc.

Fay Weldon's *The Cloning of Joanna May* accordingly illustrates the sexual and biopolitics of male control over reproduction, namely, as far as the work of male scientists and big corporations are concerned.[7] As Rosi Braidotti, among other commentators, remarks, "Where does this leave the woman, however? She is not only reduced to maternal power, but that power is also displaced on to technology-based, corporate-owned reproductive production systems. In some ways, the corporations are the real moral monsters in all the popular science fiction and cyberpunk films: they corrupt, corrode, exploit and destroy ruthlessly" (*Metamorphoses: Towards a Materialist Theory of Becoming*, 207), words that seem tailor made to address Carl May's actions.

Sandra Harding's statement that "the epistemologies, metaphysics, ethics, and politics of the dominant forms of science . . . many of its applications and technologies, its modes of defining research problems and designing experiments, its ways of constructing and conferring meanings" (*The Science Question in Feminism*, 9) result in the kind of science that has been "highly incorporated into the projects of a bourgeois, racist, and masculine-dominant state, military and industrial complex" (138) is a forceful reminder of the masculinist-inflected situation in science and medical studies. Her words seem perfectly suited to comment on the situation described in *The Cloning of Joanna May*, with the unethical, irresponsible, and arrogant way of behaving that characterizes Carl May. For Janice Raymond, in turn, with the advent of new reproductive technologies, "The myth of patriarchal creation is re-fabricated. Once more, man fashions women in his own image" ("Feminist Ethics, Ecology, and Vision," 435). As Raymond insists, in the midst of "such a man-made environment, it is important for women not only to see the whole picture, but to persist in asking woman-centered ecological questions" (435), as well as asserting "our own ecological vision" (435).

A related theme that comes in for severe criticism in Weldon's novel is the fetishization of youth and the concomitant misogynistic discourse that considers older women as dispensable, almost as inconsequential nonentities, as Carl May insinuates when he declares to his sixty-year-old ex-wife, Joanna, "You should have died twenty years ago, what use to the world are you? A woman without youth, without children, without interest, a woman without a husband" (*The Cloning of Joanna May*, 106). Indeed, Carl May's interconnected dream of creating a "perfect woman" (78), who, according to him, would be one "who looked, listened, understood and was faithful" (78) is also inextricably linked with his fantasy of an ideal of womanhood—youthful, passive, and compliant—existing only to fulfill his wishes, growing progressively worthless with age. This egocentric daydream is closely intertwined with Carl May's illusory hope that by creating four younger versions of his wife he would be able to recover her youth and benefit again from the favors of those youthful copies of Joanna May, a totally fallacious expectation, as he comes to realize. His anticipation that they would "love him as Joanna had," since "they were Joanna" (241) is proven false, as well as his assertion that when "he multiplied her he had not so much tried to multiply perfection . . . he had done it to multiply her love for him, . . . multiply it fourfold" (241). As Lana Faulks observes in relation to Carl May's actions, "Control and power are the objectives of those who try to create a world of sameness: identical beliefs and behavior sell commodities fueling the marketplace. Corporations feast on feeble minds that digest the carefully crafted ideology of progress and production" (*Fay Weldon*, 61), words that aptly synthesize Carl May's motives and corporate behavior.

The rigidly stratified and hierarchical organizational system epitomized by Carl May's leadership is a concrete example of the types of society that come in for criticism from Luce Irigaray: "All the systems of exchange that organize patriarchal societies and all the modalities of productive work that are recognized, valued, and rewarded in these society's are men's business. The production of women, signs, and commodities is always referred back to men" (*This Sex Which Is Not One*, 170). Going on from this, Irigaray notes, again in words that are particularly relevant to Weldon's novel, that

> *the feminine occurs only within models and laws devised by male subjects.* Which implies that there are not really two sexes, but only one. A single practice and representation of the sexual. . . . This model, a *phallic* one, shares the values promulgated by patriarchal society and culture, values inscribed in the philosophical corpus: property, production, order, form, unity, visibility . . . and erection. (*This Sex Which Is Not One*, 86)

These are insights that can be usefully extrapolated to offer a pertinent comment on the relative positions of Carl May, his wife, and her clones.

Metaphors of phallic mastery over the surrounding world and other people are recurrently attached to Carl May. The building where his office is situated, in a tower block, "had been designed to dominate the city skyscape" (*The Cloning of Joanna May*, 75), even though no sooner was it finished than "all around arose the thrusting towers of usurping empires—leaner, taller, glassier" (75). He is variously described as a "vampire" (38) and a wolf whose "teeth were fangs and growing as long as the wolf's were in 'Red Riding Hood'" (108). With his selfish dreams of being "master of mortality" (109), Carl May, with the help of the scientific expertise of Dr Holly, epitomizes the paradigmatic "mad male cloning scientist" (109), another "Frankenstein" (109) with powers over life and death, the "devil" (109), as Joanna May also calls him.

All these images of exploitation of others conspire to emphasize Carl May's omnipotence in the realm of work and the personal sphere. He considers himself invincible, godlike. As he explains to Joanna May, "I can make a thousand thousand of you if I choose, fragment all living things and re-create them. I can splice a gene or two, can make you walk with a monkey's head or run on a bitch's legs or see through the eyes of a newt: I can entertain myself by making you whatever I feel like, and as I feel like so shall I do" (*The Cloning of Joanna May*, 109). He implicitly brings nature and woman together in his Manichaean discourse about good and evil: Both woman and nature need to be controlled and manipulated, since both are found faulty. He further elucidates his grand designs: "I would perfect nature's universe, because nature is blind, and obsessive, and absurd . . . and has no judgement, only insists on our survival, somehow, any old how: nature is only chance, not good or bad. All I want is the any old how properly under control, directed . . . I, man, want to teach nature a thing or two, in particular the difference between good and bad; for who else is there to do it? But how can I, because woman makes man bad" (111). The text suggests that one of the principal factors that led to Carl May's becoming this monstrous other may be rooted in his childhood, in the lack of parental care, a situation that also links him with Frankenstein's monstrous creature. Indeed, we learn that "Carl's mother had kept him much of the time, when he was little and hungry and stole, chained up in the dog's kennel in the yard, to teach him a lesson. His father was a dead dog; his mother was a bitch" (14). It is thus pertinent to conjecture, from a psychoanalytic point of view, that Carl May's treatment of women as objects may betray his repressed wish to punish his mother for the heartless way she behaved toward him.

Carl May's close connections with the discourse and politics of biopower are severely condemned in the book, and his arrogance will be heavily punished in the end. Not only do Joanna May's clones band together against him, finding strength in their very close bond, but his policies of lack of transparency and monopoly in decision making in Britnuc, as far as the Chernobyl crisis is concerned, also bring about his downfall. In order to prove to the public how safe the cooling ponds in the nuclear power stations that belonged to Britnuc are, Carl May decides to call the press to witness him swimming in one of them, Britnuc B, in the Welsh hill country, whose wild beauty, as well as "the overwhelming presence of nature unorganized and unconfirmed" (*The Cloning of Joanna May*, 257), poses a sharp contrast to Britnuc B, whose public relations team assures the media that Chernobyl could not possibly happen there. At the end of the book, Carl May is dying from the consequences of this fatal swim. After he had come out of the supposedly nonradioactive cooling pond, "The meters—the ones put in to reassure the visitors—had started to chatter" (261).

Joanna May, in turn, realizes, "When I acknowledged my sisters, my twins, my clones, my children, when I stood out against Carl May, I found myself: pop! I was out. He thought he could diminish me: he couldn't: he made me" (*The Cloning of Joanna May*, 246). In spite of their origin in the authoritative behavior of Joanna's husband, who harvests her eggs without her knowledge, the result hints at a positive outcome for Joanna and her clones: "They could see now that was the trouble—they'd been lonely. . . . Now they had each other, nothing need be the same" (251). They find strength and support in the community of women they form, in their common bond, like the women in the female-only worlds I have mentioned, a more radical instance of that kind of connection, a powerfully enabling feeling of "sisterhood."[8] Joanna May, who becomes in this narrative yet another version of a "virgin mother" without ever even giving birth, becomes aware of her four cloned children only when they are already grown-up, a circumstance for which her husband and his male supremacist politics are to blame.

The final twist, however, is that Carl May will leave his own progeny, in spite of having had a vasectomy when he was eighteen. Just before dying, he pleads with Joanna May to "remake him" (*The Cloning of Joanna May*, 262), to which she accedes. Alice, one of her clones, gives birth to a clone of Carl May, a baby ironically brought up by Joanna May herself who, in her function as substitute mother and educator, now has full power over him, although presumably she will use that power to make a better human being of him and possibly at the same stroke providing ammunition to the nature-versus-nurture debate, this time on the nurture side. A caring environment and family education will, the novel suggests, go a long way toward at least partially reversing the influence of genes.

Carl May's whole attitude toward life reveals the at times dangerous consequences of too great a concentration of power in individuals inside organizations, while the novel as a whole questions the pitfalls of institutionalized networks of power enmeshed in political power games with practices that may lead to abuses of authority and, potentially, public disasters. Weldon's novel also implicitly engages with the conventional view that links masculinity with individualization, an assumption taken to extremes in Carl May's character, as compared with a femininity that equals collectivism, a notion that is given fictional representation in the strength found by Joanna May and her clones in coming together and joining forces against Carl May. These conjectures are problematized in the novel, with a narrative drive that implicitly suggests that the preponderant embeddedness of masculinity within bureaucracy should be offset by a greater presence of women within organizations, which would provide for a more collective form of decision making, which in turn would create a more eco-amiable organization. In *The Cloning of Joanna May* woman is depicted, from Carl May's perspective, as still very much the other of the same, to use Luce Irigaray's terminology, that is, woman in a patriarchal world, conceived in terms of a male normative logic, reduced to *"the economy of the Same"* (*This Sex Which Is Not One*, 74). Woman's "otherness," then, becomes to a great extent dissolved in order to be incorporated and regulated to the familiar standard of masculinity, a standpoint the novel goes a long way toward subverting, proposing a more egalitarian perspective where the otherness of the other, man or woman, would be partially included in the other sex, taking into account differences in order to bring about a more balanced organizational order.

In *The Feminist Case against Bureaucracy*, K. E. Ferguson offers a critique of contemporary bureaucracy and argues that women can provide an alternative perspective on organizational structures, one characterized by nonhierarchical and nonbureaucratic arrangements. Ferguson argues for a greater integration of the public sphere, where men have been the principal actors, with the private one, the domestic one, which women have considered as one of the chief areas of their subordination, a policy that would pave the way for a less aggressive and more cooperative organizational life. As Ferguson remarks, "In their role as subordinates, women's experience sheds considerable light on the nature of bureaucratic domination; in their role as caretakers, women's experience offers grounds for envisioning a nonbureaucratic collective life" (26). This is the plea that both Fay Weldon's *The Cloning of Joanna May* and Joanna Russ's *The Female Man* implicitly articulate.

THE FEMALE MAN

Joanna Russ's *The Female Man* radically calls into question the patriarchal and hierarchical functioning of Western society, dismantling the cultural structures that regulate the construction of genders and gender-related functions. As in *The Cloning of Joanna May*, in *The Female Man* male stereotypes and masculinist representational structures similarly come under heavy scrutiny and satire.

I see the two novels as engaged in a critical dialogue with each other and would like to suggest here that *The Cloning of Joanna May* can be read as in part a response to *The Female Man*. Both contain acute sardonic criticism of patriarchy and an invocation of the power of close female bonds to change society's mostly male organizations and their practices. Indeed, Jane, Julia, Gina, and Alice, the clones of Joanna May, are almost the equivalents of Joanna Russ's four Js, Joanna, Janet, Jeannine, and Jael, who is also called Alice Reasoner (maybe in a veiled reference to Alice, one of Joanna May's clones). In addition, both novels' protagonists share the same name: Joanna.[9]

Donna Haraway considers Joanna Russ's *The Female Man* "the founding text in anglophone feminist SF" (*Modest_Witness@Second_Millenium. FemaleMan©_Meets_Oncomouse™: Feminism and Technoscience*, 75). Russ's novel deals with the intersections and meetings of four genetically identical women—Joanna, Jeaninne, Janet, and Jael—who live in alternate worlds and come together in a time warp in Joanna's time, the 1970s, in New York. The four clone sisters, who inhabit different chronotopes, share the same genotype and are described as four versions of the same woman, like the four clones of Joanna May.[10]

Joanna, a radical feminist, is a contemporary version of and spokesperson for Russ herself, while Jeaninne works as a librarian in New York City in 1969. In Jeannine's world, however, World War II never took place and the Great Depression still shapes quotidian life. Jael, a cyborg warrior woman, named after the biblical Jael in the book of Judges, lives in Womanland, a radical version of Janet's Whileaway, a utopian country where only women reside. In Jael's world, the men inhabit Manland and the women Womanland; the former are the "Haves" and the latter the "Have-nots," an asymmetrical pattern that has led to a war between Manlanders and Womanlanders that has been raging for forty years.[11] Indeed, characterized by a warring mood between men and women, the state of affairs in Jael's country can be seen as a more radical version of that prevailing in Joanna May's world, where women are victims and prey in Carl May's hands and are often depicted in a rebellious mood against the "Haves," that is, the likes of Joanna's husband. In *The Female Man* men come in for relentless criticism and satire, depicted as arrogant, violent, and mindless of women's feelings and aspirations, in many ways like Dr Holly and Carl May in Weldon's novel.

Here I wish to concentrate on Whileaway, an all-female society, another alternative world where Janet, who comes from the far future, "but not my future or yours" (*The Female Man*, 161), lives. The societal and institutional organizations in Whileaway differ considerably from Western models and are depicted not as straightforward examples to be followed but as a tongue-in-cheek, satirical vision of a speculative world made up only of women who nevertheless exhibit no so-called feminine traits and have developed along lines that deviate from the characteristics a Western observer would expect from an all-female society. In other examples of women-only worlds, such as Mary E. Bradley Lane's *Mizora* (1890) and Charlotte Perkins Gilman's *Herland* (1915), the societal organization conforms to conventional expectations of a pastoral, profoundly ecologically minded, peaceful, and fulfilling world. While some of these traits apply to Russ's utopian society, there is a certain amount of violence, which undermines such essentialist notions as that of women's supposedly inherently peaceful propensities, which would make for a totally placid and unruffled world.

The male population in Whileaway was all wiped out by a plague that came in P. C. 17 (Preceding Catastrophe) and ended in A. C. 03 (After Catastrophe). In the third century A. C. genetic engineering became widely available, although the merging of ova had been used much earlier. In Whileaway women have their children around the age of thirty, "singletons or twins as the demographic pressures require. These children have as one genotypic parent the biological mother (the 'body mother') while the nonbearing parent contributes the other ovum ('other mother')" (*The Female Man*, 49), a process similar to that used by Dr Holly to create the clones of Joanna May, who coincidentally was also thirty when she had her mock pregnancy and her eggs were taken from her body without her knowledge. Reproductive technology in Whileaway stands in stark contrast to the violations of the protagonist's bodily integrity in *The Cloning of Joanna May*. Indeed, women in Whileaway have total control over their reproductive systems and their own bodies. They choose when to have their children, usually at thirty, and do not have to depend on sperm to produce their offspring. *The Female Man* thus provides another graphic illustration of a women-only society where babies are created through a version of cloning, one that many lesbian couples would no doubt welcome. As in "Houston, Houston, Do You Read?" in Whileaway, women still carry their genetically created babies in their wombs. In Whileaway women look forward to the time when they give birth. Having a child is considered as "a vacation" (14), a time when the women can "pursue whatever interests [they] have been forced to neglect previously, and the only leisure they have ever had—or will have again until old age" (49), a situation strongly reminiscent of a similar practice in Gilman's *Herland*, where "each mother had her year of Glory; the time to love and learn, living closely with her child, nursing it proudly, often for two years or more" (*Herland*, 104). The description of a mother in Whileaway, as "someone on vacation, someone with leisure, someone who's close to the information network and full of intellectual curiosity" (*The Female Man*, 23) also presents a sharp contrast to the conventional image of mothers in traditional Western societies, overworked and overburdened, with no leisure time to devote to intellectual or other pleasurable activities, totally absorbed in the raising of their children. Unlike this prevailing representation of mothers in the Western world, mothers in Whileaway would constitute a "top class" (23), if there were one in this nonhierarchical society. Motherhood, then, is highly valued in Whileaway, as in *Mizora* and *Herland*, as a potent source of personal fulfillment, bypassing, however, the male agency, in a recreation of Firestone's instigation to separate biology from destiny, with the means to "create life independently of sex" (*The Dialectic of Sex: The Case for Feminist Revolution*, 241).

While Joanna May finds her justification and fulfillment as a woman in being Carl May's wife, in making him feel good, as well as in her desire to be a mother, women in Whileaway do not have to hold a mirror up to men, to borrow Virginia Woolf's resonant expression, being thus able to concentrate on their own and the community's development and well-being. As Woolf puts it, in a satiric vein, "Women have served all these centuries as looking glasses possessing the magic and delicious power of reflecting the figure of man at twice its natural size. . . . How is he to go on giving judgments, civilizing natives, making laws, writing books, dressing up and speechifying at banquets, unless he can see himself at breakfast and dinner at least the size he really is?" (*A Room of One's Own*, 37–38). This insight aptly defines Carl May's behavior and often appears dramatized in *The Female Man*. As the narrator puts it, carrying out a deconstruction of the clichés of a characteristically "romantic" situation that portrays Jeannine in a pleasure boat with a suitor: "From shore it must really look quite good, the canoe, the pretty girl, the puffy summer clouds, Jeannine's sun-shade. . . . His contribution is *Make me feel good*; her contribution is *Make me exist*" (*The Female Man*, 120).

In *This Sex Which Is Not One*, Irigaray forcefully warns women against a deleterious homogenization with androcentric models: "Woman could be *man's equal*. In this case she would enjoy, in a more or less near future, the same economic, social, political rights as men. She would be a *potential man*" (84; emphasis mine). These words can usefully shed light on some of the issues raised by *The Female Man*, that is, the woman forced to fit male models who thus "becomes" a female man. Joanna Russ's project can be defined in part as revealing the woman hidden behind the masquerade of femininity. As Irigaray pertinently remarks, under the heading "How She Became Not-He," and in words that seem to offer a pointed commentary on Russ's "female man," "Instead of remaining a different gender, the feminine has become, in our languages, the non-masculine, that is to say an abstract nonexistent reality" (*Je, Tu, Nous: Toward a Culture of Difference*, 20). Irigaray's forceful exhortation that "above all . . . we must not turn completely into men" ("Women-Mothers: The Silent Substratum of the Social Order," 51) finds its fictional dramatization in Janet Evason, the emissary from Whileaway and the female man of Russ's novel.

In Whileaway there are no true cities, and social life is based on a complex clan organizational structure that at its core consists of families of thirty to thirty-five persons, the farms being the only family units. As Janet explains, there is no government in Whileaway in the traditional sense, and in addition, "There is no one place from which to control the entire activity of Whileaway, that is, the economy" (*The Female Man*, 91). Whileawayans have consistently striven to achieve a nonhierarchical, egalitarian society, a model *The Cloning of Joanna May* similarly endorses in its severe criticism of institutionalized hierarchical practices. Bringing feminist and organizational theory into productive conjunction, Kathleen P. Ianello considers the possibility of alternative models that privilege nonhierarchical, more participatory forms of organization, like that put forward in Whileaway,[12] where society is communal, children move at ease all over the planet past puberty, and "the kinship web . . . is world-wide" (81).

The invention of the induction helmet, which "makes it possible for one workwoman to have not only the brute force but also the flexibility and control of thousands" (*The Female Man*, 14) brings about a radical transformation in Whileawayan industry, turning it "upside down" (14). Indeed, Whileaway is "engaged in the reorganization of industry consequent to the discovery of the induction principle" (56). Since induction helmets make it feasible "to operate dozens of waldoes at just about any distance you please" (51), it is possible to introduce a drastically reduced workweek of sixteen hours.

Ecological concerns are of momentous significance to Whileawayans, who are farmers, scientists, police officers, artists, and so on. Whileaway is "so pastoral that at times one wonders whether the ultimate sofistication [*sic*] may not take us all back to a kind of Pre-Paleolithic dawn age, a garden without any artifacts. . . . Meanwhile, the ecological housekeeping is enormous" (*The Female Man*, 14). Waste is a taboo in Whileaway, which is "inhabited by the pervasive spirit of underpopu-lation" (100). Whileawayans are characterized by "the cast of mind that makes industrial area into gardens and ha-has, that supports wells of wilderness where nobody ever lives for long, that strews across a planet sceneries, mountains, glider preserves" (54) and so on. By means of a self-consciously satirical discourse that subverts what are often considered typical and unchanging feminine distinguishing attributes, Whileaway, although clearly not a serious utopian blueprint for a realistic future society, nevertheless provides an effective criticism of patriarchal institutions through parodical revision of some of its organizations, such as the army and institutionalized industry.

While *The Cloning of Joanna May* does not advocate such a radical scenario as the abolition of men, the narrative drive strongly suggests the urgent need to rethink hierarchical practices in organizations in order to minimize gender inequalities, as well as a pressing need, so powerfully expressed in *The Female Man*, to protect the environment.

Toward the end of both novels, family reunions that bring together the four Js and Joanna May's four clones are staged. In *The Female Man* the narrator declares very early on that "eventually we will all come together" (18), Jael later deciding to find "her other selves" (160), whereas in *The Cloning of Joanna May*, Joanna takes the initiative to look for her clones, whom she calls "my sisters, my twins, my clones, my children" (246) and her "sisters and daughters both" (127). In both cases, the experience is an empowering one, producing a feeling of strength derived from the emotional network of support provided by Joanna's clones, while a similar sentiment of reliance on the sisterhood of the four Js applies. In Joanna's case, she finds that "all of a sudden there was more of me left . . . reinforcements came racing over the hill; Joanna May was now Alice, Julie, Gina, Jane as well. Absurd but wonderful!" (247). Indeed, the family reunions in both books suggest a gathering of forces on the women's side to fight against the abusive authority of the "Manlanders," the "Haves," in Russ's case, and the ruthless domination exerted by Carl May in his managerial and personal affairs.

Jael's description of the origin of the four Js is strikingly reminiscent of the creation of Joanna May's four clones, both narratives actively participating in the debate that pits against each other the theories of biological determinism and social constructionism—nature versus nurture. As Jael explains, musing about her "other selves" (*The Female*

Man, 160), in spite of all the differences among the four of them, "We started the same. . . . If you discount the wombs that bore us, our prenatal nourishment, and our deliveries (none of which differ essentially) we ought to have started out with the same autonomic nervous system, the same adrenals, the same hair and teeth and eyes, the same circulatory system, and the same innocence. We ought to think alike and feel alike and act alike, but of course we don't. . . . I can hardly believe that I am looking at three other myselves" (162). Jael further remarks, prompting Joanna, Jeannine, and Janet to observe themselves, "Look in each other's faces. What you see is essentially the same genotype, modified by age, by circumstances, by education, by diet, by learning" (161), an insight repeated in *The Cloning of Joanna May*. Indeed, in the case of Joanna May's clones, when they first come together without knowing each other, Dr Holly's secretary assesses them in a detached way: "Reared separately, no doubt, though they'd end up much the same in the end. The impact of the rearing environment wore away with time" (*The Cloning of Joanna May*, 225–226). Like the four Js in *The Female Man*, who, as Jael explains, "started the same" (162), the four clones of Joanna May also look different, although they can be recognized as belonging to the same genotype. The underlying implication, in both cases, suggests that in spite of the superficial differences due to their distinct upbringing, the empathy experienced by the cloned women and the strength they derived from that feeling was a feature they had in common that could not be easily erased. The reunion of the cloned women brought about by Jael and Joanna May thus turned out to be a site of empowerment for each of them individually and as a group, providing added impetus to their future lines of action against patriarchy, as well as stressing the intensity of the biological link that unites them. Indeed, the endings of the two novels also bear some similarities. While Jael's purpose is to enlist the other Js help in the war between the Manlanders and the Womanlanders, Joanna May and her four clones unite against Carl May and Dr Holly's tyrannical, manipulative, and secretive ways within May's institutional empire.

CYBORG MONSTERS AND INAPPROPRIATE/D OTHERS

Donna Haraway has often acknowledged her debt to writers of "science fiction," specifically Joanna Russ, considering that "the cyborgs populating feminist science fiction make very problematic the statuses of man or woman, human, artefact, member of a race, individual entity, or body" ("A Cyborg Manifesto: Science, Technology and Socialist-Feminism in the Late Twentieth Century," 178). Haraway located in science fiction texts some of the sources of her socialist-feminist ironic cyborg myth, a political manifesto that fits particularly well Russ's characters in *The Female Man*. As Haraway asserts, "Cyborg monsters in feminist science fiction define quite different political possibilities and limits from those proposed by the mundane fiction of Man and Woman" (180), a description that seems tailor made for defining Joanna, Janet, and Jael in *The Female Man*. Indeed, these "cyborg monsters" in Russ's conception might become fundamental actors in the kind of "regenerative politics" Haraway envisions as essential grounding for the concretization of her dream of a future socialist-feminist structure where these "inappropriate/d" others will play a fitting and active role ("The Promises of Monsters: A Regenerative Politics for Innappropriate/d Others," 320). Characters like the four Js in *The Female Man*, who refuse to be classified according to the prevalent masculinist norm, and, less radically but nonetheless in an effective way, Joanna May and her four clones, who gravitate away from the centripetal pull of Carl May's empire with its tentacles extending in many directions, constitute apt examples of "others" who have not been appropriated and shaped by restrictive patriarchal rules. The kind of "cyborg subject position" ("The Promises of Monsters: A Regenerative Politics for Innappropriate/d Others," 320) advocated by Haraway constitutes a powerful strategic stance from which to argue an effective disconnection from the "sacred image of the same," that is, the masculinist norm enforced on the women characters in *The Female Man*[13] and *The Cloning of Joanna May* who might then, in Haraway's terms, become "inappropriated," their own selves, not mirror images of the male narcissistic will to power.[14]

The Cloning of Joanna May interrogates such central issues as identity, narcissistic drives, and aging,[15] concerns crucially implicated in the human cloning debate. In its extremely satirical portrayal of Carl May, *The Cloning of Joanna May* thematizes the old stereotype of a greater masculine distancing from and control over feminized nature, as well as interference in the female reproductive sphere. Carl May's ultimate punishment and fate forcefully suggest that a new model of organizational theory and practice, based on ecologically grounded procedures, is urgently needed.

Val Plumwood concisely sums up an ecofeminist position that "accepts the undesirability of the domination of nature associated with the masculine" ("Women, Humanity and Nature," 22) without falling into the traps of essentializing the supposed closer connnection of the (biological) feminine with nature or unproblematically associating the masculine with aggression and violence toward the natural world. Plumwood believes that only a degendered model, one that transcends the masculine and the feminine, "could provide some sort of basis on which to mount a revised ecofeminist argument" (24).

One of the principal thrusts of Weldon's book is an exhortation for women to become intervening organizational agents at all levels of societal and government structures, especially as far as reproductive technologies and environmental issues are concerned, to become actively implicated in decision criteria within organizations. As Haraway maintains, "If we learn how to read these webs of power and social life, we might learn new couplings, new coalitions" ("A Cyborg Manifesto: Science, Technology and Socialist-Feminism in the Late Twentieth Century," 170). Traditionally perceived as passive and susceptible to exploitation and subjugation, both women and nature have to become intervening actors in the struggle for more just organizational patterns, more interactive and earth and women friendly, but also more politically engaged in the defense and promotion of women's rights, as well as the preservation and improvement of the environment.

The Female Man similarly constitutes an eloquent plea for a peaceful world where women are granted due respect and ecological values are implemented, where "women in the integrated circuit" ("A Cyborg Manifesto: Science, Technology and Socialist-Feminism in the Late Twentieth Century," 170), at all institutional levels, are the rule and not the exception. Haraway's appeal for strategic alterations, for women to become dynamic actors within organizations, can be seen as an integral part of the process of trying to change existing legislation and widely accepted practices in organizational management. As Egri and Pinfield conclude, "While the dominant social paradigm emphasizes humans' dominionistic and utilitarian relations with nature, the radical environmentalism perspective emphasizes humans' emotional, aesthetic and spiritual connections with the natural environment" ("Organizations and the Biosphere: Ecologies and Environments," 227). They suggest, in words that suitably describe Carl May's actions, that "focusing solely on the material value and benefits to be derived from the natural environment . . . informs environmentally unsustainable actions" (227). On the other hand, "preserving the natural environment purely for its aesthetic value . . . to the exclusion of other relations with the natural environment denies the development of material relations necessary for human physical existence. In the end, there is a need for a balance among these disparate and sometimes conflicting relationships with the natural environment—not a static final balance, but a dynamic balancing between evolving human and natural systems of existence" (227). While Egri and Pinfield do not address the question of sexual politics within organizational dynamics, I want to suggest here that *The Cloning of Joanna May* plays an intervening role in introducing the question of gender into the whole larger subject of the running of organizations and institutional life, a debate in which Donna Haraway's cyborg politics participates in a fundamental way. In this chapter I have tried to bring into political play and productive convergence the intersections and contributions of these various discourses, which can be seen fictionally engaged in a dialectical struggle in both Russ's and Weldon's novels. In the disentangling of these variegated webs of power, with the help of organizational theory, several strands of environmental discourses, as well as gender politics, suggest potentially productive new directions for change implicit in *The Female Man* and *The Cloning of Joanna May*. These texts thus become political manifestoes for action along the lines put forward by Donna Haraway and her ironic cyborg myth, as well as exhortations for the development of a discourse of organization theory and practice that are both environmentally responsible and respectful of gender asymmetries.

Gender asymmetries leading to inequality in the family and at work, as well as possible ways in which new reproductive technologies, namely, cloning, might contribute to a more just society where women would not be treated as subaltern and, because of their biology, disadvantaged both in the workplace and as mothers, inserted into the sexual politics of power of traditional family structures, will be addressed in the next chapter, along with some literary representations of alternative scenarios.

NOTES

1. These novels include Joanna Russ, *The Female Man* (1975); Naomi Mitchison, *Solution Three* (1975); Pamela Sargent, *Cloned Lives* (1976); Kate Wilhelm, *Where Late the Sweet Birds Sang* (1976); Marge Piercy, *Woman on the Edge of Time* (1976); Sally Miller Gearhart, *The Wanderground* (1979); and Suzy McKee Charnas, *Motherlines* (1978), to cite only some of the most representative.

2. As Miller, Hickson, and Wilson maintain, "The lone decision-maker making choices about his or her own interests might be thought to act rationally (although psychologists may argue the evidence here) but the complexities of managerial decision-making in concert with others have been well documented" ("Decision-Making in Organizations," 45).

3. See, for instance, J. Hearn and W. Parkin, "Gender and Organizations: A Selected Review and Critique of a Neglected Area"; K. E. Ferguson, *The Feminist Case against Bureaucracy*; J. Hearn, D. L. Sheppard, P. Tancred-Sheriff, and G. Burell, eds., *The Sexuality of Organization*; A. J. Mills and S. J. Murgatroyd, *Organizational Rules: A Framework for Understanding Organizations*; A. J. Mills and P. Tancred, eds., *Gendering Organizational Analysis*; M. Savage and A. Witz, *Gender and Bureaucracy*; Albert J. Mills, "Organizational Discourse and the Gendering of Identity"; Sylvia Gherardi, *Gender, Symbolism and Organizational Cultures*.

4. As Hillel Schwartz notes, referring to male scientists who create test-tube babies, "Men are at work here, as it had been Saint Jerome who propagated the doctrine of the Virgin Birth and medieval churchmen who conceived of the male seed as the active force in a passive womb" (*The Culture of the Copy: Striking Likenesses, Unreasonable Facsimiles*, 341).

5. Carl May's actions can also be characterized with recourse to repressed feelings of womb envy. According to Eva Feder Kittay, in order to "defend against womb envy . . . the boy may devalue himself or devalue life-giving. He may maintain the high valuation on creation but displace the creative capacities from woman, appropriating them to man and to his exclusive domains. Such appropriation, in its many manifestations, is probably the most likely outcome of womb envy" ("Womb Envy: An Explanatory Concept," 121). As Kittay pertinently observes, "For the child, this means forming ambitions in which he can outdo women in their productivity, accepting ideologies in which men play the 'really' important part in procreation and the family and, in general, fiercely adopting the stance of male superiority" (121), words that help shed light on Carl May's behavior.

6. However, I believe there is a countermythology to this one which takes such forms as Mother Earth, the Mother of God, and in such widespread expressions as "the mother of all . . . ," suggesting the biggest and most encompassing of them all, but also carrying profoundly disturbing and violent connotations, such as "the mother of all battles," an elocution often used during the recent war against Iraq with implications I would like to explore but that fall outside the scope of this book.

7. As Mary Daly, in a critical and ironic vein, states, "It is impossible to miss symptoms of this male fertility syndrome in the multiple technological 'creations' (artificial wombs) of the Fathers—such as . . . corporate offices, airplanes, spaceships—which they inhabit and control. Moreover, these male-constructed artificial wombs are ultimately more tomb-like than womb-like, manifesting the profoundly necrophilic tendencies of technocracy" (*Gyn/Ecology: The Metaethics of Radical Feminism*, 61).

8. Jeanne Beckwith wrote a play dealing with cloning, which she kindly let me read, about three women clones who did not know they existed and finally meet, reminiscent of the cloned women in Weldon's *The Cloning of Joanna May*.

9. In an e-mail message to the author, Fay Weldon confirmed that it was quite probable that she had read Joanna Russ's *The Female Man* "and then forgot it, consciously but not unconsciously."

10. I have always thought of the four Js as clones, contrary to the opinion of some critics. I was therefore pleased to find that Haraway also calls them the "four clone sisters" (*Modest_Witness@Second_Millenium.FemaleMan©_Meets_OncomouseTM: Feminism and Technoscience*, 69).

11. For a full discussion of the character of Jael, see Jeanne Cortiel, *Demand My Writing: Joanna Russ/Feminism/Science Fiction*.

12. Having studied three different forms of feminist organizations, Ianello identifies the "modified consensual model" (*Decisions without Hierarchy: Feminist Interventions in Organization Theory and Practice*, 122) as an organizational structure that closely corrresponds to a more cooperative, less hierarchical organizational form that "needs to be tested in large-scale organizations" (122).

13. As Jeanne Cortiel points out, in spite of the thorough defiance of normative gender structures the novel carries out, "It is the impossibility of Joanna transcending gender and representing universal humanness that forms a central dilemma in the novel" ("Determinate Politics of Indeterminacy: Reading Joanna Russ's Recent Work in Light of Her Early Short Fiction," 221).

14. While we have to be cautious of considering the cyborg figure as unproblematically liberatory for women characters, since science fiction works are crowded with female cyborgs that have been configured as fantasy figures, playing the role of objects of pleasure for the voyeuristic male gaze, there are other elements to the cyborg woman that help break down deeply entrenched dichotomies that have contributed to the stratification of women's position in society. It is these transgressive characteristics that I am interested in here, traits that Joanna, Janet, and Jael strongly evince.

15. Kathleen Woodward, in *Aging and Its Discontents: Freud and Other Fictions* (1991), argues that the "profound gerontophobia in our culture should be extirpated" (193), suggesting that "one of the ways to begin that process is to examine critically our representations of aging and to work to produce new ones" (193). Furthermore, she maintains that "Freudian psychoanalysis is complicit with our culture's repression of aging" (192).

7

The Sexual Politics of Human Cloning: Mothering and Fathering in the New Millennium

Recent developments in reproductive technologies could open the way to radically different future social and family arrangements, new psychological configurations and a reorganization of sexual power structures. As Donna Haraway notes, "reproductive politics are at the heart of questions about citizenship, liberty, family, and nation" (*Modest_Witness@Second_Millenium.FemaleMan©_Meets_Oncomouse™: Feminism and Technoscience*, 189). In analogous vein, Gabriele Schwab remarks that "fantasies about alien impregnation and cross-species offspring, genetic engineering and freak mutations, embryo swapping and cloning, sperm banks and cryonic preservation abound at century's end, bespeaking a cultural imaginary simultaneously fascinated and horrified by reproductive technologies" ("Steps Toward a Millennial Imaginary: Reproductive Fantasies in Cussins's 'Confessions of a Bioterrorist,'" 232).

The widespread fascination, often laced with horror, elicited by human cloning articulates the strength of the cultural imaginary Schwab refers to. Human cloning, indeed, could bring about not only the fulfillment of the female fantasy of bearing children without men, through spontaneous parthenogenesis or cloning, but also the male dream of producing children without the help of women.[1] This scenario could potentially cause an even greater imbalance than that of our still predominantly patriarchal world as far as sexual power structures are concerned. On the one hand, it might decisively pave the road to operative changes as far as women's lives are concerned in our contemporary world by granting them effective reproductive autonomy. On the other hand, unless women reclaim their rightful place as active agents in the decision-making processes and play an effective participatory role in the way these new advances are applied, they might find the potentialities for a gradual movement toward greater gender equality thwarted. If men retain control over sexual politics, as they may be newly shaped by these novel reproductive scenarios, then it is possible that, to a certain extent, the feasibility of human cloning and the introduction of ectogenetic births might be to further a masculinist political agenda and not be put at the service of women's goals and aspirations. As Leon Kass observes, "These new technologies have the potential to take the reproduction away from women's bodies. If women do not act now, we may soon be marching for our *right* to bear children and give birth if we want to. . . . The advent of these new powers for human engineering means that some men may be destined to play God, to re-create other men in their own image."[2] Although Kass's prediction dates from 1972, it is still largely correct. As Robin Rowland bluntly maintains, "Ultimately the new reproductive technology will be used for the benefit of men to the detriment of women" ("Reproductive Technologies: The Final Solution to the Woman Question?" 356), a scenario dramatized in, for instance, Fay Weldon's *The Cloning of Joanna May* (1989) and Anna Wilson's *Hatching Stones* (1991).

In this chapter I wish to consider the sexual politics of human cloning in relation to alternative visions of motherhood and fatherhood and analyze some of the ways in which the advent of cloning might help change women's lives. Cloning will be seen to possess the potential to contribute to the encouragement of social equality and to the destabilization of the long-standing patriarchal "economy of the Same," as Luce Irigaray describes it (*This Sex Which Is Not One*, 74), or, alternatively, to become a genetic weapon of further oppression. The question thus becomes this: Will the availability of human cloning benefit women or, on the contrary, contribute to the perpetuation of their subordination to a still male-dominated medical and scientific establishment?

While some feminists see the prospect of the implementation of human cloning as a threat to women in the sense that it might rob them, as they see it, of their only source of power, the unique gift of motherhood,[3] cloning might, on the other hand, enable both men and women to have their cloned offspring independently of either, thus potentially contributing to a greater equality as far as sex roles are concerned.[4] The still-dominant perception that women are fundamentally childcarers would undergo a gradual change, since with the introduction of cloning and the development of artificial wombs men and women would be equally able to have children with or without a companion of the opposite or the same sex. I believe these alterations, which would be operative at both a biological and a psychological level, would slowly create the conditions for similar job and career opportunities for both sexes, since women would not be limited by their anatomy to becoming mothers only by following the traditional modes of reproduction, unless they so chose. The gradual implementation of these new techniques would inevitably create a very different perception of parenting potential, which, in turn, would lead to a wholly distinct psychological map for humanity. This new context would no longer be articulated in terms of Freudian explanatory principles as far as human traumas, drives, perversions, and taboos are concerned but would come to reflect radically altered family, social, political, and urban circumstances.

I would like to argue thus that recent developments in reproductive technologies might, provided that women struggle to attain their legitimate place as active agents in the decision-making processes, where they are glaringly underrepresented, decisively pave the road to concrete and far-reaching changes to women's lives. These new techniques, among which I will stress human cloning, have opened up the prospect of a revolutionary change in the way we consider sex roles and gendered conceptions of individuality, although a future society that included cloning technology would foreseeably look very different if envisioned by a woman or by a man. Gena Corea suggests that if it ever became possible, cloning might be predominantly used to promote male urges to self-generate, circumventing the woman's participation, in what she regards as "the classic patriarchal myth of single parenthood by the male" (*The Mother Machine: Reproductive Technologies from Artificial Insemination to Artificial Wombs*, 260), a scenario given fictional illustration in, for instance, Fay Weldon's *The Cloning of Joanna May* (1989) and Anna Wilson's *Hatching Stones* (1991). I wish to contest this view by offering and critically examining instances of women's reflections with respect to such visions. I have to acknowledge, however, that it seems likely that if men retain control of sexual politics, given that they will inevitably be reshaped by these novel reproductive conditions, the feasibility of human cloning and the introduction of ectogenetic births might be used in the furthering of a masculinist political agenda and not be put at the service of women's goals and aspirations.[5]

Since my emphasis in this chapter falls on the potential benefits the participation of women in the technological arena may bring to women's lives, I situate myself clearly on the side of like "technophiles" (Nancy Lublin, *Pandora's Box: Feminism Confronts Reproductive Technology*, 23), to borrow Nancy Lublin's word, such as Shulamith Firstone and Donna Haraway, whose theories I will use to help buttress my argument. Lublin defines feminist technophilia as a "veneration for technology because of the belief that it will free women from the burden of reproduction, the primary source of our oppression" (23) and "technophiles" as those feminist thinkers "who are enthusiastic about the supposedly emancipatory nature of technology" (23). Opposed to these technophiles stand the technophobes, who believe that intervention in reproductive technologies is inherently antiwomen. Feminists like Susan Griffin, Mary Daly, and Adrienne Rich emphasize woman's arguably closer connection with nature and celebrate women's bodies as the source of pleasure and not of oppression, indeed as weapons in the struggle for liberation, a position I consider reductionist and essentialist,[6] in direct opposition to Firestone and Haraway's argumentation.

SHULAMITH FIRESTONE'S UTOPIAN VISION

In her radical feminist book *The Dialectic of Sex: The Case for Feminist Revolution* (1970), significantly dedicated to Simone de Beauvoir, for whom woman's biology and reproductive capacities were also the main causes of her oppression, Shulamith Firestone put forward the general outlines for a future society where women would have the same privileges and prerogatives as men, not being "slaves" to their biological destiny, passive vessels, and "two-legged wombs," to use Margaret Atwood's haunting description of women in *The Handmaid's Tale* (146). Not surprisingly, Firestone advocated as an unavoidable cornerstone of her vision the absolute necessity of freeing women "from the tyranny of their biology by any means available, and the diffusion of the childbearing and childrearing role to the society as a whole, to men and other children as well as women" (*The Dialectic of Sex: The Case for Feminist Revolution*, 238), in what would amount to a radical rewriting of Freud's script according to which "anatomy is destiny." Firestone's book recognizes a causal relation between woman's biology, her reproductive capabilities, and the sexual division of labor. In their political subordination to men, women can be equated to the working class in capitalist society. Using Marx and Engels's dialectical materialist method to analyze the "dynamics of sex war" (2) and the conditions necessary to effect a feminist revolution, Firestone at the same time criticizes what she considers the shortcomings of communist theory as far as the oppression of women as a group in the arena of the class struggle is concerned. As she remarks, "An economic diagnosis traced to ownership of the means of production, even of the means of reproduction, does not explain everything. There is a level of reality that does not stem directly from economics" (5). As an alternative, Firestone suggests developing a "materialist view of history based on sex itself" (5), performing an analysis "in which biology itself—procreation—is at the origin of the dualism" (8), the sex dualism that is represented by the two categories of woman and man. As Firestone pertinently argues, "Unlike economic class, sex class sprang directly from a biological reality: men and women were created different, and not equally privileged" (8). As she goes on to stress, "Although, as de Beauvoir points out, this difference of itself did not necessitate the development of a class system—the domination of one group by another—the reproductive *functions* of these differences did" (8). In this context, drawing on biblical imagery, Firestone praises the liberating potential of technology, stressing that the double curse "that man would toil by the sweat of his brow in order to live, and woman would bear children in pain and travail" (242) would be lifted through technology. As Nancy Chodorow, in a related vein, argues, "Women's mothering is central to the sexual division of labor. Women's maternal role has profound effects on women's lives, on ideology about women, on the reproduction of masculinity and sexual inequality, and on the reproduction of particular forms of labor power. Women as mothers are pivotal actors in the sphere of social reproduction" (*The Reproduction of Mothering: Psychoanalysis and the Sociology of Gender*, 11).

In spite of her antitechnological stance, Adrienne Rich similarly engages with Marxist rethoric in terms strongly reminiscent of Firestone's when she argues,

> The repossession by women of our bodies will bring far more essential change to human society than the means of production by workers. The female body has been both territory and machine, virgin wilderness to be exploited and assembly-line turning out life. We need to imagine a world in which every woman is the presiding genius of her own body. (*Of Woman Born*, 285)

Unlike de Beauvoir and Firestone, however, Rich considers the experience of motherhood as a fundamental reservoir of pleasure and power. According to Rich, not only do women's biology and reproductive capacities not necessarily lead to oppression, but they can be a potent source of jouissance, of libidinal pleasure, an aspect that is not taken into account by either de Beauvoir or Firestone.

For women to repossess their bodies, Firestone argues, a formidable upheaval in societal structures and engagement with technology is fundamental. Pushing the parallels with Marxism she has been using even further, Firestone forcefully declares that

> just as to assure elimination of economic classes requires the revolt of the underclass (the proletariat) and, in a temporary dictatorship, their seizure of the means of *production*, so to assure the elimination of sexual classes requires the revolt of the underclass (women) and the seizure of control of *reproduction*: not only the full restoration to women of ownership of their own bodies, but also their (temporary) seizure of control of human fertility—the new popula-

tion biology as well as all the social institutions of childbearing and childrearing. (*The Dialectic of Sex: The Case for Feminist Revolution*, 10–11)[7]

Firestone further elaborates on this revolutionary vision, putting forward what she considers the desirable measures that would have to be implemented in order to achieve her blueprint for an egalitarian, socialist-feminist society. Her insistence on the possibility of divorcing motherhood from being solely attached to woman is translated into her anticipatory fantasy, according to which,

> The reproduction of the species by one sex for the benefit of both would be replaced by (at least the option of) artificial reproduction: children would be born to both sexes equally, or independently of either, however one chooses to look at it; the dependence of the child on the mother (and vice versa) would give way to a greatly shortened dependence on a small group of others in general, and any remaining inferiority to adults in physical strength would be compensated for culturally. The division of labor would be ended by the elimination of labor altogether (cybernation). The tyranny of the biological family would be broken. (*The Dialectic of Sex: The Case for Feminist Revolution*, 11)

As Firestone observes, the end of the tyranny of the biological family would also spell the cessation of the "psychology of power" (*The Dialectic of Sex: The Case for Feminist Revolution*, 11) on which it is grounded. Female biology, then, would no longer mean motherhood as the only destiny open to most women.

In turn, Christine Battersby's notion of a feminist metaphysics, which include "an emphasis on birth," articulated in *The Phenomenal Woman: Feminist Metaphysics and the Patterns of Identity* (1998), works as a partial corrective to Simone de Beauvoir's and Firestone's demand for a radical alteration of and alternative to the vision and representation of woman's body as exclusively devoted to pregnancy and maternity. What I wish to stress here, in response to Christine Battersby's feminist metaphysics, is the need to start theorizing the patterns of individuation and identity of a being who might literally not be born,[8] but rather develop inside an artificial womb.[9]

The fact that there are two virtually opposed views about women's reproductive capacities and how society should deal with them, however, does not mean that Firestone's radically new concept of woman's role in society cannot coexist with more traditional feminine participation in procreation and childrearing.[10] On the one hand, there are those feminist thinkers who, like Adrienne Rich, see a woman's biology and her childbearing responsibilities as conducive to a sense of empowerment over her own body and, to a certain extent, over men. However, in this respect we need to ask why, then, in spite of the traditional sacralization of motherhood, women are very often drastically restricted in their societal expectations precisely because of those very reproductive capabilities. As Rosalind Pollack Petchesky pertinently asks, "Can feminism reconstruct a joyful sense of childbearing and maternity without capitulating to ideologies that reduce women to a maternal essence? Can we talk about morality in reproductive decision-making without invoking the spectre of maternal duty?" ("Foetal Images: The Power of Visual Culture in the Politics of Reproduction," 79). It is these problems that both Firestone and Haraway, in their different theoretical strategies and practical suggestions, try to solve. As Donna J. Haraway, in tune with Firestone, argues,

> A concept of a coherent inner self, achieved (cultural) or innate (biological), is a regulatory fiction that is unnecessary—indeed, inhibitory—for feminist projects of producing and affirming complex agency and responsibility. A related "regulatory fiction" basic to Western concepts of gender insists that motherhood is natural and fatherhood is cultural: mothers make babies naturally, biologically. ("A Cyborg Manifesto: Science, Technology, and Socialist-Feminism in the Late Twentieth Century," 135)

In an analogous vein, Angela Carter argues that the widespread theory of "maternal superiority is one of the most damaging of all consolatory fictions and women themselves cannot leave it alone" (*The Sadeian Woman*, 106), although this situation may drastically change with the implementation of human cloning and artificial wombs.

> Not of Woman Born
>
> Donna Haraway, "A Cyborg Manifesto: Science, Technology, and Socialist-Feminism in the Late Twentieth Century," 177

Donna J. Haraway's "ironic political myth" ("A Cyborg Manifesto: Science, Technology, and Socialist-Feminism in the Late Twentieth Century," 149) in which the cyborg is the main protagonist, provides a series of helpful ideas that can be said to roughly work in the same direction as some of the possibilities that human cloning holds for women in terms of a political strategy to help furthering their position in a masculinist society.

According to Haraway, the cyborg eludes conventional humanist concepts of woman as childbearer and raiser, destabilizing the kind of dualistic thinking that privileges man over woman, self over other, culture over nature, dualisms that have been "systemic to the logics and practices of domination of women, people of colour, nature, workers, animals—in short, domination of all constituted as others" ("A Cyborg Manifesto: Science, Technology, and Socialist-Feminism in the Late Twentieth Century," 177). The cyborg also evades traditional notions of the nuclear family, calling into question such long-established psychoanalytic concepts as the Oedipus and Electra complexes. With the gradual dissolution of the traditional nuclear and biological families, relations between parents and putatively cloned children might start developing dynamics distinct from those laid down by Freud. What genetic relation would a cloned child have with his or her parent? Would he or she be a son or daughter, twin brother, or twin sister of the parent? These and other questions will inevitably arise.[11]

For Haraway, cyborgs[12] "have more to do with regeneration and are suspicious of the reproductive matrix of most birthing" ("A Cyborg Manifesto: Science, Technology, and Socialist-Feminism in the Late Twentieth Century," 181), a scenario that is dramatized in, for instance, Marge Piercy's Mattapoisett section in *Woman on the Edge of Time* and Lisa Tuttle's "World of Strangers," where "the reproductive matrix of most birthing" is deconstructed. I thus situate my argument about the potentially empowering consequences for women of human cloning along the lines laid down by Haraway, who describes her cyborg myth as being about "transgressed boundaries, potent fusions, and dangerous possibilities which progressive people might explore as one part of needed political work" ("A Cyborg Manifesto: Science, Technology, and Socialist-Feminism in the Late Twentieth Century," 154).

Donna Haraway analyzes scientific discourses as both constructed and as "instruments for enforcing meanings" ("A Cyborg Manifesto: Science, Technology, and Socialist-Feminism in the Late Twentieth Century," 164). In tune with Firestone, Haraway argues that "one important route for reconstructing socialist-feminist politics is through theory and practice addressed to the social relations of science and technology, including crucially the systems of myth and meanings structuring our imagination" (163). The relations between science and technology constitute a material reality that women need to be aware of—not fear or disparage. These relations are "rearranging" categories of race, sex, and class; feminism needs to take this into account. Indeed, Haraway's analysis of "women in the integrated circuit" tries to suggest, without relying too much on the category of "woman" (as a natural category), that as technologies radically restructure "life" on earth, "women" do not, and are not, through education and training, learning to control these technologies, to "read these webs of power" (170). A socialist-feminist politics must therefore address these restructurings. As Haraway points out, "Who controls the interpretation of bodily boundaries in medical hermeneutics is a major feminist issue" (160). Her often-reiterated exhortation for women to participate in the making of science is inextricably linked with the control of webs of power, since it is "the production of science and technology that constructs scientific-technical discourses, processes, and objects" (169) and is instrumental in the creation of "the new world, just as it has participated in maintaining the old one" (*Simians, Cyborgs, and Women: The Reinvention of Natire*, 68).

In related vein, Haraway addresses the issue of women's victimization, remarking that the traditional plots that shape Western culture

> are ruled by a reproductive politics—rebirth without flaw, perfection, abstraction. In this plot women are imagined either better or worse off, but all agree they have less selfhood, weaker individuation, more fusion to the oral, to Mother, less at stake in masculine autonomy, a route that does not pass through Woman, Primitive, Zero, the Mirror Stage and its imaginary. It passes through women and other present-tense, illegitimate cyborgs, *not of Woman born*, who refuse the ideological resources of victimization so as to have a real life. ("A Cyborg Manifesto: Science, Technology, and Socialist-Feminism in the Late Twentieth Century," 177; emphasis mine)

Haraway's discussion of cyborg politics, although not referring to human cloning in particular, can be said to pertinently apply and buttress the argument I have been developing here. She states, "Sexual reproduction is one kind of reproductive strategy among many, with costs and benefits as a function of the system environment. Ideologies of sexual reproduction can no longer reasonably call on notions of sex and sex role as organic aspects in natural objects like organisms and families" ("A Cyborg Manifesto: Science, Technology, and Socialist-Feminism in the Late Twentieth Century," 162). From this perspective, then, human cloning, I believe, can be seen as a potentially liberating alternative to the rigid boundaries imposed on women by those ideologies of sexual reproduction.

Firestone's "cybernetic feminism" (*The Dialectic of Sex: The Case for Feminist Revolution*, 238) and Haraway's vision of a "socialist-feminist culture" ("A Cyborg Manifesto: Science, Technology, and Socialist-Feminism in the Late Twentieth Century," 150) can then be brought together in a productive symbiosis that will help us reflect on the fictional works previously named that through prophetically anticipating many medical and technological procedures and their repercussions on human life will provide an invaluable imaginative blueprint with which to assess critically imminent developments and their implications as far as the sexual politics of the near future are concerned.

WOMEN'S SCIENCE, WOMEN'S BODIES

The discourse of science has been an object of scrutiny on the part of many feminist critics, who have seen it as heavily male gendered and catering mostly to male political agendas. Evelyn Fox Keller and Sandra Harding are among the most influential critics of what they see as the sexist way science has operated so far. Drawing principally on the work of Nancy Chodorow and Dorothy Dinnerstein, one of Evelyn Fox Keller's main arguments is that the predominant philosophy of science practices and techniques is strongly masculine and individualistic. In order to counterbalance that attitude, Keller, like Haraway, calls for a greater participation of women in the research and practice of science so that those sexist paradigms can be changed.[13]

Sandra Harding, for her part, argues for the need "to produce a feminist science—one that better reflects the world around us than the incomplete and distorting accounts provided by traditional social science" ("Conclusion: Epistemological Questions," 318), taking into account the specifics of gender, race, and class, which inevitably give form to our experience of the world around us. Following on from this insight, Harding defends a type of scientific practice based on an ethics of care, of greater relational habits not grounded on exploitation, and, like Keller, alerts women to the need for a more active agency in the construction and practice of science.

Keller's notion of a gender-neutral science found a strongly receptive echo in many feminist activists who demand the kind of policy that will be sympathetic to women's needs and aspirations. Indeed, as Catherine MacKinnon contends, in order for women to achieve greater control of their bodies and hence a greater political power to subvert male domination, they should be able to start exerting that control earlier. As she insists, "If women are not socially accorded control over sexual access to their bodies, they cannot control much else about them" ("Toward Feminist Jurisprudence," 616).

In this context, I believe the possibility of having their own cloned children would be a potential way for women to possess greater control over their bodies and reproductive choices, as I have suggested. Michelle Stanworth, in a related vein, notes that

> The thrust of feminist analysis has been to rescue pregnancy from the status of the "natural"—to establish pregnancy and childbirth not as a natural condition, the parameters of which are set in advance, but as an accomplishment which we can actively shape according to our own ends. . . . In the feminist critique of reproductive technologies, it is not technology as an *artificial* invasion of the human body that is at issue—but whether we can create the political and cultural conditions in which such technologies can be employed by women to shape the experience of reproduction according to their own definitions. ("The Deconstruction of Motherhood," 34–35)

Indeed, it is never redundant to stress that cloning would be fully empowering for women only if it went on a par with social and economic independence, as well as political parity, so as to ensure that women's problems would receive the adequate amount of attention that would go with appropriate representation in the institutional organs with power of decision. This aspect is equally stressed by Michele Barrett who, in *Women's Oppression Today* (1980), observes that "the way in which the biology of human reproduction is integrated into social relations is not a biological question: it is a political issue" (76).

Similar concerns are articulated by Firestone, as we have seen, who acknowledges that

> Though the sex class system may have originated in fundamental biological conditions, this does not guarantee, once the biological basis of their oppression has been swept away, that women and children will be freed. On the contrary, the new technology, especially fertility control, may be used against them to reinforce the entrenched system of exploitation. (*The Dialectic of Sex: The Case for Feminist Revolution*, 10)

This powerful call for women's effective participation in medicine, biology, and all stages of the decision-making process is one of the most forcefully articulated requirements put forward by feminists in their struggle for social equity and justice. Hilary Rose is similarly critical of the "pervasive conservatism [which] lies at the heart of the debate about the so-called new challenges to ethics posed by the new science and technology of reproduction" ("Victorian Values in the Test-tube: The Politics of Reproductive Science and Technology," 172). As she goes on to observe, the "problem for feminists is that we want to resist specific oppressive technologies while at the same time working to change nothing less than the values and structures of science. Thus our debates must be located within an understanding of the biologically determinist direction of modern science and medicine which contain within them fixed notions of woman's and man's natures" (172).

Donna Haraway's prophetically cautionary words carry a crucial message that I want to expand on in this book:

> A socialist-feminist science will have to be developed in the process of constructing different lives in interaction with the world. Only material struggle can end the logic of domination. . . . I do not know what life science would be like if the historical structure of our lives minimized domination. I do know that the history of biology convinces me that basic knowledge would reflect and reproduce the new world, just as it has participated in maintaining an old one. (*Simians, Cyborgs, and Women: The Reinvention of Nature*, 68)

Haraway's call for a much more significant representation of women in the effective shaping and evolution of the sciences of life so that the balance of domination can be evened out and the new medical practices might not reinscribe patriarchal ascendancy and control has to be heeded and put into practice so that future society does not perpetuate deeply questionable political power asymmetries.

In an analogous vein, Luce Irigaray, whose trajectory includes a longstanding involvement with the Italian feminist communists and the women of the Italian left-wing parties, as well as with issues related with women's bodily experiences, is worried that instead of helping to free women from their subordination to patriarchy, the recent reproductive technologies might further accentuate the traditional view that "the framework for women's existence is exclusively maternal" (*Je, Tu, Nous: Toward a Culture of Difference*, 135). Irigaray further remarks that "there's a real risk that some women, who call themselves freed from their nature such as it was defined by patriarchy, will once again subject themselves body and 'soul' to this variant on their fate called artificial procreation" (135). Irigaray's doubts find powerful vocalization:

> Test-tube mothers, surrogate mothers, men engendering futuristically (in their intestines): what next? Will all this help us get away from the pressure to have children, our sole sexual "vocation" according to the patriarchs, so as to get to know ourselves, to love and create ourselves in accordance to our bodily differences? (*Je, Tu, Nous: Toward a Culture of Difference*, 135)

Cloning, I believe, would provide a possible answer to Irigaray's distrust of some new reproductive technologies, which she fears might go on reproducing the same male webs of power.[14] Indeed, in "So When Are We to Become Women?" (in *Je, Tu, Nous: Toward a Culture of Difference*) Irigaray observes, "Today's scientists poring over their test tubes to decide a woman's fertility or fertilization very much resemble theologians speculating about the possibility of a female soul or about the point at which the fetus' soul comes into existence. The approach is similar, perhaps worse. And if need be, some of these scientists will be women" (134–135).

Irigaray further suggests that having children without men, contrary to some women's expectations, will not release them from their collusion with the male, with patriarchy. She argues that it "still amounts to defining oneself in relation to the other sex rather than oneself; it amounts to thinking of oneself without the other and not to thinking one's self, thinking about oneself, about myself as a woman (*à moi-elle*), about ourselves by our selves as women (*à nous et avec nous-elles*)" (*Je, Tu, Nous: Toward a Culture of Difference*, 133). However, Irigaray has not considered the possibility of human cloning and the inevitable changes in perspective and societal dynamics it will necessarily bring about. Indeed, the first item in Irigaray's list that describes the three ways in which man "is not absent from artificial procreation" (134), which reads, "First, a child conceived without a male sexual partner nevertheless depends upon male semen" (134), is no longer the case with cloning, a technique which, when fully developed, will enable a woman to gestate her own baby, parthenogenetically, as it were. She might then give birth to her own self as a baby, in what would be the apex of narcissism, of "amour de soi," or to some other woman's child, having another woman's embryo implanted in her womb, amid several other potential scenarios.

The prospect of human cloning has effectively forced renewed attention on the sexual politics of reproductive technologies, as well as the potential social and family-related scenarios they may bring about. As Julia Kristeva remarks in *Revolt, She Said* (2002), in the context of reflecting on genetic manipulation and cloning, "We'll need male cells, and it doesn't matter which ones, but you can't do without eggs. Whatever the horrors of science fiction—not all so fanciful actually—it seems that the future of our species depends more and more on women" (69). However, as she later points out, "Women hold the key to the species on the condition they share it with men" (94), adding in jocular fashion, "at least given the present state of science" (94). As Kristeva insists, "Beyond spermatozoa and ova, since they're still indispensable, haven't we sufficiently insisted . . . on the fact that a human being, a speaking subject, emerges from a tripartite structure where the father plays an essential role?" (94).

GYNOGENESIS AND ANDROGENESIS

One variant made potentially feasible with human cloning would be a child with two genetic mothers and no father, a scenario many lesbian couples would welcome.[15] Elizabeth Sourbut, following lesbian feminist writer Ryn Edwards, called this possibility gynogenesis, although the process envisaged by her did not contemplate the hypothesis of human cloning, relying instead on two ripe eggs from each woman's ovaries and subsequently fusing their genetic material ("Gynogenesis: A Lesbian Appropriation of Reproductive Technologies").[16] A similar scenario would apply to gay male couples using androgenesis, although in their case an egg with its genetic material removed, as well as a womb, whether a real or an artificial one, would have to be used.[17]

With human cloning, then, a world of only women, like those idealized by Mary E. Bradley Lane in *Mizora*, Charlotte Perkins Gilman in *Herland*, Joanna Russ in the Whileaway section of *The Female Man*, and James Tiptree Jr. in "Houston, Houston, Do You Read?" among others, might come into being, as well as, alternatively, one of only men, or a world with an accentuated gender imbalance.[18] These are obviously radical futuristic scenarios that, however, need to be carefully considered if we are to take some of the serious implications of cloning to its ultimate and apocalyptic consequences.[19] These possibilities, indeed, serve only to alert us even more pressingly to the urgent need I have been advocating here to have an increasing number of women in decision-making positions in science and medicine, which have been traditionally gendered as masculine, taking effective charge of these reproductive technologies that will so drastically reshape women's existence. The regendering of science and technology, called for by such feminist thinkers as Shulamith Firestone, Sandra Harding, Evelyn Fox Keller, and Donna Haraway, should thus be considered not simply as a matter for philosophical or literary speculation but as a fundamental political goal, a priority to be privileged in the political agenda of present and future governments.

MOTHERHOOD DECONSTRUCTED

Woman has been for so long inextricably associated with motherhood that for many to contemplate the idea of woman as nonmother would appear as a scandal, socially unacceptable, impossibly revolutionary, and potentially deeply threatening to the patriarchal stronghold.[20] Motherhood, as has often been noted, has served throughout the ages to glorify woman as mother but at the same time to subordinate her according to the argument that as men cannot biologically fulfill the function of bearing a child, woman has to consecrate most of her time, as well as her ambitions and inclinations, to the higher good of society, its perpetuation and well-being.[21]

In *Patterns of Dissonance: A Study of Women in Contemporary Philosophy*, Rosi Braidotti points out that woman, "whether she likes it or not, only exists in her culture as a potential mother" (260), a fact that if it has been the source of one of the only avenues to empowerment women have experienced throughout the ages, has also simultaneously constituted the basis of their enslavement to anatomy. Reflecting on woman's role of mother, Julia Kristeva observes, "If it is not possible to say of a *woman* what she *is* (without running the risk of abolishing her difference), would it perhaps be different concerning the *mother*, since that is the only function of the 'other sex' to which we can definitely attribute existence? And yet, there too, we are caught in a paradox" ("Stabat Mater," 234). These difficulties, inherent in the contested site that is motherhood and the many theorizations surrounding it, are described by Patrice DiQuinzio as precisely a "paradoxical politics of mothering" (*The Impossibility of Motherhood: Feminism, Individualism, and the Problem of Mothering*, xvii), one that would "take up a wide variety of issues related to conception, pregnancy, birth, and child rearing, but it would recognize that it cannot offer a completely coherent and consistent position on these issues" (248). The advent of the new reproductive technologies we have been alluding to would bring into greater relief many of the paradoxes attendant upon these various conceptions of motherhood.

In "Stabat Mater," Julia Kristeva provides a detailed and critical look at the image of the "virginal maternal" (257), associating it with notions of essentialism and motherhood, which are problematized in the parallel columns of text and considering the manifold paradoxes attached to the concept of "virgin motherhood" (259). Drawing on Marina Warner's *Alone of All Her Sex: The Myth and the Cult of the Virgin Mary* (1976) as a privileged intertext, Kristeva ends her piece calling for a "*herethics*" (263), a notion that would fuse the kind of ethics that is specifically feminine, "hers," with the subversive notion of heresy, of transgression. In this context, it is fundamental to consider the role the Kristevan notion of the virginal maternal might come to play in a society where human cloning would be a viable and socially accepted reproductive choice, where fatherless maternity could become widespread, as well as the paradoxes that need to be negotiated when conceptualizing those new scenarios.[22]

With human cloning and ectogenesis, as well as the sharing of the reproductive capacity with men, I believe that instead of relinquishing the source of power that motherhood has been perceived in some respects as yielding, as some critics would argue,[23] women would, on the contrary, achieve a greater equality by dint of that very interchange of roles, in particular if and when ectogenesis became the norm.

With the introduction of artificial wombs and the possibility of women bearing their own children without any male intervention (semen would not be needed for a woman's ovum to start dividing into an embryo), men would see their prerogatives and claims for (legal) domination over women and their offspring severely curtailed. It goes without saying that these potentialities likewise would provide a man with the possibility of having his own child without a woman's help, thus putatively originating the development of an imbalance in the relative number of male and female births, on the one hand (as in Anna Wilson's dystopian novel *Hatching Stones*), and the appropriation or, rather, perpetuation of the power of decision over the future shape of society, on the other, if women are not equally at the center of the processes of decision making. This is obviously an extreme scenario, just as the idea that, given these tantalizing possibilities, women would immediately avail themselves of these opportunities and decide not to become pregnant any longer, leaving to the laboratory and the artificial womb the task of bearing their children, is rather far-fetched. This futuristic prospect is many decades away, which is not to say that in the relatively near future many women might not be able to decide, for medical or other reasons, to profit from these new resources, if indeed they become available.[24] These new reproductive technologies would inevitably bring about a radical rewriting and revising of gender dichotomies, opening the way to completely new sexual dynamics at work in society. Given the possibility, many women would foreseeably profit from the available technology to choose when and with whom to have a baby, for example, or to give birth to their own cloned children, completely on their own.

Cloning technology applied to human beings would thus predictably have far-reaching effects in all social and family dynamics. Men who decided to have cloned offspring (assuming that artificial wombs would be available) could thus choose to have their children completely on their own, as though there were no women in the world. Conversely, a woman could do exactly the same, and much earlier, since she would not have to depend on ectogenesis to bear and give birth to a baby, if she so chose.[25] These can be seen as very powerful motivating factors that might lead a woman to opt for having a cloned child, brought to term either inside her own womb or in an artificial one, circumstances that would potentially and gradually lead to a greater parity with men and equality of opportunities for women in the social structure, as I have been arguing. However, this relative symmetry in access to reproductive roles could lead, nonetheless, to a predominance of a masculinist political agenda, as already mentioned, perpetuating the status quo of men in an androcentric society. As J. Raymond observes, the new reproductive techniques can be seen as a powerful means for men to wrest "not only control of reproduction, but reproduction itself" ("Preface," 12) from women, a point that is also stressed by Michelle Stanworth, who notes that new reproductive technologies "are the vehicle that will turn men's illusions of reproductive power into a reality" ("The Deconstruction of Motherhood," 16).[26] The implications embedded in these developments as far as women and pregnancy are concerned are thus much more extensive than the consideration of their biological consequences might lead us to suspect. As Michelle Stanworth goes on to note, in tune with my reflections previously mentioned, "Motherhood as a unified biological process will be effectively deconstructed" (16).[27]

Stanworth summarizes the thrust of the effects of these technologies in the following terms: "Through the eventual development of artificial wombs, the capacity will arise to make biological motherhood redundant. Whether or not women are eliminated . . . the object and the effect of the emergent technologies is to deconstruct motherhood and to destroy the claim to reproduction that is the foundation of women's identity" ("The Deconstruction of Motherhood," 16).[28] These fears are similarly sounded by Julien S. Murphy, who addresses some of the issues linked with ectogenesis. Engaging with Firestone's proposals in *The Dialectic of Sex*, Murphy believes that "we are far from achieving the sort of revolution required in order for ectogenesis to be liberating" ("Is Pregnancy Necessary? Feminist Concerns about Ectogenesis," 194), going on to argue that ectogenesis "would definitely not replace pregnancy. For if it were so, that would suggest that women's bodies had been judged unfit for pregnancy. Is the best way to abolish sexism a method that downgrades a female capacity—pregnancy?" (194–195). I see two different aspects to this question that Murphy is not, from my perspective, addressing. On the one hand, I do not believe that for a woman to have the possibility to let *her own child* develop outside her womb, *if she so chose*, would in any way downgrade the female capacity to bear children. On the contrary, I consider that potentiality a very liberating prospect for a woman, who might want to have a child but be unable to for health reasons, or might prefer, for an extensive panoply of other reasons pertaining to her own inner wishes or social circumstances, not to carry a child for nine months in her womb. Related to this possibility but a separate issue is the prospect of human cloning technology becoming available before ectogenesis. In this case, women's bodies would still be indispensable for pregnancies to be carried to term, a scenario that could make women absolutely autonomous as far as conception and pregnancy are concerned, while men would still be dependent on the availability and willingness of women to carry their children. Peter Singer and Deane Wells, however, accurately represent my point of view in their defense of ectogenesis when they write,

> Can it seriously be claimed that in our present society the status of women rests entirely on their role as nurturers of embryos from conception to birth? If we argue that to break the link between women and childbearing would be to undermine the status of women in our society what are we saying about the ability of women to obtain true equality in other spheres of life? We, at least, are not nearly so pessimistic about the abilities of women to achieve equality with men across the broad range of human endeavour. For that reason, we think women will be helped rather than harmed by the development of a technology that makes it possible for them to have children without being pregnant. (*Making Babies*, 129)

Shulamith Firestone and Marge Piercy would heartily agree with this assessment.

From the perspective I have been presenting, human cloning can thus be seen to constitute a very important and empowering step forward for women, a fundamental strategic move on the way to egalitarian rights with men. Might cloning be, in fact, one of the stepping stones to enable such egalitarianism gradually to arise? I tend to believe so, as long as it is used with the necessary caution and common sense with respect to the unavoidable ethical and moral issues. As many critics have abundantly stressed, however, women have to play an actively participatory role in the forging and implementation of new technological advances in science in general and in the networks of power. There is a pressing need to reverse Haraway's pessimistic position, according to which there is no place for woman in these networks, "only geometries of difference and contradiction crucial to women's cyborg identities. If we learn how to read these webs of power and social life, we might learn new couplings, new coalitions" ("A Cyborg Manifesto: Science, Technology, and Socialist-Feminism in the Late Twentieth Century," 170). Woman's insertion into these webs of power, her greater agency in the scientific and medical arenas, constitute a fundamental step toward the implementation of a political agenda that will contemplate women's welfare and potential new reproductive scenarios in our cyberspace age.

> Feminine nonwomen conceived by male mothers.
>
> Mary Daly, *Gyn/Ecology: The Metaethics of Radical Feminism*, 53

In provocative fashion, Mary Daly muses that with technological developments what she terms "phallotechnic progress" (*Gyn/Ecology: The Metaethics of Radical Feminism*, 53) will likely advance

> toward the production of three-dimensional, perfectly re-formed "women," that is, hollow holograms. These projections, or feminine nonwomen, the replacement for female Selves, could of course eventually be projected in "solid" form—as solid waste products of technical progress, as robots. Eventually, too, the "solid" substitutes could be "flesh and blood" (not simply machines), produced by such "miraculous" techniques as total therapy . . . transsexualism, and cloning. (*Gyn/Ecology: The Metaethics of Radical Feminism*, 53)[29]

As Daly further asserts in critical and pessimistic fashion, the "march of mechanical masculinist progress is towards the elimination of female Self-centering reality. Whether or not our re-placements are materially 'hollow' or 'solid' is not the ultimate issue. These are simply different ways of describing the absence of Female Depth, of spirit, in *feminine nonwomen conceived by male mothers*" (53; emphasis mine). Although Daly goes on to note that this state, which she calls "robotitude" (53), is not exclusive of women, it seems to me her technophobia is preventing her from apprehending the advantages cloning might bring to both men and women, and that a cloned woman need not be man's slave—on the contrary. Even if a man wished to create a (cloned) woman to serve him, that woman would have her own personality and legal rights, so it would be extremely unlikely for such scenarios to materialize in the twenty-first century. In analogous fashion, males created to serve similar purposes as far as women are concerned would consent to satisfy those women's needs only if they agreed to do so and would not be transformed into robot slaves, such as the android in Joanna Russ's *The Female Man*, who exists only to supply Jael's desires.

SOME LITERARY REPRESENTATIONS

In *The Dialectic of Sex: The Case for Feminist Revolution* (1970) Firestone claimed that at the time we did not have "even a literary image of this future society" and that there was "not even a *utopian* feminist literature in existence" (227). This statement leaves out several feminist utopias written by women in the nineteenth and early twentieth centuries, probably the most salient being Mary E. Bradley Lane's *Mizora* (1880) and Charlotte Perkins Gilman's *Herland* (1915). Both of these works depict female-only societies where women reproduce parthenogenetically, giving birth only to women. These peaceful, orderly, classless, pastoral women-only societies, which do not need to take male politics into consideration, might be a version of wish-fulfillment on many women's parts. This is not, however, the kind of separatist scenario Firestone has in mind. She offers a concrete set of proposals that would effectively establish the agenda and pave the way for measurable changes in our contemporary world. Her suggestions and recommendations may sound just as utopian as Lane's or Gilman's, but they are grounded in what are now predictable short-term developments in biotechnology. What to her readers sounded like impossibly futuristic scenarios, drawn from a science fiction context, appear to us at the beginning of the third millennium almost feasible.

In "Stabat Mater," Kristeva claims that "motherhood . . . today remains, after the Virgin, without a discourse" (262). With cloning, however, the fantasy of a virginal maternal might come to be fulfilled, providing a revolutionary new vision to the concept of motherhood and potentially enabling new feminine discourses to arise, alongside Kristeva's call for a "*herethics*" (263). Kristeva notes that "what is needed in the West today is a reevaluation of the 'maternal function,' seeing it not as explosive and repressed but as a source of practices considered to be marginal (such as 'aesthetic' practices) and a source of innovation. Men as well as women are seeking this, and they are turning in particular to the women's movement" (*Julia Kristeva Interviews*, 108). As Toril Moi similarly stresses, with relation to "Stabat Mater," "There is . . . an urgent need for a 'post-virginal' discourse on maternity, one which ultimately would provide both women and men with a new ethics" (*The Kristeva Reader*, 161). This discourse can be seen tentatively at work in the projecting of new alternatives for women and men in the "*utopian* feminist literature" Firestone lamented being unable to find in 1970, which was just starting to take shape, responding to the same perceived need for change Firestone was heeding. Indeed, the 1970s saw the publication of such works as Monique Wittig's *Les Guérillères* (1969), Joanna Russ's *The Female Man* (1975), Naomi Mitchison's *Solution Three* (1975), Pamela Sargent's *Cloned Lives* (1976), Kate Wilhelm's *Where Late the Sweet Birds Sang* (1976), Marge Piercy's *Woman on the Edge of Time* (1976), Sally Miller Gearhart's *The Wanderground* (1978), and Suzy McKee Charnas's *Motherlines* (1978), to cite only some of the most representative. There is clearly a common impulse at work here, a drive toward a better, more fulfilled life for women, which goes hand in hand with a strongly operative ecological defense of the planetarian ecosystems. All these novels engage to a greater or lesser extent with Firestone's concerns, providing a fictional dramatization of her blueprint for a utopian society. Many of the narratives that these women writers went on to produce in the following decades stand in a line of continuity with the earlier ones, giving vivid expression to these ongoing issues. In order to reflect critically on these important themes, which stand out as the most pressing concerns women perceive in terms of enabling them to progress toward a more egalitarian societal organization, I have chosen to concentrate on a cluster of texts that deal in greater detail with the implications of human cloning or versions of it, namely, as far as motherhood is concerned, and draw some inevitable consequences as to how that development may come to affect our contemporary society. I will be looking at Marge Piercy's *Woman on the Edge of Time* (1976), as well as a spate of recent fiction that has reengaged with the theme of human cloning, opening up new scenarios and alternatives, such as Naomi Mitchison's "Mary and Joe," Lisa Tuttle's "World of Strangers," Felicia Ackerman's "Flourish Your Heart in This World," and Martha C. Nussbaum's "Little C," as well as two stories by Nancy Kress, "To Cuddle Amy" and "Sex Education."

"WHY DO WOMEN BOTHER TO HAVE SONS?"

In *The Redundant Male: Is Sex Irrelevant in the Modern World?* Jeremy Cherfas and John Gribbin ask the controversial question, "Why do women bother to have sons?" (1) and in the course of their argument put forward a scenario very similar to the one described in Lane's *Mizora*, Gilman's *Herland*, and Naomi Mitchison's *Memoirs of a Spacewoman*:

> Now imagine that a mutant female arises, that is, one who differs genetically from the bulk of the population. She can do without males and still have young. Her offspring will be females who, like their mother, can reproduce without the help of males by a process called parthenogenesis. . . . Because she does not produce males, such a female would have twice as many daughters as the other females; and because only daughters put much effort into raising offspring the mutation would spread very rapidly indeed. Within a few generations all the females will be asexual. . . . In such a species males would have become redundant. (*The Redundant Male: Is Sex Irrelevant in the Modern World?* 3)

As Cherfas and Gribbin go on to explain, "Females *could* reproduce parthenogenetically if the egg cells could be induced to develop on their own. But males can never do so" (6).[30]

In a related vein, Cherfas and Gribbin point out that

> doctors working on test-tube fertilisation have discovered, much to their surprise, that human eggs, all on their own and with no help from a sperm, can divide. It

requires no great leap of imagination to see that in the very near future women might be quite capable of dispensing with men altogether . . . if an adequately supported team were to put their minds to the problem it would be no time before women could do without men entirely. They would be able to clone themselves. (*The Redundant Male: Is Sex Irrelevant in the Modern World?* 177–178)

These words are one more corroboration of my own vision of the kind of empowerment cloning could grant women.

A version of this scenario is provided in Naomi Mitchison's "Mary and Joe" (1962), a story about spontaneous parthenogenesis, dealing with a woman as a sole genetic mother, thus envisaging what we might describe as "the mother as sole parent" mythology, to reverse Jane Murphy's concept of "'The Father' as the sole parent in creation" ("From Mice to Men? Implications of Progress in Cloning Research," 84). In Mitchison's "Mary and Joe," Mary, a geneticist working on surgical transplants, has to remove strips of her own skin and graft them onto the face and body of her political activist daughter Jaycie, who has been in an accident during political riots. In normal circumstances, a skin graft from mother to daughter would not take, but since Mary knew that she and Jaycie were genetically identical she was not afraid to carry out that procedure, which saved Jaycie's life. Mary had, however, to explain to the doctor and the nurse at the hospital, as well as to her husband Joe, that Jaycie did not have a father, that she was a virgin at the time she conceived Jaycie and that she could not explain what had triggered it, what had made the cells in the ovum start to divide to produce an embryo.

"Mary and Joe" can be read as a feminist revision of the biblical annunciation and nativity stories. The name of the protagonist, Mary, evokes the Virgin Mary, and biblical echoes are scattered throughout the tale. In her desire to be active and make a difference, Jaycie invokes a biblical passage where the sons of god and the daughters of men are mentioned:[31] "'I suppose, Mother, that's what it means to be a Son of God, as they used to say. You go straight to the light. You know.' And Mary had said yes and had felt something gripping at her, a rush of adrenaline no doubt! Jaycie had said: 'No daughters of God, of course!' and had laughed a little. And then she had stood up and looked straight toward her mother and said: 'But I too, I know. Directly'" (165). The implication here seems to be that in her different origin, a fatherless daughter, she can be seen as initiating another, parallel mythology to that of the son of god, in this case the daughter of woman and not of men, but with visionary and redeeming powers similar to those of Jesus Christ. Indeed, in this reworking of the virgin's story, Mary, instead of giving birth to a son, will have a girl child, in her own way a new messiah, for she will be the leader of a revolutionary political movement.

Mary's version of her mysterious conception, a subversive, epiphanic version of the Annunciation, is described in terms of the unexplainable, the inscrutable and mysterious: "It might have been imagination. It must have been. Lower than far thunder, higher than the bat's squeak, the whispering of a million leaves. Sometimes the murmur of wind-shifted leaves in summer reminded her" ("Mary and Joe," 171). In a further twist of the Virgin Mary story, Mary saves her daughter from the violent persecution of her political enemies, staying with her in the hospital during her recovery.

As Jenni Calder points out, "Mary and Joe" "carries a subversive political charge. Women can reproduce without men; there is no need for the Y chromosome; women are leaders, politically and professionally" (*The Nine Lives of Naomi Mitchison,* 271).

In "Women's Time," Julia Kristeva reflects on the question of a fatherless maternity, considering that if it "were to become the norm, it would become absolutely necessary to develop appropriate laws that could diminish the violence that might be inflicted on the child and the father" (*The Portable Kristeva,* 364). These probing questions can also pertinently be applied to meditate on potential scenarios to which the advent of human cloning could give rise, namely, a family configuration where the child is the mother's clone, genetically unrelated to the "father" or the mother's spouse, as is the case in Mitchison's "Mary and Joe," where the father feels left out and, indeed, forces himself to respond to his wife's touch after she has told him the truth about Jaycie's origins: "Deliberately and with a slow effort he made the hand respond, warmly, gently, normally. For the hand left to itself had wanted to pull away, not to touch her. Not to touch" (173).

Lisa Tuttle's short story "World of Strangers," in turn, offers a fictional illustration of a future world where both men and women can have their own, cloned children, a possibility that, in this story, has paved the way for an egalitarian society where the battle of the sexes has virtually disappeared. In this world, it is the usual practice for better-off people to have their own cloned children, a routine not so widespread in the rest of the population. Indeed, cloning is described as "the nice, modern way of continuing a species" (298). As Nick, the protagonist, explains,

> Like me, most of the children at my exclusive, very expensive school had single parents. But unlike me, they were all precisely matched. Day after day as I watched the fathers walking away with their miniature selves, the mothers so happy with their identical little daughters, I felt more and more powerfully the strangeness of my situation. Why didn't my mother have a little girl? I was a boy—so where was my father? ("World of Strangers," 297)

Nick eventually finds out that he was cloned from the man his mother loved with recourse to traces of his blood and skin. Shocked at what he perceives as his mother's treason, he turns against her, arguing that he is not after all related to her genetically, to which she retorts, "There's more to motherhood than a genetic link! Of course I'm your mother, in all the ways that matter. I love you. I've always taken care of you. I carried you for nine months before you were born" ("World of Strangers," 300). The story revolves around the discourses of mothering and fathering in a society of cloned people, where these new reproductive scenarios raise fresh ethical questions. In this society power asymmetries do not depend to a large extent on the sex of the persons. As Nick muses, "Human cloning set men—and women—free. We don't have to fight each other any more. Now men have the same, inalienable right as women. We can reproduce ourselves. We can have our own children" (305–306). This reflection comes in the wake of a train of thought that reviewed the way in which, roughly from the 1960s onward, women were often granted custody of children in cases of divorce, while fathers were usually required to assume monetary responsibility for the child.

In Nick's world, technology has gradually given men the possibility of reproduction on an almost equal footing with women. When Nick was an infant, we are told,

> Men were still dependent on women for the gestation period, although at least use of women in this way could be a relatively straightforward, commercial transaction. The development of a completely satisfactory artificial womb lagged, perhaps because there was no perceived demand. There were always plenty of women willing to rent out their wombs. . . . Another possibility was animal wombs: dairy cows were fine, and farmers were always on the lookout for extra sources of income. . . . A few men were dedicated—or mad—enough to actually give birth to their own children, but the failure rate (including parental, as well as infant) mortality was significant. ("World of Strangers," 306)[32]

In addition, most fathers suckled their babies, thanks to a course of injections.[33] After birth, the baby would be placed in "his father's waiting arms, laid on his naked, swollen breast to suck" (306), a scenario which might, after all, be not so far away.[34]

> How can men be mothers?
>
> Marge Piercy, *Woman on the Edge of Time*, 96

A similar vision is put forward in Marge Piercy's *Woman on the Edge of Time* (1976), a novel directly influenced by Firestone's ground-breaking *The Dialectic of Sex*. In Piercy's Mattapoisett, a society in the future, breast-feeding is also shared between men and women, in a powerful concretization of Firestone's vision of a society where child-bearing and child-rearing responsibilities would be shared by men and women. Furthermore, in Mattapoisett men are mothers in the sense that, as in Nick's futuristic world, they have breasts and suckle their babies. The embryos grow not in women's wombs but in special tanks, where they each have "a sac of their own inside a larger fluid receptacle. . . . All in a sluggish row, babies bobbed. . . . Languidly they drifted in a blind school" (*Woman on the Edge of Time*, 94).

Connie, a marginalized Chicana woman living in New York, whose child was taken away from her and given for adoption, manages to communicate with Mattapoisett, a society in 2137.[35] In Mattapoisett she is "the person from the past" (*Woman on the Edge of Time*, 92), the person from the age of greed and waste, their "notions of evil center around power and greed—taking from other people their food, their liberty, their health, their land, their customs, their pride" (131). Mattapoisett is a pastoral, nonsexist world, where there is no racism or violence and where children are not genetically related to their mothers (male and female). Luciente summarizes the philosophy underlying life in Mattapoisett to Connie:

> It was part of women's long revolution. When we were breaking all the old hierarchies. Finally there was that one thing we had to give up too, the only power we ever had, in return for no more power for anyone. The original production: the power to give birth. 'Cause as long as we were biologically enchained, we'd never be equal. And males never would be humanized to be loving and tender too. *So we all became mothers*. Every child has three. To break the nuclear bonding. (*Woman on the Edge of Time*, 97; my emphasis)

This remarkable dramatization of some of Firestone and Haraway's utopian visions leaves Connie horrified. As she muses,

> How could anyone know what being a mother means, who has never carried a child nine months heavy under her heart, who has never born a baby in blood and pain, who has never suckled a child. Who got that child out of a machine the way that couple, clean and white and rich and healthy, got my flesh from the State. All made up already, a canned child. . . . She hated them, the bland bottle-born monsters of the future, born without pain multicolored like a litter of puppies, without the stigmata of race and sex. (*Woman on the Edge of Time*, 98)

She also feels angry when she sees a man in the nursery breast-feeding a baby:

> How dare any man share that pleasure. These women thought they had won, but they had abandoned to men the last refuge of women. What was special about being a woman here? They had given it all up, they had let men steal from them the last remnants of ancient power, those sealed in blood and in milk. (*Woman on the Edge of Time*, 126)

The profound ambivalence Connie experiences toward these new developments is synthesized in her feelings for Barbarossa, whom she watches suckling a baby: "She could almost hate him in the peaceful joy to which he had no natural right; she could almost like him as he opened like a daisy to the baby's sucking mouth" (*Woman on the Edge of Time*, 127). The word "natural" carries the burden of Connie's mixed sentiments toward these novel scientific and medical advances. What in her world is defined as "natural" has in Mattapoisett become the object of technological intervention, so that the already fluid frontiers between natural and artificial have here undergone a radical transformation. "Natural" is no longer "essentially" good; artificial has been constructed as the normal state of events and thus aquired the veneer of intrinsic authenticity.

By releasing women from their apparently unbreakable imprisonment in their bodies, power relations in Mattapoisett are radically shifted.[36] Freud's seemingly unswerving dictum "anatomy is destiny" is rewritten, Firestone's vision is vindicated, as well as Simone de Beauvoir's, who in *The Second Sex*, however less radically, diagnosed women's subordination as located in their bodies' reproductive capabilities and advocated a greater availability in society's roles for women, a change that would open the way to greater freedom of movement for women and an added feeling of personal fulfillment. As Nancy Lublin points out, "Once the reproductive meaning of sexual difference is erased, equality (of result) can be established. Women will not become men; rather, these distinctions will be completely destroyed, and people will be recognized only as human beings. Sex as an activity might still be practiced, but it will be void of any reproductive meaning" (*Pandora's Box: Feminism Confronts Reproductive Technology*, 30). These words seem tailor made to describe the circumstances in Piercy's Mattapoisett, as well as potentially the kind of society that may come to exist given the effective development of human cloning technology and ectogenetic births. In this society, reproduction would no longer be considered as a public commodity. Pushing a socialist-feminist scenario further, the dissociation of reproduction from women's bodies, in the case of those who so wished, could come to spell the transcendence of property and, as a consequence, the deliverance of woman from the traditional ascendancy of man in the sexual and reproductive arena. In addition, discrimination based on sexual orientation might cease to exist in this kind of society, dramatized in Mattapoisett and in Firestone's socialist-feminist one. As Lublin remarks, in Firesone's utopia "The law and society more generally might finally accept the notion of love and sexual relations among individuals rather then binary opposites" (39).

Beside the potentially enabling circumstances, as far as women are concerned, that human cloning could bring about, this new technology might also generate a whole new set of novel relational configurations. In Lisa Tuttle's "World of Strangers," when Nick first confronts his father, his feeling toward him is not filial, but a lover's passion for his other half, a sentiment that is fully requited by his father. Similarly, in Martha Nussbaum's short story "Little C," the possibility of an incestuous relation between mother and son hovers over the plot. Little C is a clone of the mother's own dead husband whom she brings up as her son in the expectations that he will turn out to be like her husband. The story provides a pointed commentary on the debate between genetic determinism and social constructivism, for Little C grows up to be substantially different from his genetic father, both physically and in terms of tastes. When he is about to leave home at seventeen to pursue his studies, he asks his mother why she is not happy, for he would like to provide that missing happiness. He intuits that he is in a position similar to the foundling in George Sand's story "François de Champi," which Little C's mother would often read to him at bedtime, where Champis, having been brought up as a son, falls in love with his mother and she with him, becoming her lover and husband. Little C's mother, however, experiences only filial love for him, not a lover's desire, and hints as much to him. This confusion of boundaries brought up by these new sets of family patterns is likely to have serious consequences that need to be foreseen, as is being done in these short stories.[37]

In Felicia Ackerman's "Flourish Your Heart in This World," with action that takes place around 2006, a person had first been cloned in 2003. The protagonist, Laurel, is coming to terms with the fact that her cousin Juliana, about her age, has just had a cloned daughter, after trying to get pregnant for eight years. Laurel muses, thinking of the old Juliana and her beautiful daughter thirty years later, that "she has heard people say you have to be very egotistical to get yourself cloned. Now it strikes her the opposite is true. Generous and non-competitive, that's what you have to be—imagine bringing someone into the world who will be genetically just like you, but as you age, decades younger" (319).[38] Laurel, thirty-six, is considering all the new possibilities this technique might open for her when one of the patients at the hospice where she is a social worker, Mrs Noll, leaves in order to have herself cloned so that the doctors can subsequently grow a new liver to substitute her old, sick one (this procedure would entail killing off Mrs Noll's own embryo in order to save her life). The ethical implications of these scenarios, it scarcely needs stressing, require much greater attention but also open the way for new life-saving techniques to be developed.

Nancy Kress's "To Cuddle Amy" thematizes another chilling scenario that addresses some of the possible consequences of the widespread use of genetic engineering. "To Cuddle Amy" describes a future society where cloned children can be discarded at will if they fail to satisfy their parents, for instance when they grow up and change from submissive and obedient children to rebellious adolescents. This is the case with Amy, who at fourteen is behaving like a typical adolescent, fractious and insubordinate. Amy's distraught mother, Allison, cannot cope with Amy's mutinous, irresponsible actions and decides to get rid of her. As Allison tells her husband, "I'll throw her out tonight . . . and call the clinic in the morning" ("To Cuddle Amy," 307). Amy is one of six embryos that were the result of in vitro fertilization, stored in the fertility clinic as "standard procedure against a failure to carry to term. Or other need. Three more versions of the same embryo, the product of forced division before the first implantation. Standard procedure, yes, all over the country" (306–307).

Allison knows that the clinic will provide her with another little Amy, a delightful baby who will grow into an adorable little girl "climbing on our laps for a cuddle . . . oh, Paul, I want my little girl back!" ("To Cuddle Amy," 306). In this society, then, it is common for parents to get rid of troublesome children, who will be disposed of by fertility clinics, and have them replaced with their younger, pliable versions. In this case what we have is not children suffering from Peter Pan syndrome, unwilling to grow up, but parents who now have the means to remove their children when they grow too old for their comfort and exchange them for younger ones, parents who in effect do not want and will not allow their children to develop normally, parents whose narcissism, fueled by the technological possibilities in a posthumanist world, gives them the possibility of going back in time and endlessly, selfishly enjoying their young children.

There is, however, another chilling twist in this tale. Amy is not the first one to have been discarded by her parents. Another Amy was allowed to reach the age of sixteen and only then thrown out. Human life in this society is discardable and valueless, parents having total control over their children's bodies. Kress's "To Cuddle Amy" is to a certain extent reminiscent of the plight of David, the protagonist of Steven Spielberg's film *Artificial Intelligence* (2001), the human-machine hybrid developed by a big corporation. David, like Pinocchio, wishes to be a real boy, a son whom his "mother," Monica, could love without restrictions and whom he would be allowed to love boundlessly in return. However, like Amy, David is eventually thrown out, Monica being unable to cope with the potential danger he might pose to her biological child, Martin, but also threatened by the devotion David unconditionally bestowed upon her, the ideal child who, unlike Amy, would never grow up.

Nancy Kress's "Sex Education" is another story that presents a negative view of cloning. In "Sex Education," Mollie, a little girl, is confronted with a host of journalists who want to know what she thinks of the fact that one of her twelve clones, about whom she knew nothing, was born with a heart disease. When Mollie, ten, finds out about this sick baby, one of several "Xerox copies" ("Sex Education," 217) of her, as her mother explains, she considers it is her responsibility, since "the baby that's me" (218). She decides to steal the baby from her parents' house and brings her back to her own house, where she keeps her in the basement. Although the baby would not stop crying, Mollie had to keep walking around, trying to soothe it, "because nobody wanted this baby with the broken thing in her heart, nobody would accept the responsibility, and the baby was hers, was her, please sweet baby please sweet baby—" (224–225). This story raises the crucial question of ethical responsibility, since, as the narrator puts it, the situation did not seem to be anybody's fault: "It wasn't Mommy and Daddy's fault because they were just helping the Berringers to have a baby like Mollie. It wasn't the scientists' fault because they were just helping, too. It wasn't the Berringers' fault because they didn't ask to have a baby that wasn't perfect. It wasn't the court's fault because somebody else had to pay the court money. No one was responsible" (220).

These stories can be said to vindicate in some ways James D. Watson's prediction that with the introduction of human cloning, "The nature of the bond between parents and their children, not to mention everyone's values about the individual's uniqueness, could be changed beyond recognition" ("Moving toward the Clonal Man: Is This What We Want?" 52).

SOME FURTHER CONCERNS

> Mother the machine
>
> Marge Piercy, *Woman on the Edge of Time*, 94

Several critics have voiced their doubts and concerns about possible consequences of new reproductive technologies, and we have heard already from Adrienne Rich and Luce Irigaray.

What the texts previously examined seem to suggest, I would argue, is that human cloning, or versions thereof, can come to play a powerfully enabling role in women's struggle for an egalitarian society without power asymmetries caused by the biology of the female body and reproduction. Indeed, when what once was women's almost exclusive source of power—pregnancy and lactation—is distributed evenly between men and women, women's oppression might also be lifted and a more just society could emerge, a postpatriarchal one where phallogocentric psychoanalytic structures would no longer apply.[39]

The visions of human cloning these texts dramatize are on the whole positive, proposing an imaginative anticipation of various scenarios depicting the shape that a future society, where cloning exists as an option, might take. These include further alternatives in health care, such as developing organs for transplants grown from one's own cells so that there will not be a risk of rejection, as in Naomi Mitchison's "Mary and Joe" and Felicia Ackerman's "Flourish Your Heart in This World," as well as the emergence of new societal configurations. However, some of these narratives are also cautionary tales sounding a note of warning as far as the implications that the breakdown of the traditional nuclear family might entail, such as a parallel disintegration of the biological family, as in Piercy's *Woman on the Edge of Time*, Lisa Tuttle's "World of Strangers," Martha Nussbaum's "Little C," and Felicia Ackerman's "Flourish Your Heart in This World," where the protagonist considers having a cloned baby on her own. The structures of kinship, defined by Lévi-Strauss, as well as familial incest taboos, would also in all likelihood undergo significant transformations, as all the texts examined suggest. Overall, I get a sense from these texts that the new social arrangements cloning gives rise to have significantly contributed to improve women's status, providing a reshaping of the political and social organization where women appear on an equal footing with men while simultaneously giving rise to a new panoply of complex psychological and social problems.

Whether human cloning might contribute to bringing about these conditions is, of course, highly contentious and will remain so for many decades to come. In the end, however, the bottom line about human cloning has to do with social tolerance of ever more sophisticated aspects of choice, such as the possibility of electing a given reproductive technology that will most suit a woman's wishes and lifestyle. In this context, Michelle Stanworth argues that feminists should concentrate upon "creating the political and cultural conditions in which . . . technologies can be employed by women to shape the experience of reproduction according to their own definitions" ("The Deconstruction of Motherhood," 35). What these definitions are will, of course, differ from woman to woman, but in the important political struggle, to fight is to achieve the legal possibility to exercise freedom of choice, a choice that might one day include cloning. In tune with Nancy Lublin, who advocates a "truly socialist response to reproductive technology" (*Pandora's Box: Feminism Confronts Reproductive Technology*, 156), a "praxis feminist materialist approach [that] places the realities of sex within the scope of social justice more generally because public education, subsidized housing, and welfare benefits condition women's choices about, and the implications of, sexual reproduction" (156), I similarly situate myself in this broad socialist-feminist scenario. With particular reference to cloning, it needs to be pointed out that, as with other reproductive technologies, all women should have access to it, if they decide to follow that path, at the risk of turning that option into an elitist alternative open only to a select few. Indeed, cloning might, arguably, perpetuate class inequalities if not offered to everybody irrespective of class, sex, or ethnic background. Otherwise, only people with the appropriate acquisitive power might have access to that possibility. In a related vein, Haraway notes that

> another critical aspect of the social relations of the new technologies is the reformulation of expectations, culture, work, and reproduction for the large scientific and technical workforce. A major social and political danger is the formation of a strongly bimodal social structure, with the masses of women and men of all ethnic groups, but especially people of colour, confined to a homework economy, illiteracy of several varieties, and general redundancy and impotence, controlled by high-tech repressive apparatuses. ("A Cyborg Manifesto: Science, Technology, and Socialist-Feminism in the Late Twentieth Century," 169)

While it is incontrovertible that, as Nancy Lublin maintains, "changing the basic structures of society—education, the law, the family—must occur in order to dramatically change the political, social, and economic situation of women" (117–118), I would like to suggest that the onset of such a radically new technology as cloning might potentially help achieve the same results, as far as social equality for women is concerned, as years of tireless political struggle, promoting an effective concretization of reproductive interchangeability and balance between women and men,[40] as dramatized in Marge Piercy's *Woman on the Edge of Time* and Lisa Tuttle's "World of Strangers."

BEYOND A PHALLIC ECONOMY OF DESIRE

In the context of a discussion of the possible and predictable repercussions of the advent of human cloning in light of some of Irigaray's theories, it is significant to remember that Irigaray considers science as a privileged locus for the pursuit of a measure of truth. Irigaray, however, does not always show much patience with what she sees as "an excessive amount of interest in discoveries related to artificial methods of procreation" (*Je, Tu, Nous: Toward a Culture of Difference*, 103). On the other hand, she goes on to remark that "artificial procreation does raise a good number of scientific and ethical questions. In this respect, we can't be indifferent to it.[41] I also think that, among other things, scientists sense in this, whether consciously or not, a way of overcoming god the creator, while some women see it as a way to rid themselves of the need for men" (103–104). According to Irigaray, "All these ways of destabilizing a given social model are being carried through too hastily, without clear-sightedness and without the affirmation or establishment of better values" (104).[42]

Both Baudrillard and Irigaray can be said to be actively intent on theorizing the grounds for the implementation of "better values," to borrow Irigaray's words, attempting in their admittedly different ways, which, however, often intersect in their engagement with these issues, to extend postmodern notions of the (technologized) body and of identity. We might say, then, that while for Baudrillard it is fundamental to keep the other, for Irigaray the crucial strategy will be to preserve the other, defining this other, however, not as the other of the (male) same but an other (of man or woman) who would not be subjugated to either, whom I would call a "she-self" or a "he-self," an independent woman or man whose bodily, physical, and social configurations are constantly being refashioned and reformulated in this era of biopower.[43] It would be fair to say, then, that the stories our age of electronic media culture, virtual reality, and biotechnology originate, which so eloquently speak about it, uttering its ever-changing contours, will be increasingly populated by cyborgs and clones, eminently posthuman beings who aptly embody the at times confused and blurred outlines of our posthuman era. Indeed, in our contemporary world, increasingly shaped by biopower relations, in the sense defined by Foucault, human cloning will inevitably soon be with us.

Firestone's "'dangerously utopian' concrete proposals" (*The Dialectic of Sex*, 226), then, might be said to be at long last taking definite shape, while her vision of a "cybernetic socialism" (238) is almost with us.

NOTES

1. A man, however, would still need a woman's womb in order to have his own cloned offspring, at least until ectogenetic births become feasible. I am assuming here that the technology for cloning a human being may predictably be developed before artificial wombs become ready for use.

2. Leon R. Kass, "Making Babies—the New Biology and the 'Old' Morality," 13–56; quoted in Robyn Rowland, "Reproductive Technologies: The Final Solution to the Woman Question?"

3. This aspect is clearly articulated in Katherine Burdekin's *The End of This Day's Business* (1989), a role-reversal narrative written in the 1930s, set in 6250, where women rule the world. As the narrator explains, in spite of their technological and scientific advancement, women never wanted, "naturally, to make human life in any way but the ordinary animal procreation, because it would undermine their own power" (98).

4. The prospect of the introduction of human cloning has produced heated debate and widespread criticism. Among the most salient concerns cloning has elicited is the link between cloning and eugenic thinking, raising fears of the elimination of those deemed less valuable, as well as the creation and reproduction of certain genotypes—possibilities that could lead to a profu-

sion of designer babies and, given time, to a much more uniform population pool. The power to choose whom to clone and what characteristics designer babies should possess is inextricably linked with economic and political privilege, leaving out a large part of the world population and inevitably leading to discrimination and imbalances in terms of skin color and related ethnic issues. Human cloning also raises numerous political, economic, ethical, and religious questions, which address such notions as the inviolable uniqueness of an individual, family dynamics, and the laws that regulate the family unit, as well as what is perceived by some as the highly transgressive act of daring to create life, a gift exclusive to God. For extended discussion of these issues, see, for instance, Ruth F. Chadwick, *Ethics, Reproduction and Genetic Control*; Daniel J. Kevles, *In the Name of Eugenics: Genetics and the Uses of Human Heredity*; Martha C. Nussbaum and Cass R. Sustein, *Clones and Clones: Facts and Fantasies about Human Cloning*; John Harris, *Clones, Genes, and Immortality: Ethics and the Genetic Revolution*; Lisa Yount, *Cloning*.

5. For a full discussion of the complex questions involved in ectogenesis, see, for instance, Susan Merrill Squier, *Babies in Bottles: Twentieth-Century Visions of Reproductive Technology*; Julien S. Murphy, "Is Pregnancy Necessary? Feminist Concerns about Ectogenesis."

6. I find these technophobic arguments too dependent on essentialist notions of woman and nature, excluding culture and technology as the inevitable enemies of that implied connection with the natural world, itself unavoidably enmeshed in binary, exclusionary dualisms. As I wish to argue, science and technology can benefit women immensely if developed and applied according to ethical rules drafted by committees in which both men and women are equally represented.

7. These changes would happen only gradually. Apocalyptic visions of armies of cloned people being "fabricated" for specific purposes is clearly not a part of Firestone's project, which considers versions of cloning as enabling women to bypass the seemingly inevitable fate their biology dictates for them.

8. This much-needed theorization remains, however, beyond the scope of this book.

9. From which that baby would subsequently be "decanted," to borrow the pejorative terminology Aldous Huxley used in *Brave New World*.

10. I am here anticipating criticism to the purpose that my implied defense of cloning as potentially liberating for women would erase the body of the mother, always already a site of contradictory, vexed feelings and readings, from this scenario, but that criticism would only apply in the case of the widespread use of ectogenesis, that is to say, artificial wombs.

11. In this respect, see Ronald M. Green, "I, Clone."

12. I am here assimilating clones with cyborgs, since the former are the result of laboratory manipulation of cells.

13. See, for instance, Evelyn Fox Keller, *Reflections on Gender and Science* (1984); idem, "From Secrets of Life to Secrets of Death" (1990).

14. In a private conversation with me in June 2001, however, Luce Irigaray considered the prospect of cloning as "sad" in terms of social and family life.

15. For a discussion of these and other possibilities that might become available with the advent of human cloning, see William N. Eskridge Jr. and Edward Stein, "Queer Clones."

16. Charis Thompson Cussins, in her short story "Confessions of a Bioterrorist: Subject Position and Reproductive Technologies," offers some interesting speculative visions of virgin birth fantasies. Eva, an embryologist and senior lab technician at an infertility clinic, describes her "Virgin fantasy" to her two friends, Gabriela and Mary, in the following terms, clearly reminiscent of Ryn Edwards's scenario: "You know the micromanipulation techniques we use for severe male infertility that I taught you, Mary? If you think about it, you manually put the sperm inside the egg for fertilization. So we don't need sperm morphology for fertilization anymore; it's just the chromosomes and a few activating proteins, right? What I would really love to do is to take two eggs and fertilize one with the other. Not as the only way to get pregnant; not as parthenogenesis; no cloning; fully normal meiotic recombinant reproduction; just combining two eggs, rather than having to involve sperm" (205–206).

17. Until, that is, such time when ectogenesis or male pregnancy is technically possible.

18. The Whileaway section of Russ's *The Female Man* and James Tiptree Jr.'s "Houston, Houston, Do You Read?" might be seen as corresponding to the kind of vision put forward by Adrienne Rich, who considers lesbian separatism as a possible societal scenario to circumvent the oppression of gender. Rich declares, "Lesbian existence comprises both the breaking of a taboo and the rejection of a compulsory way of life. . . . We may first begin to perceive it as a form of nay-saying to patriarchy, a form of resistance" ("Compulsory Heterosexuality and Lesbian Experience," 645).

19. If cloning technology became widespread, however, some of its potentially dangerous applications would have to be addressed. From a Western point of view, the sex of his or her child might not make much difference to a prospective mother or father, or at least not in terms of the possible later discrimination against the female offspring. A mother or father in India or China, however, might decide to have only boys, since girls are still considered less important by society in general and would necessarily have disadvantaged lives, a decision that might involve neglect or mistreatment of female children or, even more drastically, infanticide. For dicussions of the issues involved in this thorny problem, see Helen Bequaert Holmes, "Sex Preselection: Eugenics for Everyone?"; Mary Anne Warren, "The Ethics of Sex Preselection"; Gregory Stock, *Redesigning Humans: Choosing Our Children's Genes*.

20. In "Motherhood: The Annihilation of Women," Jeffner Allen rejects motherhood "on the grounds that motherhood is dangerous to women" (315) and argues the case for a "philosophy of evacuation" (315), which "proposes women's collective removal of ourselves from all forms of motherhood" (315). Allen explains the rationale behind her statement: "Freedom is never achieved by the mere inversion of an oppressive construct, that is, by seeing motherhood in a 'new' light. Freedom is achieved when an oppressive construct, motherhood, is vacated by its members and thereby rendered null and void" (315).

21. Thus implicitly suggesting that in spite of their alleged superiority men are often found deficient in providing a smoothly running environment for their home and children. Dorothy Dinnerstein in *The Mermaid and the Minotaur: Sexual Arrangements and Human Malaise* (1976), Nancy Chodorow in *The Reproduction of Mothering: Psychoanalysis and the Sociology of Gender* (1978), and Jane Gallop in *The Daughter's Seduction: Feminism and Psychoanalysis* (1982) offer a

pertinent critique of patriarchal patterns of socialization in the traditional nuclear family. Chodorow, in particular, forcefully argues the case for a much greater participation of the father in the raising of the children and the running of the household.

22. Some fictional accounts that illustrate different versions of the potential scenarios human cloning might give rise to include, among others, Mary E. Bradley Lane, *Mizora* (1890); Charlotte Perkins Gilman, *Herland* (1915); Gwyneth Jones, *Divine Endurance* (1984); Joanna Russ, *The Female Man* (1975); Pamela Sargent, *Cloned Lives* (1976); Kate Wilhelm, *Where Late the Sweet Birds Sang* (1977); James Tiptree Jr. "Houston, Houston, Do You Read?" (1976); Sally Miller Gearhart, *The Wanderground* (1978); Suzy McKee Charnas, *Motherlines* (1978); Joan Slonczewski, *A Door into Ocean* (1986); Sheri S. Tepper, *The Gate to Women's Country* (1988); Anna Wilson, *Hatching Stones* (1991); Fay Weldon, *The Cloning of Joanna May* (1989); Lisa Tuttle, "World of Strangers" (1998).

23. Among these could be cited Renate Duelli Klein, Gena Corea, Roberta Steinbacher, Helen Holmes, Rita Arditti, and others.

24. Susan Merrill Squier addresses the different feelings attached to the attraction to ectogenesis by men and women. While referring to the fascination that ectogenesis held for Charlotte Haldane, wife of scientist and geneticist J.B.S. Haldane and author of, among many other books, *Man's World* (1927), Squier argues that "if for male writers in the modern period . . . the image of ectogenesis fed a fantasy of male agency and autonomy, acquired through denigration of the image of the gestating woman, for women writers the image meant something rather different" (*Babies in Bottles: Twentieth-Century Visions of Reproductive Technology*, 131). As Squier goes on to point out, Charlotte Haldane "saw implications in ectogenesis diametrically opposed to those envisioned by her male friends and colleagues. Her goal as a modern woman writer was to empower women, particularly mothers" (131). This ambivalence felt by women toward extrauterine gestation is still very strong in our contemporary world, but like Haldane a growing number of women see in ectogenesis a strategic and empowering technique that enables them to carry on with their lives, as their male colleagues do, without creating potential inequalities in the workplace and in domestic life. For an account of the fascinating debate on ectogenesis in the 1920s and 1930s in England and on the divided responses of contemporary feminists, see Squier, *Babies in Bottles: Twentieth-Century Visions of Reproductive Technology*.

25. I am assuming here that probably the technology for cloning a human being will be developed before artificial wombs become ready for use, indispensable for men to have their cloned babies. On the other hand, however, it is not medically impossible for a man to get pregnant, according to Robert Winston, who, in *The IVF Revolution: The Definitive Guide to Assisted Reproductive Techniques*, maintains that the techniques already exist to achieve a male pregnancy, although there are so many risks that it would be ethically unacceptable to try the procedure at the present time. On the other hand, as Patrick D. Hopkins pertinently remarks, in relation to the prospect of men getting pregnant, "The possibility of this technology leads us to ask: Should women be the only ones *allowed* to bear children? Are women as a class *obligated* to bear children? Does the state have a compelling interest in preventing men

from reproducing as they see fit? Should technology be judged wrong or made illegal *because* it upsets traditional biological sex roles?" ("[Re]locating Fetuses," 172). See also Dick Teresi and Kathleen Mcauliffe, "Male Pregnancy" (1998).

26. This scenario is given fictional illustration in, for example, Maureen Duffy, *Gor Saga* (1981) and Fay Weldon, The *Cloning of Joanna May* (1989).

27. In this context, surrogate motherhood is a case in point. In her discussion of surrogacy Lori B. Andrews considers that the arguments brought to bear on this question are predominantly political and have to do with the fundamental issue of whether the government should have ascendency over women's bodies and regulate them, as is the case with surrogacy. As Andrews synthesizes her position, "Some feminists have criticized surrogacy as turning participating women, albeit with their consent, into reproductive vessels. I see the danger of the antisurrogacy arguments as potentially turning *all* women into reproductive vessels, without their consent, by providing government oversight for women's decisions and creating a disparate legal category for gestation. Moreover, by breathing life into arguments that feminists have put to rest in other contexts, the current rationales opposing surrogacy could undermine a larger feminist agenda" ("Surrogate Motherhood: The Challenge for Feminists," 168).

28. I have some problems with the claim that "reproduction is the foundation of women's identity." There is a plurality of women, including those who do not wish to become mothers. Arguing that women are defined through motherhood is a very debatable issue, an argument that, for me, comes very close to inserting women into the rigid binary dualisms that have contributed so decisively to inflexibly placing woman into unyielding boundaries and strictly defined sex roles.

29. Here I am reminded of Bryan Forbes's 1975 film *The Stepford Wives*, where the wives are substituted by copies of themselves who are programmed to behave according to men's wishes and fantasies.

30. Professor Robert Winston, in *The IVF Revolution: The Definitive Guide to Assisted Reproductive Techniques*, corroborates this view. He remarks, "It is possible, under some circumstances, to activate a human egg without actually placing a sperm near it. Certain forms of physical injury will do this, as will a small electrical shock. With activation, the egg may start to divide spontaneously and form a structure which, using routine microscopy, is indistinguishable from a normal embryo. . . . It has been suggested that it might be possible, eventually, to produce a complete human being by parthenogenesis" (204). However, as he goes on to maintain, "Evidence strongly suggests that lethal genes would prevent further development beyond a few cells. It seems, therefore, that at least for the time being humans will not reproduce like ants, termites or bees" (204).

31. God, of course, does not have daughters, only sons.

32. In 1924, philosopher Anthony Ludovici was already putting forward and criticizing similar scenarios. For Ludovici, intraspecies uterine tranfers would be utterly debasing and humiliating for humankind: "It is probable . . . that in the early days of extracorporeal gestation, the fertilized human ovum will be transferred to the uterus of a cow or an ass, and left to mature as a parasite on the animal's tissues. . . . And with this innovation, we shall prob-

ably suffer increased besotment, and intensified bovinity or asininity, according to the nature of the quadruped chosen" (*Lysistrata, or Woman's Future and Future Woman*, 92).

33. In this context Chris Hables Gray cites an article by Jared Diamond entitled "Father's Milk," published in *Discover* in 1995. As Gray explains, it "described in loving detail how and why human fathers can be easily modified with a few hormone shots and some maula nipple stimulation so that they can nurse their young" (*Cyborg Citizen: Politics in the Posthuman Age*, 148). Gray quotes Diamond: "Experience may tell you that producing milk and nursing youngsters is a job for the female mammal, not the male. But your experience is probably limited, and the potential of biology—and medical technology—is vast" (148). As Gray further notes in words that could apply to Tuttle's story, "Slightly more difficult, but just as likely, is the modification of men to carry a baby, as in the Arnold Schwarzenegger movie *Junior*" (148).

34. As Dick Teresi and Kathleen Mcauliffe note, in "Male Pregnancy," men are "already lining up. . . . when the time comes for the first embryo transfer into a man, there will be no shortage of volunteers" (180–181).

35. Issues of class and ethnicity are very important in this novel. Connie's working-class Chicana background is to a great extent responsible for her being pressured to withdraw into a mental hospital, where patients are in many respects guinea pigs of the medical establishment, with women's bodies being used by doctors to carry on experiments, such as fertility treatments.

36. Alice E. Adams suggests that Mattapoisett provides an example of a communist society, stressing the link between women's reproductive capabilities and their exploitation in a capitalist system: "In Mattapoisett women keep their wombs empty so they will no longer serve as the guarantors of reproductive capitalism. Luciente defines childbirth as 'the original production.' Only when the means of reproduction are communally owned can true communism arise" (*Reproducing the Womb: Images of Childbirth in Science, Feminist Theory, and Literature*, 150), in a version of the utopian dream envisioned by Firestone.

37. In her polemical vision of the future, Firestone addresses the question of incest. According to her, if a child "should choose to relate sexually to adults, even if he should happen to pick his own genetic mother, there would be no *a priori* reasons for her to reject his sexual advances, because the incest taboo would have lost its function. The 'household,' a transient social form, would not be subject to the dangers of inbreeding. . . . Thus, without the incest taboo, adults might return within a few generations to a more natural polymorphous sexuality" (*The Dialectic of Sex: The Case for Feminist Revolution*, 240).

38. This scenario prompts reference to Fay Weldon's novel *The Cloning of Joanna May* (1989), where Joanna May is confronted, at age sixty, by her four clones, her sisters, daughters, twins, she is not sure how to describe them, who are thirty years younger than she and show her the contrast between what she looks like now and what she used to be like.

39. The abolition of private property in Mattapoisett can be read as arguably a logical consequence that goes in tandem with the discontinuance of the monopoly of reproduction as pertaining solely to woman. In a Marxist register, Luciente, one of the inhabitants of Mattapoisett, explains to Connie that

the reason everything belongs to the government and that they do not keep things for themselves has to do with passing "along the pleasure" ("Woman on the Edge of Time," 168) those things give them, with sharing that pleasure with others.

40. I am aware that this sweeping statement has to take into account the fact that, as Nancy Lublin succintly puts it, "because technology must be completely comprehensive to eradicate the social meaning of sexual difference . . . the possibility of completely artificial creation or independent replication is still very much in the distant scientific future" (*Pandora's Box: Feminism Confronts Reproductive Technology*, 35). In spite of Lublin's caveat, I believe those changes might come to happen in the not so distant future.

41. This is a point also stressed by Adrienne Rich, who calls attention to the need for women "to become well informed about current developments in genetics, cloning, and extrauterine reproduction" (*Of Woman Born: Motherhood as Experience and Institution*, 282).

42. Adrienne Rich perceptively comments on the social and philosophical changes that a potential future scenario of equality between men and women would necessarily entail, suggesting, as I do, that that equality would depend to a great extent on the implementation of a whole different set of new reproductive technologies, which would make conception and birth not directly dependent on women's bodies, as in the case of asexual reproduction and ectogenesis (ectogenetic wombs). According to Rich, "The repossession by women of our bodies will bring far more essential change to human society than the seizing of the means of production by workers. The female body has been both territory and machine, virgin wilderness to be exploited and assembly-line turning out life" (*Of Woman Born: Motherhood as Experience and Institution*, 285–286). Rich's profoundly disturbing image of the female body turned into an assembly line to produce babies and exploited as a commodity inescapably brings to mind Baudrillard's perspicacious comments on the facile reproducibility of our consumer society.

43. I borrowed the nomenclature of the "she-self" from Rosi Braidotti (*Nomadic Subjects: Embodiment and Sexual Difference in Contemporary Feminist Theory*, 199) and then extended it by analogy to "he-self."

Conclusion

> Cloning was neither good nor bad; only its purpose was important. And that purpose should not be one that was trivial or selfish.
>
> Arthur C. Clarke, *Imperial Earth*, 209

"Human cloning—the creation of exact replicas of another individual from any cell in the body except the sex cells—had been achieved early in the twenty-first century" (Arthur C. Clarke, *Imperial Earth*, 9–10). Arthur C. Clarke's prognostications may yet prove to be more accurate than he could ever have imagined.[1] However, as Clarke's narrator in *Imperial Earth* (1976) goes on to observe, "Even when the technology had been perfected, it had never become widespread, partly because there were few circumstances that could ever justify it" (10). That is precisely one of the points I have been arguing for here: that all the feelings of horror and misgiving to which the prospect of human cloning have given rise would only ever be justified if large numbers of identical people were genetically engineered in order to fulfil specific tasks, as was the case of such nightmarish scenarios as those put forward in Aldous Huxley's *Brave New World*. When used in specific cases, where the benefits outweigh the disadvantages, there seems to be no reason why new technologies should not be tested and tried, if they can improve people's lives.

In Clarke's *Imperial Earth*, cloning is regarded as just another reproductive technology, which was justified if the end purpose was deemed righteous, not serving selfish interests or ethically wrong objectives. It is employed, in this light, to enable a dynasty of space pioneers to carry on. Malcolm, an engineer-administrator who was unable to have children due to the effect of radiation during his space travels, has a clone son, Colin, "his identical twin—but half a century younger" (10). As the narrator further explains,

> When Colin grew up, there was no way in which he could be distinguished from his clone father at the same age. Physically, he was an exact duplicate in every respect. But Malcolm was no narcissus, interested in creating a mere carbon copy of himself; he wanted a partner as well as a successor. So Colin's educational program concentrated on the weak points of Malcolm's. (*Imperial Earth*, 10)

Imperial Earth addresses a number of issues that impact on the problems of human cloning and a future world where cloning might have become as acceptable as any other fertility treatment. These issues include the formation of identity, narcissism, family structures, genetic determinism, the feasibility and desiribility of the "mechanical womb" (203), and the necessity and justification for the use of cloning.

As Daniel J. Kevles notes, J.B.S. Haldane in *Daedalus* had observed that "to humanity biological innovation is initially abhorrent, a perversion, an offense not to some god but to man himself. History suggested to Haldane that, led by the scientist with his songs of deicide, man might slay his inner demons, come to terms with the seeming perversions, and transform unnatural innovations into natural, humanly advantageous customs" (*In The Name of Eugenics: Genetics and the Use of Human Heredity*, 298). The same might occur with the fantasy of human cloning. Within a few decades from now, the whole idea may have become reality and cloned babies accepted without a second thought, just as babies created by means of in vitro fertilization are accepted as completely "normal."

Looking forward, it is not too far-fetched to imagine a time when one could foreseeably download one's brain into a computer, then have it then reinstalled in another, younger brain, clean of memories, or even to transplant one's brain into a cloned body, our own body as younger, or older, a body that might putatively have been kept preserved in some form of cryogenic suspension or any other technique. In this way cloning would enable people to extend their lifespans almost indefinitely, moving from body to younger body, provided the brain could be transferred in good condition.[2]

For Slavoj Zizek, in analogous fashion, the future will bring "not the replacement of the human mind by the computer, but a combination of the two" ("Bring Me My Philips Mental Jacket," 5). As Janeen Webb similarly observes, since we are already comfortably, if vicariously, living "with replicated simulacra of experience, perhaps we are also ready to accept a simulacrum of the subjectivity that does the experiencing" ("Bone of My Bones, Flesh of My Flesh: A Brief History of the Clone in Science Fiction," 163).

Reflecting on the "prospect of biogenetic intervention" ("Bring Me My Philips Mental Jacket," 3), Slavoj Zizek considers that it "effectively emancipates humankind from the constraints of a finite species" (3), although it might radically undermine the notion that we are "natural" beings, not artificially enhanced with recourse to technology. Indeed, as Zizek observes, "ultimately, biogenetic intervention could render the idea of education meaningless" (3), since traits deemed desirable could be manipulated or added to the human genome so as to produce more intelligent and skillful individuals. In addition, as Zizek adds, "Such interventions will give rise to asymmetrical relations between those who are 'spontaneously' human and those whose characters have been manipulated; some individuals will be the privileged 'creators' of others" (3). Furthermore, access to these technologies will be limited to an elite, those with financial means, a situation that will also apply to cloning, as it already does to in vitro fertilization and other fertility techniques.

Biologist Lee M. Silver provides a vision of Zizek's speculations. Silver imagines a futuristic New York in 2350, where "the extreme polarization of society that began during the 1980s has now reached its logical conclusion, with all people belonging to one of two classes. The people of one class are referred to as *Naturals*, while those in the second class are called the *Gene-enriched* or simply the *GenRich*. . . . The GenRich—who account for 10 percent of the American population—all carry synthetic genes. . . . The genRich are a modern-day hereditary class of genetic aristocrats" (*Remaking Eden: Cloning and Beyond in a Brave New World*, 4). Silver further predicts that there will be almost complete social polarization, with the GenRich controlling every section of the economic, political, scientific, and cultural arenas.

Still, according to Silver's forecasts, in 2050 having your own cloned child might be a common event. He describes the case of Melissa, about to give birth to her cloned daughter, who, unlike many other mothers-to-be, does not need to see a computer printout of her about-to-be-born child, since she

> already has thousands of pictures that show her future daughter's likeness, and they're all real, not virtual. For the fetus inside Jennifer is her identical twin sister—her clone—who will be born thirty-six years after she and Jennifer were both conceived within the same single-cell embryo. As Jennifer's daughter grows up, she will constantly behold a glimpse of the future simply

by looking at her mother's photo album and her mother. (*Remaking Eden: Cloning and Beyond in a Brave New World*, 3–4)

In this kind of scenario, the psychological consequences for the daughter, but also for the mother, can be devastating and hard to gauge. From the daughter's point of view, this particularly strong mother–daughter bond can be perceived as more difficult to disentangle in order for the daughter to be able to achieve individuation, a sense of selfhood that is not contiguous or undistinguishable from that of the mother.

This scenario, where a woman gives birth to a younger replica of herself, her own clone, can, of course, be read as an instantiation of the acme of selfishness, of narcissism. This is also the case in Eva Hoffman's novel *The Secret: A Fable for Our Time* (2001), where Elizabeth, an investment consultant, decides to leave the hectic pace of New York life, with its pollution and noise, and retreat to the Midwest, where nature was still available to be enjoyed. A willful, determined woman, Elizabeth decides to have a cloned daughter manufactured at a Manhattan laboratory—Iris, who becomes her project, her enterprise. In remaking herself in her own image, Elizabeth may be said to be reenacting the creation of humankind by god—a Judeo–Christian mythical framework, but with an important twist, since she becomes goddess the mother in this rewriting of Genesis.

Iris, who comes to see herself as a monster, a freak, muses on whether she is real, an authentic child, and whether a copy can have desires of her own. These haunting doubts, reminiscent of those experienced by Frankenstein's creature ("Who was I? What was I? Whence did I come? What was my destination?" [Mary Shelley, *Frankenstein, or, the Modern Prometheus*, 128]), will propel Iris to look for the rest of her family, estranged from Elizabeth because of what they perceived as her tinkering with life, in search for a solution to her predicament. Following on from Iris's therapist's diagnosis of a pathological version of the mother–daughter bond that only separation can heal, Iris eventually forges her own self distinct from her mother's through the tender and accepting gaze of a lover who sees her as herself, not as her mother's copy. Since, as Iris believes, "to be is to be perceived" (*The Secret*, 256) and one "cannot be a subject without being an object, visible to others" (177) her lover's perceptive and kind gaze legitimizes her and proves to her that a "monster who is loved" (255) ceases to be one and becomes a human being deserving respect. Unlike Frankenstein's monster, who is denied this love, Iris manages to negotiate the perilous waters of her life and come through mostly unscathed, but Hoffman's novel sounds a cautionary note in relation to the psychological wounds such scenarios involving cloned daughters and sons may involve.[3] In this Buildungsroman obsessed with mirrors and reflecting surfaces, with portraits and frames, with images and representation, it is as if Iris cannot access her own image, which, paradoxically, pursues her with absolute clarity in all its contours (inasmuch as one can only always see ourself in a mediated fashion) in the shape of her mother. Like the man in Magritte's painting, *La Réproduction Interdite*, staring at the mirror and seeing his back instead of his face and chest, Iris wishes in many ways to deny her own image and concept of selfhood attached to it in order to forge a sense of personhood and individuality distinct from her mother's, thus symbolically refusing to confront her own mirror image.

LA RÉPRODUCTION INTERDITE

René Magritte's painting *La Réproduction Interdite* (1937), portraying a man with his back to the viewer looking into a mirror which, instead of faithfully returning his image reflects his back, so that there is no reciprocity in the gaze, encapsulates many of the themes addressed in this book. I read the painting as a reflection on identity, suggesting such questions as these: Is it ever possible to know oneself? Am I who I think I am? Can mirrors as identity creators and promotors be trusted? Mirrors, eyes, the questioning of the visible and invisible and the subversion of scopic rules, substituted by a logic derived from the power of thought to envisage scenes not regulated by the tenets of reason constitute recurring leitmotifs in Magritte's oeuvre. The mirror's refusal to reproduce the man's front can also be interpreted, in the context of my reflections on the philosophical and psychological underpinnings of human cloning, as a reminder that identity, a sense of the self, cannot be copied or duplicated. The uncanny resonances of this painting evoke the disquieting moment of panic upon finding the loss of one's reflected self, or its inversion, in many ways comparable to that experienced by the two pairs of lovers in Dante Gabriel Rossetti's painting, *How They Met Themselves* (1851–1860). Meeting one's double or realizing that one's reflection no longer appears on a reflecting surface, as a mirror, or appears truncated or deficient, as in Magritte's *La Réproduction Interdite*, can be seen as intricately interrelated experiences, leading to profoundly disquieting interrogations pertaining to the nature of the self, perceived as diluted or in risk of decaying.[4] Here again we are faced with the contradictions attendant upon the contemporary drive to an increasingly individualized society, some of whose members, to an extent paradoxically, wish to duplicate their uniqueness, which, however, by virtue of this very duplication, will necessarily lead to the decrease of that very uniqueness, for there would be at least another one of them.

"IF YOU ARE ME, WHO AM I?"[5]

Crucially, but also paradoxically, the very search for origins that is one of the main drives fueling the fantasy of human cloning can also pave the way for gradually increasing indeterminacy, translated into a generalized indistinctness or vagueness that spreads inside the clone groups, whose members seem to grow less and less preoccupied about questions of identity and their place in society. What matters for them is remaining within the closeness and comfort of the presence of their brothers and sisters, as in Wilhelm's *Where Late the Sweet Birds Sang*, Ursula K. Le Guin's "Nine Lives," and Damon Knight's "Mary."

In "The Clone or the Degree Xerox of the Species," Baudrillard yet again forcefully reflects on the consequences and ramifications of living in a society where metaphorical and mental cloning has become the dominant cultural practice. As Baudrillard remarks, in words that shed light on the social dynamics at work in for example Huxley's *Brave New World*,

> What the system produces and reproduces are virtually matching beings, beings substitutable for each other, already mentally cloned. In the end, this whole cloning business is not so very new; we have experience of it in all areas of life—intellectual, cultural and operational, not to mention the fields of work and technology, where the system has long trained us to be clones of ourselves or of each other. ("The Clone or the Degree Xerox of the Species," 199)

In turn, the drive to singularity, which appears with renewed vigor in a society dominated by copies and simulacra, constitutes an effort to restore the kind of individual wholeness that can be attained only through difference from the other individuals in a given society (an aspect that does not exclude cloned human beings who, unlike some of the characters in the stories analyzed in this book, do not have to be just numbers in a series of endless duplications of the same, as we have seen dramatized in such characters as Bernard in *Brave New World*, Mary in Damon Knight's "Mary," Mark in *Where Late the Sweet Birds Sang*, and the five clones in Pamela Sargent's *Cloned Lives*.

Addressing the vexed question of sameness and psychic cloning, translated into a wish for community, for safety in sameness, while simultaneously concealing a mistrust of that very same similarity, Adam Phillips observes that "there is, for some people, a deep fear of not being a clone, of not being identical to someone, or identical to someone's wishes for oneself . . . as though we can only work out what or who we are like from the foundational belief—the unconscious assumption—that there is someone else that we are exactly like" ("Sameness Is All," 92). It is indeed the paradoxical wish to be the same, therefrom drawing comfort and a sense of ontological justification for one's existence, while at the same time vying to be different, that stands at the core of the difficulties and conflicts inherent in human development from early childhood, which to a great extent also informs the ambiguities attendant upon the drive to clone or be cloned. Although, as Adam Phillips remarks, people "can never be identical to each other," he insightfully concludes that "perhaps this relentless wish for absolute identity—that even real cloning cannot satisfy—conceals, tries to talk us out of, a profound doubt about our being the same as anything" (94). Through the elimination of otherness, discord and hostility might gradually disappear, but so might attraction and seduction by the same token be abolished, although this would only apply given a radical scenario where mass cloning was the norm.

This is also Baudrillard's position. In spite of the allure that the notions of duality, twinship, and cloning exert on him, Baudrillard forcefully argues for the preservation of alterity and against the multiplication of the same and "his" [*sic*] proliferated figures: "incest, autism, twinning, cloning" (*Figures de l'altérité*, 174; my translation). Only by means of keeping this alterity, this differentiation between the sexes and among men and women which is, for Baudrillard, the source of seduction, can the disappearance of the other be averted. As he forcefully reminds us, "One must not be reconciled with our own bodies or with oneself. One must not be reconciled with the Other, one must not be reconciled with nature.... There lies the secret to a strange attraction" (175; my translation). In the last analysis, then, Baudrillard's theory of seduction arguably constitutes his most important argument against cloning.

In spite of the strategic exclusion of men from many utopian texts written by women, envisaging female-only worlds where women-centered policies and a way of life could be developed, most texts ultimately advocate a harmonious existence between men and women, often envisaging a gradual "femalization of man," as Herbert Marcuse postulated (*Counterrevolution and Revolt*, 74). Shulamith Firestone considers that we now "have the knowledge to create a paradise on earth anew" (*The Dialectic of Sex: The Case for Feminist Revolution*, 242), although she goes on to warn that "the alternative is our own suicide through that knowledge, the creation of a hell on earth, followed by oblivion" (242). Firestone's solution to the question of women's subordinate position in society is, among other strategic moves, the "freeing of women from the tyranny of their biology by any means available" (238), which would sever the seemingly unbreakable bond between woman and motherhood. It is sad to consider that some women still believe that motherhood may be their only source of "power" and control in an androcentric society, which bespeaks the need for a much greater emphasis on education and financial independence for those women. If motherhood becomes a *real* choice, through whatever means (normal conception and pregnancy, in vitro fertilization, cloning, extrauterine developing of the fetus following traditional or artificial insemination, or cloning) then the concept of women as wombs will likely gradually disappear. Could it be, then, that cloning might be the means for women to achieve a *for-herself*, a definition without recourse to male norms, so often dreamed of and dramatized in so many of the texts looked at in this book? There seem to be good reasons to entertain some measure of optimism, even though laced with caution and the need to monitor closely developments in the sexual politics being applied to decisions to implement new reproductive technologies, namely, who will have access to them and what they are being used for.

In polemical fashion, Adrienne Rich puts forward an anticipatory utopian fantasy that appears to me to sum up my own vision as well as that of many of the texts analyzed in this book. According to Rich,

> Ideally, of course, women would choose not only whether, when, and where to bear children, and the circumstances of labor, but also between biological and artificial reproduction. But I do not think we can project any such idea onto the future—and hope to realize it—without examining the shadow images we carry in us, the magical thinking of Eve's curse, the social victimization of women-as-mothers. . . . If motherhood and sexuality were not wedged resolutely apart by male culture, if we could *choose* both the forms of our sexuality and the terms of our motherhood or non-motherhood freely, women might achieve sexual autonomy (as opposed to "sexual liberation"). ("The Theft of Childbirth," 26)

As Rich goes on to state, as if addressing the scenarios visited in some of the fictional narratives examined in the previous chapters,

> The mother should be able to choose the means of conception (biological, artificial, or even parthenogenetic), the place of birth, her own style of giving birth, and her birth attendants. Birth might then become one event in the unfolding of our diverse and polymorphous sexuality—not a necessary *consequence* of sex, but one aspect of liberating ourselves from fear and the loathing of our bodies. (29–30)

As I have been arguing, human cloning technology, if applied according to well-regulated principles (coupled or not with the use of artificial wombs), could significantlty contribute to ameliorate women's lives, giving them a wider range of choices as far as reproductive technologies are concerned, which would impact on their lifestyles and on promoting a more egalitarian status quo between women and men.

Maybe in a society in the not-so-distant future women who so wish will be able to find a fulfilling sense of maternity without having to feel obliged by their biological clocks or social expectations, and maybe men will be given a similar opportunity—that of bearing children themselves or seeing their own children develop inside artificial wombs, should they choose to have children on their own.

In *Of Woman Born: Motherhood as Experience and Institution*, Rich stresses the need to "seek visions, to dream dreams . . . to try new ways of living, to make room for serious experimentation" (282). At the same time, however, she introduces a note of caution when she remarks that "in the light of most women's lives as they are now having to be lived, it can seem naïve and self-indulgent to spin forth matriarchal utopias, to 'demand' that the technologies of contraception and genetics be 'turned over' to women (by whom, and under what kinds of pressure?)" (282). These are indeed very important questions, but they no longer sound so naïve with an ever growing number of women in the scientific and medical professions, even though there is still a long way to go before a modicum of balance between men and women in the seats of political decisions is reached, the feminist project that is the most difficult but also the most strategically important to attain.

As Haraway, in a related vein, stresses, "Making connections is the kind of physiology in feminist science studies that I want to foster. I want feminists to be enrolled more tightly in the meaning-making process of technoscientific world-building" (*Modest_Witness@Second_Millenium. FemaleMan©_Meets_Oncomouse™: Feminism and Technoscience*, 127). N. Katherine Hayles, in turn, concurs with Donna Haraway in stressing the potential pleasures and potentialities of our posthuman state. As Hayles remarks, "For some people, including me, the posthuman evokes the exhilarating prospect of getting out of the old boxes and opening up new ways of thinking about what being human means" (*How We Became Posthuman: Virtual Bodies in Cybernetics, Literature, and Informatics*, 285). What being human means lies at the very center of the psychological and philosophical preoccupations raised by human cloning.

I wish to finish with Kira, a character in Pamela Sargent's *Cloned Lives*, whose utopian vision for the future of humanity, which includes immortality, freedom from most diseases, and what she sees as the benefits of a judicious use of genetic engineering, bears many resemblances to my own utopian fantasies:

> She saw another humanity on Earth, freed from the determinants of genetic disabilities, of aging bodies, of unbalanced minds, and of death. She saw them freed from the tyranny of time and the roulette of reproduction, able to deliberate, to consider, to enjoy each moment. She saw a people freed from the necessity to change everything around them because they could instead change themselves. (*Cloned Lives*, 308–309)

The vision is further elaborated in the following terms, which again can be seen as engaging with and articulating Shulamith Firestone's blueprint for utopia in *The Dialectic of Sex: The Case for Feminist Revolution*:

People might at last walk a peaceful path, themselves whole in body and mind, able to turn from the problems that had always beset human beings to the more important ones of purpose and discovery. They might even learn to treasure the Earth. . . . Some would venture off the earth and in confronting other life forms confront their own hidden desires. . . . She saw a world that might finally achieve the Marxist dream of the withering away of the state, the libertarian dream of freedom for each person. . . . It would not be a utopia, of that she was sure. There would be new problems, perhaps more threatening than the old. . . . But there would be a new pattern for human existence, enough time, hopefully, for anyone to succeed, a chance for everyone to explore all possible alternatives, unlimited by time. (309)

I consider this fantasy a strategic outline of what a more peaceful, just, egalitarian future might look like, where both women and men would lead their lives free from structures of dominance related to their sex. Moreover, given the state of technological development, it does not seem too far-fetched to surmise that many of these conditions might in fact materialize in the next few decades. It is toward such a vision that the drive of my argumentation and reflections on human cloning and related issues have been leading. Indeed, this is also very much my vision.

NOTES

1. As I am writing this, Raelians (a sect led by Rael, the founder of Clonaid), claim they have cloned several babies, the first of whom was born on December 26, 2002, and was given the symbolic name of Eva, the initiator of a new age for humanity. Prof Panagiotis Zavos (who works at the University of Kentucky) and the Italian doctor Severino Antinori have also been busy attempting to clone the first baby.

2. While George Johnson discusses the potentiality of cloning a brain, concluding that it is impossible ("Soul Searching," 67), what I am suggesting here is brain transference.

3. The problematic relations between mother and cloned daughter, who is simultaneously her twin sister and could have been her mother, are also at the center of John Varley's short story "Lollipop and the Tar Baby" (1977). As Elizabeth in Hoffman's *The Secret*, who has a cloned daughter who serves in many ways as a challenge to her but also as someone to raise as a companion, in Varley's tale Zoe, whose job is to catch black holes and who has to spend extended periods of time in her spaceship, decides to clone herself at a time when she craves for company. As Zoe puts it, "And what better company than herself? With the medical facilities aboard *Shirley* she had grown a copy of

herself and raised the little girl as her daughter" (359). As Xanthia, the cloned daughter, muses, "In her whole life she had seen and talked to only one other human being, and that was Zoe, who was one hundred and thirty-five years old and her identical twin" (358). This claustrophobic atmosphere between mother and daughter, confined to the same spaceship, and compounded by the ambiguity and intersection of their family ties, contributes to create an atmosphere of suspicion on the part of Xantia, who believes she is not safe and that her mother/sister's goal is eventually to kill her. She feels like a disposable object, serving her mother's selfish and narcissistic impulses, a pawn in her mother's hands. Eventually Zoe gives Xanthia her own spaceship, *Lollipop*, telling her that she will be able to raise "little copies" (363) of herself if she wants to. As Xanthia realizes, she "was recapitulating the growth Zoe had already been through a hundred years ago" (362) and her life "had been a process of growing slowly into the mold Zoe represented" (365). Indeed, as Xanthia recognizes, in the society of two where she had grown up, "The only other person she knew had her own face" (366), and unsurprisingly she grows afraid of males and mirrors. In Varley's future world, where people could live almost forever, "Clone Control Regulations" (370) had been instituted in order to limit population, for "even if everyone had only one child—the Birthright—population would still grow. For a while, clones had been a loophole" (370), but that had come to an end with the introduction of the genetic statutes, which dictated that "only one person had the right to any one set of genes. If two possessed them, one was excess, and was summarily executed" (371). Xanthia rebels, refusing to emulate her mother's actions, suspecting she was not the only clone her "sister-mother" (364) had raised, as well as conjecturing that like the other clones she would probably be killed in the end, since she is "excess" according to the Clone Control Regulations. In order to save herself she allows her mother to disappear into a black hole.

4. Walter Benjamin's "The Work of Art in the Age of Mechanical Reproduction" (first published in 1936 in *Zeitschrift für Sozialforschung* 5, 1), a year before Magritte's painting, touches on some of the issues raised by *La Réproduction Interdite,* such as the consequences of reproducibility in terms of the cultural and artistic value of a work of art, whose copying would cause its aura to decrease and consequently its worth to decline.

5. Wendy Doniger, "Sex and the Mythological Clone," 136.

Works Cited

Ackerman, Felicia. "Flourish Your Heart in This World." In *Clones and Clones: Facts and Fantasies about Human Cloning*, edited by Martha Nussbaum and Cass R. Sustein, 310–331. New York: W. W. Norton, 1998.

Ackroyd, Peter. *The House of Doctor Dee*. Harmondsworth, UK: Penguin, 1993.

Adam, Villiers de L'Isle. *L'Eve Future* (1886); *Eve of the Future Eden*, translated by Marilyn Gaddis Rose. Lawrence, KS: Coronado Press, 1981.

Adams, Alice E. *Reproducing the Womb: Images of Childbirth in Science, Feminist Theory, and Literature*. Ithaca and London: Cornell University Press, 1994.

Aeschylus. *Eumenides*, translated by Philip Vellacott. Harmondsworth, UK: Penguin, 1956.

Albinski, Nan Bowman. "Utopia Reconsidered: Women Novelists and Nineteenth-Century Utopian Visions." *Signs* 13 (Summer 1988): 830–841.

———. *Women's Utopias in British and American Fiction*. New York: Routledge, 1988.

Aldrich, H. E., and Marsden, P. V. "Environments and Organizations." In *The Handbook of Sociology*, edited by N. J. Smelser, 361–392. Newbury Park, CA: Sage, 1988.

Alford, C. Fred. *Narcissism: Socrates, the Frankfurt School, and Psychoanalytic Theory*. New Haven and London: Yale University Press, 1988.

Allen, Jeffner. "Motherhood: The Annihilation of Women." In *Mothering: Essays in Feminist Theory*, edited by Joyce Trebilcot, 315–330. Totowa, NJ: Rowman and Littlefield, 1984.

Allen, S. G., and J. Hubbs. "Outrunning Atalanta: Destiny in Alchemical Transmutation." *Signs* 6, 2 (1980): 210–229.

Anderson, Poul. *Virgin Planet*. New York: Warner Books, 1959.

Andrews, Lori B. "Surrogate Motherhood: The Challenge for Feminists." In *Sex/Machine: Readings in Culture, Gender, and Technology*, edited by Patrick D. Hopkins, 157–170. Bloomington and Indianapolis: Indiana University Press, 1998.

Annas, George. "Scientific Discoveries and Cloning: Challenges for Public Policy." In *Flesh of My Flesh: The Ethics of Cloning Humans*, A Reader, edited by Gregory E. Pence, 77–83. Lanham, MD: Rowman & Littlefield, 1998.

Apollinaire, Guillaume. *L'Enchanteur Pourrissant, suivi de "Les Mamelles de Tirésias" et de "Couleur du temps."* Paris: Gallimard, 1972.

Arditti, Rita, Renate Duelli Klein, and Shelley Minden, eds. *Test-Tube Women: What Future for Motherhood?* London: Pandora, 1984.

Arnzen, Michael. "The Return of the Uncanny." *Paradoxa* 3, 3–4 (1997): 315–320.

Attebery, Brian. "Women Alone, Men Alone: Single-Sex Utopias." *FEMSPEC* 1, 2 (2000): 4–15.

Atwood, Margaret. *The Handmaid's Tale* (1985). London: Virago, 1991.

———. *Oryx and Crake*. London: Bloomsbury, 2003.

Auerbach, Nina. *Communities of Women: An Idea in Fiction*. Cambridge and London: Harvard University Press, 1978.

Bakhtin, Mikhail. *Problems of Dostoevsky's Poetics*, translated by R. W. Rotsel. Ann Arbor, MI: Ardis, 1973.

Barr, Marleen S. *Feminist Fabulation: Space/Postmodern Fiction*. Iowa City: University of Iowa Press, 1992.

———. *Alien to Femininity: Speculative Fiction and Feminist Theory*. Westport, CT: Greenwood Press, 1987.

———. "Post-Phallic Culture: Reality Now Resembles Utopian Feminist Science Fiction." In *Future Females, The Next Generation: New Voices and Velocities in Feminist Science Fiction Criticism*, edited by Marleen S. Barr, 67–84. Lanham, MD: Rowman and Littlefield, 2000.

———. "'We're at the Start of a New Ball Game and That's Why We're All Real Nervous': Or, Cloning—Technological Cognition Reflects Estrangement from Women." In *Learning from Other Worlds: Estrangement, Cognition, and the Politics of Science Fiction and Utopia*, edited by Patrick Parrinder, 193–207. Durham, NC: Duke University Press, 2001.

Barrett, Michele. *Women's Oppression Today*. London: Verso, 1980.

Barthes, Roland. *Roland Barthes by Roland Barthes*, translated by Richard Howard. Berkeley and Los Angeles: University of California Press, 1994.

Bartkowski, Frances. *Feminist Utopias*. Lincoln and London: University of Nebraska Press, 1989.

Battersby, Christine. *The Phenomenal Woman: Feminist Metaphysics and the Patterns of Identity*. New York: Routledge, 1998.

Baudrillard, Jean. "Clone Story ou L'enfant prothèse." *Traverses* 14–15 (1979): 143–148.

———. "Clone Story." In *Simulacra and Simulation*, translated by Sheila Faria Glaser, 95–103. Ann Arbor: University of Michigan Press, 1981.

———. *Simulacra and Simulation*, translated by Sheila Faria Glaser. Ann Arbor: University of Michigan Press, 1981.

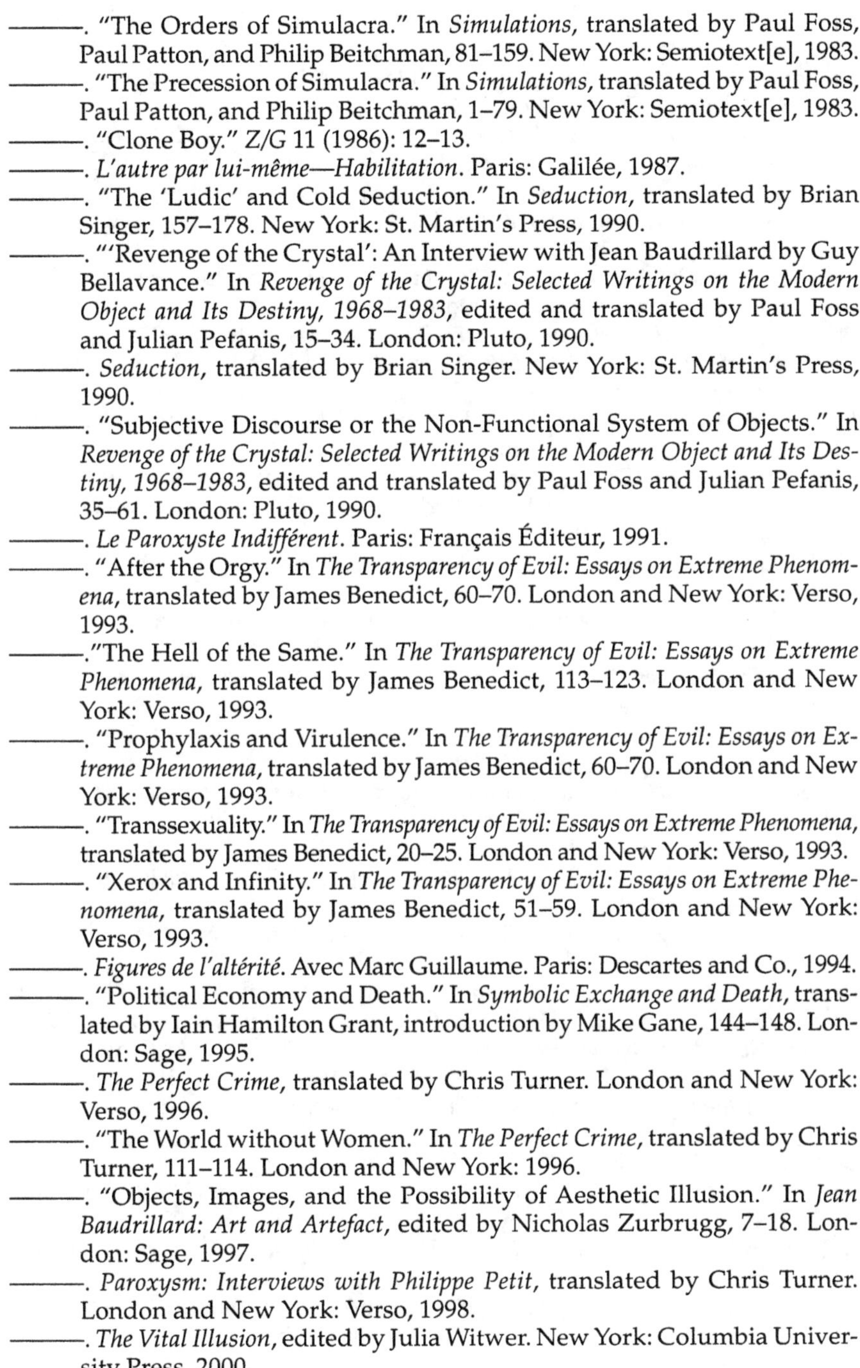

———. "The Orders of Simulacra." In *Simulations*, translated by Paul Foss, Paul Patton, and Philip Beitchman, 81–159. New York: Semiotext[e], 1983.

———. "The Precession of Simulacra." In *Simulations*, translated by Paul Foss, Paul Patton, and Philip Beitchman, 1–79. New York: Semiotext[e], 1983.

———. "Clone Boy." *Z/G* 11 (1986): 12–13.

———. *L'autre par lui-même—Habilitation*. Paris: Galilée, 1987.

———. "The 'Ludic' and Cold Seduction." In *Seduction*, translated by Brian Singer, 157–178. New York: St. Martin's Press, 1990.

———. "'Revenge of the Crystal': An Interview with Jean Baudrillard by Guy Bellavance." In *Revenge of the Crystal: Selected Writings on the Modern Object and Its Destiny, 1968–1983*, edited and translated by Paul Foss and Julian Pefanis, 15–34. London: Pluto, 1990.

———. *Seduction*, translated by Brian Singer. New York: St. Martin's Press, 1990.

———. "Subjective Discourse or the Non-Functional System of Objects." In *Revenge of the Crystal: Selected Writings on the Modern Object and Its Destiny, 1968–1983*, edited and translated by Paul Foss and Julian Pefanis, 35–61. London: Pluto, 1990.

———. *Le Paroxyste Indifférent*. Paris: Français Éditeur, 1991.

———. "After the Orgy." In *The Transparency of Evil: Essays on Extreme Phenomena*, translated by James Benedict, 60–70. London and New York: Verso, 1993.

———."The Hell of the Same." In *The Transparency of Evil: Essays on Extreme Phenomena*, translated by James Benedict, 113–123. London and New York: Verso, 1993.

———. "Prophylaxis and Virulence." In *The Transparency of Evil: Essays on Extreme Phenomena*, translated by James Benedict, 60–70. London and New York: Verso, 1993.

———. "Transsexuality." In *The Transparency of Evil: Essays on Extreme Phenomena*, translated by James Benedict, 20–25. London and New York: Verso, 1993.

———. "Xerox and Infinity." In *The Transparency of Evil: Essays on Extreme Phenomena*, translated by James Benedict, 51–59. London and New York: Verso, 1993.

———. *Figures de l'altérité*. Avec Marc Guillaume. Paris: Descartes and Co., 1994.

———. "Political Economy and Death." In *Symbolic Exchange and Death*, translated by Iain Hamilton Grant, introduction by Mike Gane, 144–148. London: Sage, 1995.

———. *The Perfect Crime*, translated by Chris Turner. London and New York: Verso, 1996.

———. "The World without Women." In *The Perfect Crime*, translated by Chris Turner, 111–114. London and New York: 1996.

———. "Objects, Images, and the Possibility of Aesthetic Illusion." In *Jean Baudrillard: Art and Artefact*, edited by Nicholas Zurbrugg, 7–18. London: Sage, 1997.

———. *Paroxysm: Interviews with Philippe Petit*, translated by Chris Turner. London and New York: Verso, 1998.

———. *The Vital Illusion*, edited by Julia Witwer. New York: Columbia University Press, 2000.

———. "The Clone or the Degree Xerox of the Species." In *Screened Out*, translated by Chris Turner, 196–202. London and New York: Verso, 2002.
Becket, Samuel. *Malone Dies*. London: Calder, 1994.
Benjamin, Walter. "The Work of Art in the Age of Mechanical Reproduction." In *Illuminations*, edited with an introduction by Hannah Arendt, translated by Harry Zohn, 219–253. London: Collins/Fontana, 1973.
———. "The Image of Proust." In *Illuminations*, edited with an introduction by Hanna Arendt, translated by Harry Zohn, 203–217. London: Collins/Fontana Books, 1973.
———. *Charles Baudelaire: A Lyric Poet in the Era of High Capitalism*, translated by Harry Zohn. London: Verso, 1997.
Benton, Jill. *Naomi Mitchison: A Biography*. London: Pandora, 1992.
Berman, Jeffrey. *Narcissism and the Novel*. New York and London: New York University Press, 1990.
Bettelheim, Bruno. *Symbolic Wounds: Puberty Rites and the Envious Male*, rev. ed. New York: Collier Books, 1962.
Blake, William. *The Complete Poetry and Prose of William Blake*, newly rev. ed., edited by David V. Erdman. Berkeley and Los Angeles: University of California Press, 1982.
Bleiler, Everett. *Science-Fiction: The Early Years*. Kent, OH: Kent State University Press, 1990.
Bloch, Ernst. *The Principle of Hope*, vol. 1., translated by Neville Plaice, Stephen Plaice, and Paul Knight. Cambridge, MA: MIT Press, 1996.
———. *The Spirit of Utopia*, translated by Anthony A. Nassar. Stanford, CA: Stanford University Press, 2000.
Bloom, Harold. *Blake's Apocalypse: A Study in Poetic Argument*. Ithaca, NY: Cornell University Press, 1970.
Bogue, Ronald. *Deleuze and Guattari*. London and New York: Routledge, 1990.
Bova, Ben. *Immortality: How Science Is Extending Your Life Span—And Changing the World*. New York: Avon Books, 1998.
Bowen-Moore, Patricia. *Hanna Arendt's Philosophy of Natality*. Basingstoke, UK: Macmillan, 1989.
Bradley, Marion Zimmer. *The Ruins of Isis*. New York: Simon and Schuster, 1978.
Braidotti, Rosi. *Patterns of Dissonance: A Study of Women in Contemporary Philosophy*, translated by Elizabeth Guild. Cambridge: Polity Press, 1991.
———. *Nomadic Subjects: Embodiment and Sexual Difference in Contemporary Feminist Theory*. New York: Columbia University Press, 1994.
———. *Metamorphoses: Towards a Materialist Theory of Becoming*. Cambridge, UK: Polity Press, 2002.
Brin, David. *Glory Season*. New York: Bantam Books, 1993.
Broderick, Damien. *The Last Mortal Generation: How Science Will Alter Our Lives in the 21st Century*. Sydney: New Holland Publishers, 1999.
Bukatman, Scott. *Terminal Identity: The Virtual Subject in Postmodern Science Fiction*. Durham and London: Duke University Press, 1993.
Burdekin, Katharine. *Quiet Ways*. London: Thornton Butterworth, 1930.
———. *Swastika Night* (1937), introduction by Daphne Patai. Old Westbury, NY: Feminist Press, 1985.

———. *The End of This Day's Business*. Afterword by Daphne Patai. New York: Feminist Press, 1989.

Butler, Octavia. "Bloodchild." In *Bloodchild and Other Stories*, 3–32. New York: Seven Stories Press, 1996.

Calder, Jenni. *The Nine Lives of Naomi Mitchison*. London: Virago, 1997.

Callahan, Daniel. "Cloning: Then and Now." In *Cloning: Responsible Science or Technomadness?* edited by Michael Ruse and Aryne Sheppard, 309–313. Amherst, NY: Prometheus Books, 2001.

Carey, John, ed. *The Faber Book of Utopias*. London: Faber and Faber, 1999.

Carroll, Noël. *A Philosophy of Mass Art*. Oxford: Clarendon Press, 1998.

Carter, Angela. *The Passion of New Eve* (1977). London: Virago, 1992.

———. *The SadeianWoman: An Exercise in Cultural History*. London: Virago, 1979.

Cavarero, Adriana. *In Spite of Plato: A Feminist Rewriting of Ancient Philosophy*, foreword by Rosi Braidotti, translated by Anderlini-D'Onofrio and Áine O'Healy. Cambridge, UK: Polity Press, 1995.

Chadwick, Ruth F., ed. *Ethics, Reproduction and Genetic Control*. London and New York: Croom Helm, 1987.

Charnas, Suzy McKee. *Walk to the End of the World* and *Motherlines* (1978). London: The Women's Press, 1989.

———. "A Woman Appeared." In *Future Females: A Critical Anthology*, edited by Marleen S. Barr. Bowling Green, OH: Bowling Green University Popular Press, 1981.

Cherfas, Jeremy, and John Gribbin. *The Redundant Male: Is Sex Irrelevant in the Modern World?* London: Bodley Head, 1984.

Cherry, C. J. *Cyteen* (1988). New York: Warner Books, 1995.

Chodorow, Nancy. *The Reproduction of Mothering: Psychoanalysis and the Sociology of Gender*. Berkeley and Los Angeles: University of California Press, 1978.

Cixous, Hélène. *With, Ou L'Art de L'Innocence*. Paris: des femmes, 1981.

Cixous, Hélène, and Catherine Clément. *The Newly Born Woman*, translated by Betsy Wing; introduction by Sandra M. Gilbert. Manchester: Manchester University Press, 1987.

Clark, I. F. *The Pattern of Expectation 1644–2001*. New York: Basic Books, 1979.

Clark, R. *The Life and Work of J.B.S. Haldane*. London: Hodder and Stoughton, 1968.

Clark, Stephen R. L. *Biology and Christian Ethics*. Cambridge: Cambridge University Press, 2001.

Clarke, Arthur C. *Imperial Earth*. New York: Ballantyne Books, 1976.

Clayes, Gregory, and Lyman Tower Sargent, eds. *The Utopia Reader*. New York and London: New York University Press, 1999.

Clyde, Irene (pseud. of Thomas Baty). *Beatrice the Sixteenth*. London: George Bell, 1909.

———. *Alone in Japan: The Reminiscenses of an International Jurist Resident in Japan, 1916–1954*, edited by Motokichi Hasegawa. Tokyo: Maruzen, 1959.

Cohn, Carol. "Sex and Death in the Rational World of Defense Intellectuals." In *Exposing Nuclear Phallacies*, edited by Diana E. H. Russell, 127–159. New York: Pergamon Press, 1989.

Colker, Ruth. *Pregnant Men: Practice, Theory, and the Law*. Bloomington and Indianapolis: Indiana University Press, 1994.

Cook, Robin. *Mutation*. New York: Berkley Books, 1990.

Cooper, Edmund. *Five to Twelve*. New York: G. P. Putnam's Sons, 1968.

———. *Who Needs Men?* (1972). London: Coronet Books, 1974.

Corbett, Mrs George. *New Amazonia: A Foretaste of the Future*. London: Tower Publishing, 1889.

Corea, Gena. *The Mother Machine: Reproductive Technologies from Artificial Insemination to Artificial Wombs*. London: The Women's Press, 1985.

———. "The Reproductive Brothel." In Gena Corea, Renate Duelli Klein, Jalna Hanmer, Helen B. Holmes, Betty Hoskins, Madhu Kishwar, Janice Raymond, Robyn Rowland, and Roberta Steinbacher, *Man-Made Women: How New Reproductive Technologies Affect Women*, 38–51. Bloomington and Indianapolis: Indiana University Press, 1985.

Corea, Gena, Renate Duelli Klein, Jalna Hanmer, Helen B. Holmes, Betty Hoskins, Madhu Kishwar, Janice Raymond, Robyn Rowland, and Roberta Steinbacher, *Man-Made Women: How New Reproductive Technologies Affect Women*. Bloomington: and Indianapolis: Indiana University Press, 1985.

Cortiel, Jeanne. *Demand My Writing: Joanna Russ/Feminism/Science Fiction*. Liverpool, UK: Liverpool University Press, 1999.

———. "Determinate Politics of Indeterminacy: Reading Joanna Russ's Recent Work in Light of Her Early Short Fiction." In *Future Females, the Next Generation: New Voices and Velocities in Feminist Science Fiction Criticism*, edited by Marleen S. Barr, 219–236. Lanham, MD: Rowman and Littlefield, 2000.

Cranny-Francis, Anne. *The Body in the Text*. Carlton South, Victoria: Melbourne University Press, 1995.

Creed, Barbara. *The Monstrous-Feminine: Film, Feminism, Psychoanalysis*. London and New York: Routledge, 1993.

Cussins, Charis Thompson. "Confessions of a Bioterrorist: Subject Position and Reproductive Technologies." In *Playing Dolly: Technocultural Formations, Fantasies, and Fictions of Assisted Reproduction*, edited by E. Ann Kaplan and Susan Squier, 189–219. New Brunswick, NJ: Rutgers University Press, 1999.

Daly, Mary. *Gyn/Ecology: The Metaethics of Radical Feminism*, with a New Intergalactic Introduction by the Author. Boston: Beacon Press, 1990.

———. *Quintessence: Realizing the Archaic Future: A Radical Elemental Feminist Manifesto*. London: The Women's Press, 1999.

Dann, Jack, and Gardner Dozois, eds. *Clones*. New York: Ace Books, 1998.

de Beauvoir, Simone. *The Second Sex* (1949), translated and edited by H. M. Parshley. Harmondsworth: Penguin, 1977.

De Lauretis, Teresa. *Alice Doesn't: Feminism, Semiotics, Cinema*. Basingstoke, UK: Macmillan, 1987.

Delblanc, Sven. *Homunculus: A Magic Tale* (1965), translated by Verne Moberg. Englewood Cliffs, NJ: Prentice-Hall, 1969.

Deleuze, Gilles. *Nietzsche and Philosophy*, translated by Hugh Tomlinson. New York: Columbia University Press, 1983.

———. "The Simulacrum and Ancient Philosophy." In *The Logic of Sense*, translated by Mark Lester with Charles Stivale; edited by Constantin V. Boundas. New York: Columbia University Press, 1990.

———. *Difference and Repetition*, translated by Paul Patton. New York: Columbia University Press, 1994.

Derrida, Jacques. *Of Grammatology*, translated by Gayatri Chakravorty Spivak. Baltimore and London: Johns Hopkins University Press, 1976.

———. *Dissemination*, translated with an introduction and additional notes by Barbara Johnson. Chicago: University of Chicago Press, 1981.

———. *Of Spirit: Heidegger and the Question*, translated by Geoffrey Bennington and Rachel Bowlby. Chicago and London: University of Chicago Press, 1989.

———. "The Double Session." In *A Derrida Reader: Between the Blinds*, edited with an introduction and notes by Peggy Kamuf, 171–199. New York: Harvester Wheatsheaf, 1991.

Derrida, Jacques, and Elisabeth Roudinesco. *De quoi demain . . . Dialogue*. Paris: Galilée, 2001.

Diagnostic and Statistical Manual of Mental Disorder, 3d ed. (DSM-III). Arlington, Va.: American Psychiatric Association, 1980.

Dick, Philip K. *Do Androids Dream of Electric Sheep?* London: Millenium Books, 1968.

Dinnerstein, Dorothy. *The Mermaid and the Minotaur: Sexual Arrangements and Human Malaise*. New York: Harper and Row, 1976.

DiQuinzio, Patrice. *The Impossibility of Motherhood: Feminism, Individualism, and the Problem of Mothering*. New York and London: Routledge, 1999.

Dolar, Mladen. "I Shall Be with You on Your Wedding Night." *October* 58 (1991): 5–23.

Donawerth, Jane L. "Science Fiction by Women in the Early Pulps, 1926–1930." In *Utopian and Science Fiction by Women: Worlds of Difference*, edited by Jane L. Donawerth and Carol A. Kolmarten; foreword by Susan Gubar, 137–152. Liverpool, UK: Liverpool University Press, 1994.

Doniger, Wendy. "Sex and the Mythological Clone." In *Clones and Clones: Facts and Fantasies about Human Cloning*, edited by Martha C. Nussbaum and Cass R. Sustein, 114–138. New York: W. W. Norton, 1998.

Duffy, Maureen. *Gor Saga* (1981). London: Methuen, 1988.

Durham, Scott. *Phantom Communities: The Simulacrum and the Limits of Postmodernism*. Stanford, CA: Stanford University Press, 1998.

Eagleton, Terry. *Walter Benjamin, or, Towards a Revolutionary Criticism*. London: Verso, 1981.

Egri, Carolyn P., and Lawrence T. Pinfield. "Organizations and the Biosphere: Ecologies and Environments." In *Managing Organizations: Current Issues*, edited by Stewart R. Clegg, Cynthia Hardy, and Walter R. Nord, 208–233. London: Sage, 1999.

Elliott, David. "Uniqueness, Individuality, and Human Cloning." In *Cloning: Responsible Science or Technomadness?* edited by Michael Ruse and Aryne Sheppard, 103–121. Amherst, NY: Prometheus Books, 2001.

Erdrich, Louise. *The Antelope Wife*. New York: HarperPerennial, 1998.

Ertz, Susan. *Woman Alive*. London: Hodder and Stoughton, 1935.

Eskridge, William N., Jr., and Edward Stein. "Queer Clones." In *Clones and Clones: Facts and Fantasies about Human Cloning*, edited by Martha Nussbaum and Cass R. Sustein, 95–113. New York: W. W. Norton, 1998.

Euripides. *Hippolytos*, translated by Robert Bagg. New York: Oxford University Press, 1973.

———. *Medea and Other Plays*, translated with an introduction by Philip Vellacott. Harmondsworth: Penguin, 1986.

Farquahr, Dion. *The Other Machine: Discourse and Reproductive Technologies*. New York and London: Routledge, 1996.

Faulks, Lana. *Fay Weldon*. New York: Twayne Publishers, 1998.

Ferguson, K. E. *The Feminist Case against Bureaucracy*. Philadelphia: Temple University Press, 1984.

Ferreira, Maria Aline. "'Glad Wombs' and 'Friendly Tombs': Reembodiments in D. H. Lawrence's Late Works." In *Writing the Body in D. H. Lawrence: Essays on Language, Representation, and Sexuality*, edited by Paul Poplawski, 163–175. Westport, CT: Greenwood Press, 2001.

Firestone, Shulamith. *The Dialectic of Sex: The Case for Feminist Revolution* (1970). New York: Bantam, 1972.

Fitting, Peter. "So We All Became Mothers: New Roles for Men in Recent Utopian Fiction." *Science Fiction Studies* 12, 2 (1985): 156–183.

———. "For Men Only: A Guide to Reading Single-Sex Worlds." *Women's Studies* 14, 2 (1987): 101–117.

———. "Reconsiderations of the Separatist Paradigm in Recent Feminist Science Fiction." *Science Fiction Studies* 19 (1992): 32–48.

Foster, Hal, ed. *Postmodern Culture*. London: Pluto Press, 1985.

———. *Compulsive Beauty*. Cambridge, MA: MIT Press, 1997.

Foucault, Michel. *The Order of Things: An Archeology of the Human Sciences*. New York: Vintage, 1973.

Francavilla, Joseph. "The Concept of the Divided Self in Harlan Ellison's 'I Have No Mouth and I Must Scream' and 'Shatterday.'" *Journal of the Fantastic in the Arts* 6, 2 & 3 (1994): 107–125.

Freedman, Carl. *Critical Theory and Science Fiction*. Hanover and London: Wesleyan University Press, 2000.

———. "Science Fiction and the Triumph of Feminism." *Science Fiction Studies* 27 (2000): 278–289.

Freedman, Estelle. "Separatism as Strategy: Female Institution Building and American Feminism, 1870–1930." In *Feminism and Community*, edited by Penny A. Weiss and Marilyn Friedman, 85–104. Philadelphia: Temple University Press, 1995.

Freedman, Nancy. *Joshua Son of None*. New York: Delacorte Press, 1973.

Freud, Sigmund. "Twenty-Sixth Lecture." In *The Standard Edition of the Complete Psychological Works of Sigmund Freud*, vol. 14, translated and edited by James Strachey. London (1953–1974).

———. *The Three Essays on the Theory of Sexuality* (1905), edited and translated by James Strachey. New York: Basic Books, 1962.

———. "Little Hans" (1909). In *Case Histories I*, translated from the German under the general editorship of James Strachey. Pelican Freud Library, vol. 8. Harmondsworth, UK: Penguin, 1977.

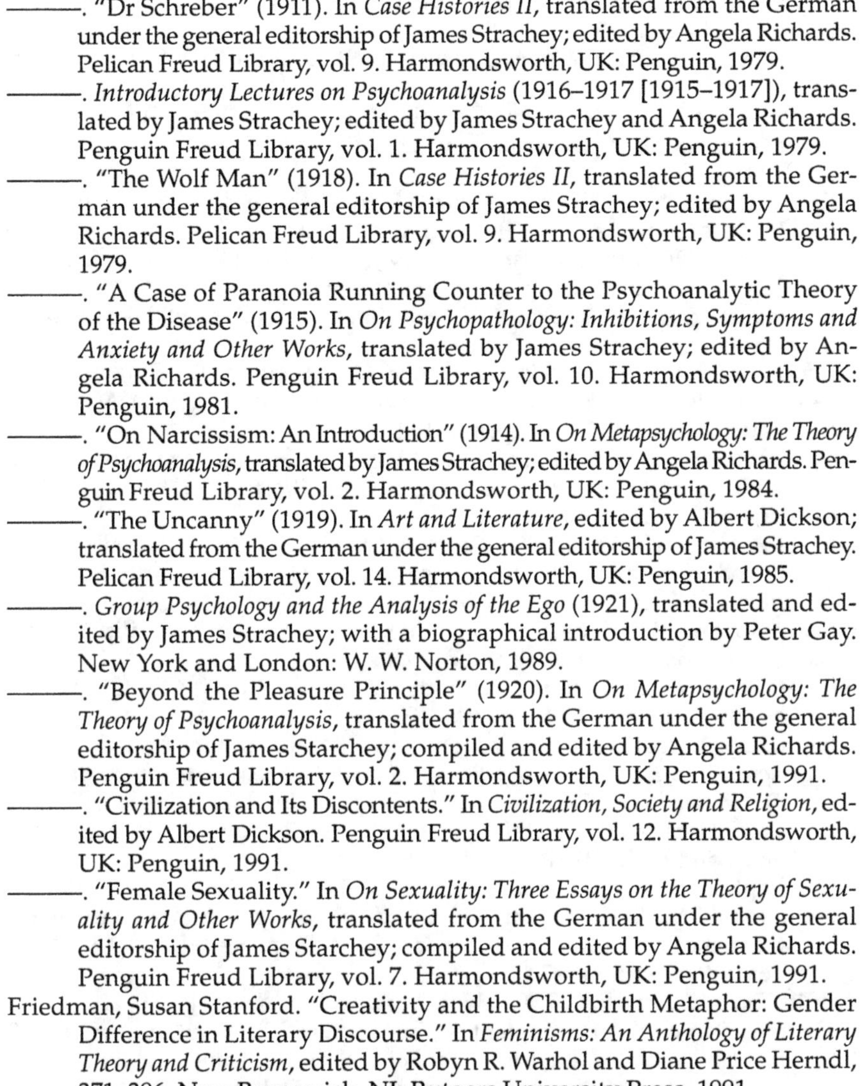

———. "Dr Schreber" (1911). In *Case Histories II*, translated from the German under the general editorship of James Strachey; edited by Angela Richards. Pelican Freud Library, vol. 9. Harmondsworth, UK: Penguin, 1979.

———. *Introductory Lectures on Psychoanalysis* (1916–1917 [1915–1917]), translated by James Strachey; edited by James Strachey and Angela Richards. Penguin Freud Library, vol. 1. Harmondsworth, UK: Penguin, 1979.

———. "The Wolf Man" (1918). In *Case Histories II*, translated from the German under the general editorship of James Strachey; edited by Angela Richards. Pelican Freud Library, vol. 9. Harmondsworth, UK: Penguin, 1979.

———. "A Case of Paranoia Running Counter to the Psychoanalytic Theory of the Disease" (1915). In *On Psychopathology: Inhibitions, Symptoms and Anxiety and Other Works*, translated by James Strachey; edited by Angela Richards. Penguin Freud Library, vol. 10. Harmondsworth, UK: Penguin, 1981.

———. "On Narcissism: An Introduction" (1914). In *On Metapsychology: The Theory of Psychoanalysis*, translated by James Strachey; edited by Angela Richards. Penguin Freud Library, vol. 2. Harmondsworth, UK: Penguin, 1984.

———. "The Uncanny" (1919). In *Art and Literature*, edited by Albert Dickson; translated from the German under the general editorship of James Strachey. Pelican Freud Library, vol. 14. Harmondsworth, UK: Penguin, 1985.

———. *Group Psychology and the Analysis of the Ego* (1921), translated and edited by James Strachey; with a biographical introduction by Peter Gay. New York and London: W. W. Norton, 1989.

———. "Beyond the Pleasure Principle" (1920). In *On Metapsychology: The Theory of Psychoanalysis*, translated from the German under the general editorship of James Starchey; compiled and edited by Angela Richards. Penguin Freud Library, vol. 2. Harmondsworth, UK: Penguin, 1991.

———. "Civilization and Its Discontents." In *Civilization, Society and Religion*, edited by Albert Dickson. Penguin Freud Library, vol. 12. Harmondsworth, UK: Penguin, 1991.

———. "Female Sexuality." In *On Sexuality: Three Essays on the Theory of Sexuality and Other Works*, translated from the German under the general editorship of James Starchey; compiled and edited by Angela Richards. Penguin Freud Library, vol. 7. Harmondsworth, UK: Penguin, 1991.

Friedman, Susan Stanford. "Creativity and the Childbirth Metaphor: Gender Difference in Literary Discourse." In *Feminisms: An Anthology of Literary Theory and Criticism*, edited by Robyn R. Warhol and Diane Price Herndl, 371–396. New Brunswick, NJ: Rutgers University Press, 1991.

Gablik, Suzi. *The Reenchantment of Art*. New York: Thames and Hudson, 1995.

Gallop, Jane. *The Daughter's Seduction: Feminism and Psychoanalysis*. Basingstoke, UK, and London: Macmillan, 1982.

Gardner, Thomas S. "The Last Woman" (1932). In *When Women Rule*, edited by Sam Moskowitz, 131–148. New York: Walker, 1972.

Gasché, Rodolphe. *The Tain of the Mirror: Derrida and the Philosophy of Reflection* (1986). Cambridge and London: Harvard University Press, 1994.

Gatens, Moira. *Imaginary Bodies: Ethics, Power and Corporeality*. London and New York: Routledge, 1996.

Gaylin, William. "Frankenstein Myth Becomes a Reality." *New York Times Magazine*, 5 March 1972.

Gearhart, Sally Miller. *The Wanderground* (1979). London: The Women's Press, 1985.

Genette, Gérard. "The Narcissus Complex." In *Figures I*. Paris: Seuil, 1966.

Gherardi, Sylvia. *Gender, Symbolism and Organizational Cultures*. London: Sage, 1995.

Gilbert, Sandra M., and Susan Gubar. "'Fecundate! Discriminate!' Charlotte Perkins Gilman and the Theologizing of Maternity." In *Charlotte Perkins Gilman: Optimist Reformer*, edited by Jill Rudd and Val Gough, 200–216. Iowa City: University of Iowa Press, 1999.

Gilchrist, Cherry. *The Elements of Alchemy*. Shaftesbury, Dorset, UK: Element, 1991.

Gilman, Charlotte Perkins. *His Religion and Hers*. London: T. F. Unwin, 1924.

———. *The Man-Made World: Or, Our Androcentric Culture* (1911). New York and London: Johnson Reprint, 1971.

———. *With Her in Ourland*, edited by Mary Jo Deegan and Michael R. Hill, with an introduction by Mary Jo Deegan. Westport, CT and London: Praeger, 1997.

———. *Herland, The Yellow Wall-Paper, and Selected Writings*, edited with an introduction and notes by Denise Knight. Harmondsworth, UK: Penguin, 1999.

Goethe, Johann Wolfgang von. *Faust*. Part I. Translated by Philip Wayne. Harmondsworth, UK: Penguin, 1986.

———. *Faust*. Part II. Translated with an introduction by Philip Wayne. Harmondsworth, UK: Penguin, 1979.

Gough, Val. "Lesbians and Virgins: The New Motherhood in *Herland*." In *Anticipations: Essays on Early Science Fiction and Its Precursors*, edited by David Seed, 195–215. Liverpool: Liverpool University Press, 1995.

Gray, Chris Hables. *Cyborg Citizen: Politics in the Posthuman Age*. New York and London: Routledge, 2001.

Gray, Stephen. *Born of Man*. London: GMP, 1989.

Green, Ronald M. "I, Clone." In *Understanding Cloning*, compiled with an introduction by Sandy Fritz; foreword by William Haseltine, 89–96. New York: Warner Books, 2002.

Griffith, Nicola. *Ammonite* (1992). New York: Ballantine Books, 2002.

Habermas, Jürgen. "An Argument against Human Cloning: Three Replies." In *The Postnational Constellation: Political Essays*, translated and edited with an introduction by Max Pensky, 163–172. London: Polity, 2001.

———. *The Future of Human Nature*. London: Polity, 2003.

Halberstam, Judith, and Ira Livingston, eds. *Posthuman Bodies*. Bloomington and Indianapolis: Indiana University Press, 1995.

Haldane, Charlotte. *Man's World*. New York: George H. Doran, 1927.

Haldane, J.B.S. *Daedalus, or Science and the Future*. New York: E. P. Dutton, 1924.

———. "Biological Possibilities for the Human Species in the Next Ten Thousand Years." In *Man and His Future*, edited by Gordon Wolstenholme, 337–361. Boston: Little, Brown, 1963.

———. *The Man with Two Memories*. London: Merlin Press, 1976.

Haraway, Donna. *Primate Visions: Gender, Race, and Nature in the World of Modern Science.* New York and London: Routledge, 1989.

———. "A Cyborg Manifesto: Science, Technology and Socialist-Feminism in the Late Twentieth Century." In *Simians, Cyborgs, and Women: The Reinvention of Nature,* 149–181. London: Free Association Books, 1991.

———. *Simians, Cyborgs, and Women: The Reinvention of Nature.* London: Free Association Books, 1991.

———. Interview with Gary A. Olson. "Writing, Literacy and Technology: Toward a Cyborg Writing." In *Women Writing Culture,* edited by Gary A. Olson and Elizabeth Hirsh; foreword by Donna Haraway; afterword Henry A. Giroux, 45–77. Albany: State University of New York Press, 1995.

———. *Modest_Witness@Second_Millenium.FemaleMan©_Meets_Oncomouse™: Feminism and Technoscience.* New York and London: Routledge, 1997.

———. "The Actors Are Cyborg, Nature Is Coyote, and the Geography Is Elsewhere: Postscript to 'Cyborgs at Large.'" In *Technoculture,* edited by Constance Penley and Andrew Ross, 21–26. Minneapolis: University of Minnesota Press, 1997.

———. "Cyborgs at Large: Interview with Donna Haraway," conducted by Constance Penley and Andrew Ross. In *Technoculture,* edited by Constance Penley and Andrew Ross, 1–20. Minneapolis: University of Minnesota Press, 1997.

———. "The Promises of Monsters: A Regenerative Politics for Innappropriate/d Others." In *Cybersexualities: A Reader on Feminist Theory, Cyborgs and Cyberspace,* edited with an introduction by Jenny Wolmark, 314–366. Edinburgh: Edinburgh University Press, 1999.

———. *How Like a Leaf! An Interview with Thyrza Nichols Goodeve.* New York: Routledge, 2000.

Harding, Sandra. *The Science Question in Feminism.* Ithaca, NY: Cornell University Press, 1986.

———. "Conclusion: Epistemological Questions" (1987). In *Feminisms: A Reader,* edited with an introduction by Maggie Humm, 319–322. New York and London: Harvester Wheatsheaf, 1992.

Harris, John. *Clones, Genes, and Immortality: Ethics and the Genetic Revolution.* Oxford: Oxford University Press, 1998.

Harrison, Jane. *Mythology.* New York: Harcourt Brace, 1963.

Hayles, N. Katherine. "The Life Cycle of Cyborgs: Writing the Posthuman." In *Cybersexualities: A Reader on Feminist Theory, Cyborgs and Cyberspace,* edited with an introduction by Jenny Wolmark, 156–173. Edinburgh: Edinburgh University Press, 1999.

———. *How We Became Posthuman: Virtual Bodies in Cybernetics, Literature, and Informatics.* Chicago and London: University of Chicago Press, 1999.

Hearn, J., and W. Parkin. "Gender and Organizations: A Selected Review and Critique of a Neglected Area." *Organisation Studies* 4, 3 (1983): 219–242.

Hearn, J., D. L. Sheppard, P. Tancred-Sheriff, and G. Burrell, eds. *The Sexuality of Organization.* London: Sage, 1989.

Hillman, David, and Carla Mazzio, eds. "Introduction: Individual Parts." In *The Body in Parts: Fantasies of Corporeality in Early Modern Europe,* xi–xxix. New York and London: Routledge, 1997.

Hoffman, E.T.A. "The Sandman." In *The Golden Pot and Other Stories*, translated and edited by Ritchie Robertson, 85–118. Oxford: Oxford University Press, 1992.

Hoffman, Eva. *The Secret: A Fable for Our Time*. London: Secker and Warburg, 2001.

Holland, Cecilia. *Floating Worlds* (1971). London: Gollacz, 2000.

Hollinger, Veronica. "(Re)reading Queerly: Science Fiction, Feminism, and the Defamiliarzation of Gender." In *Future Females, The Next Generation: New Voices and Velocities in Feminist Science Fiction Criticism*, edited by Marleen S. Barr, 197–215. Lanham, MD: Rowman and Littlefield, 2000.

Holmes, Helen Bequaert. "Sex Preselection: Eugenics for Everyone?" In *Sex/Machine: Readings in Culture, Gender, and Technology*, edited by Patrick D. Hopkins, 116–142. Bloomington and Indianapolis: Indiana University Press, 1998.

Hopkins, Patrick D."(Re)Locating Fetuses." In *Sex/Machine: Readings in Culture, Gender, and Technology*, edited by Patrick D. Hopkins, 172–174. Bloomington and Indianapolis: Indiana University Press, 1998.

———., ed. *Sex/Machine: Readings in Culture, Gender, and Technology*. Bloomington and Indianapolis: Indiana University Press, 1998.

Horney, Karen. "The Dread of Woman." *International Journal of Psychoanalysis* 13 (1932): 348–360.

Horrocks, Christopher. *Baudrillard and the Millennium*. Cambridge: Icon Books, 1999.

Houellebecq, Michel. *Elementary Particles*, translated from the French by Frank Wynne. New York: Vintage International, 2000.

Hunter, Dianne. "Doubling, Mythic Difference, and the Scapegoating of Female Power in *Macbeth*." *Psychoanalytic Review* 75, 1 (1988): 129–152.

Huxley, Aldous. *Brave New World* (1932). New York: HarperPerennial, 1998.

———. *Island* (1962). Harmondsworth, UK: Penguin, 1972.

Huxley, Julian. *What Dare I Think?* New York: Harper & Brothers, 1931.

Ianello, Kathleen P. *Decisions without Hierarchy: Feminist Interventions in Organization Theory and Practice*. New York and London: Routledge, 1992.

Ingram, Angela, and Daphne Patai. "Fantasy and Identity: The Double Life of a Victorian Sexual Radical." In *Rediscovering Forgotten Radicals: British Women Writers, 1889–1939*, 265–302. Chapel Hill and London: University of North Carolina Press, 1993.

Irigaray, Luce. *Speculum of the Other Woman*, translated by Gillian C. Gill. Ithaca, NY: Cornell University Press, 1989.

———. *This Sex Which Is Not One*, translated by Catherine Porter with Carolyn Burke. Ithaca, NY: Cornell University Press, 1990.

———. "The Bodily Encounter with the Mother." In *The Irigaray Reader*, edited with an introduction by Margaret Whitford, 34–46. Oxford: Basil Blackwell, 1991.

———. "How to Define Sexuate Rights?" In *The Irigaray Reader*, edited with an introduction by Margaret Whitford, 204–212. Oxford: Basil Blackwell, 1991.

———. *Marine Lover of Friedrich Nietzsche*, translated by Gillian C. Gill. New York: Columbia University Press, 1991.

———. *The Irigaray Reader*, edited with an introduction by Margaret Whitford. Oxford: Basil Blackwell, 1991.

———. "Women-Mothers: The Silent Stratum of the Social Order." In *The Irigaray Reader*, edited with an introduction by Margaret Whitford, 47–52. Oxford: Basil Blackwell, 1991.

———. *Elemental Passions*, translated from the French by Joanne Collie and Judith Still. London: Athlone Press, 1992.

———. *Sexes and Genealogies*, translated by Gillian C. Gill. New York: Columbia University Press, 1993.

———. "Divine Women." In *Sexes and Genealogies*, translated by Gillian C. Gill, 55–72. New York: Columbia University Press, 1993.

———. *An Ethics of Sexual Difference*, translated by Carolyn Burke and Gillian C. Gill. London: Athlone Press, 1993.

———. *Je, Tu, Nous: Toward a Culture of Difference*, translated from the French by Alison Martin. New York and London: Routledge, 1993.

———. "Love of Same, Love of Other." In *An Ethics of Sexual Difference*, translated by Carolyn Burke and Gillian C. Gill, 97–115. London: Athlone Press, 1993.

———. "Women, the Sacred, Money." In *Sexes and Genealogies*, translated by Gillian C. Gill. New York: Columbia University Press, 1993.

———. "Your Health: What, or Who, Is It?" In *Je, Tu, Nous: Toward a Culture of Difference*, translated from the French by Alison Martin, 101–105. New York and London: Routledge, 1993.

———. *Thinking the Difference: For a Peaceful Revolution*, translated from the French by Karin Montin. London: Athlone Press, 1994.

———. "'Je-Luce Irigaray': A Meeting with Luce Irigaray" (Interview conducted by Elizabeth Hirsch and Gary A. Olson). In *Women Writing Culture*, edited by Gary A. Olson and Elizabeth Hirsh; foreword by Donna Haraway; afterword by Henry A. Giroux, 141–166. Albany: State University of New York Press, 1995.

———. *Democracy Begins between Two*, translated by Kirsteen Anderson. London: Athlone Press, 2000.

———. *To Be Two*, translated by Monique M. Rhodes and Marco F. Cocito-Monoc. London and New Brunswick, NJ: Athlone Press, 2000.

Jackson, Rosemary. "Narcissism and Beyond: A Psychoanalytic Reading of *Frankenstein* and Fantasies of the Double." In *Aspects of Fantasy*, edited by William Coyle, 43–53. Westport, CT: Greenwood Press, 1986.

———. *Fantasy: The Literature of Subversion*. London and New York: Routledge, 1988.

Jacobs, Naomi. "The Frozen Landscape in Women's Utopian and Science Fiction." In *Utopian and Science Fiction by Women: Worlds of Difference*, edited by Jane L. Donawerth and Carol A. Kolmarten; foreword by Susan Gubar, 190–202. Liverpool: Liverpool University Press, 1994.

Jacobus, Mary. *First Things: The Maternal Imaginary in Literature, Art, and Psychoanalysis*. New York and London: Routledge, 1995.

James, Henry. *The Sense of the Past*. London: W. Collins Sons, 1917.

Jameson, Fredric. "Postmodernism, or The Cultural Logic of Late Capitalism." *New Left Review* 146 (1984): 53–92.

———. "Postmodernism and Consumer Society." In *Postmodern Culture*, edited by Hal Foster, 111–125. London, 1985.

Jantzen, Grace. *Becoming Divine: Towards a Feminist Philosophy of Religion*. Manchester: Manchester University Press, 1998.

Jennings, Herbert Spencer. *Prometheus, or Biology and the Advancement of Man*. London: E. P. Dutton, 1925.

Johnson, George. "Soul Searching." In *Clones and Clones: Facts and Fantasies about Human Cloning*, edited by Martha C. Nussbaum and Cass R. Sunstein, 67–70. New York and London: W. W. Norton.

Jones, Gwyneth. *Divine Endurance*. London: Unwin, 1984.

———. *Flower Dust*. New York: Tor, 1993.

Josipovici, Gabriel. "Scratch and Change." *TLS*, April 27, 2001, 6.

Kass, Leon R. "Making Babies—The New Biology and the 'Old' Morality." *Public Interest* 26 (Winter 1972): 13–56.

———. "Human Cloning Should Be Banned." In *Cloning*, edited by Paul A. Winters, 26–48. San Diego, CA: Greenhaven Press, 1998.

Keller, David. "The Feminine Metamorphosis," edited by Hugo Geinsback. *Science Wonder Stories* 3 (August 1929): 149–197.

Keller, Evelyn Fox. *Reflections on Gender and Science*. New Haven: Yale University Press, 1984.

———. "From Secrets of Life to Secrets of Death." In *Body/Politics: Women and the Discourse of Science*, edited by Mary Jacobus, Evelyn Fox Keller, and Sally Shuttleworth, 177–191. New York: Routledge, 1990.

Kellner, Douglas. *Jean Baudrillard: From Marxism to Postmodernism and Beyond*. Cambridge: Polity Press, 1989.

Kestner, Joseph. "Narcissism as Symptom and Structure: The Case of Mary Shelley's *Frankenstein*." In Mary Shelley, *Frankenstein*, edited by Fred Botting, 68–80. Houndmills, UK: Macmillan, 1995.

Kevles, Daniel J. *In the Name of Eugenics: Genetics and the Uses of Human Heredity*, with a new preface by the author. Cambridge: Harvard University Press, 1997.

Kittay, Eva Feder. "Womb Envy: An Explanatory Concept." In *Mothering: Essays in Feminist Theory*, edited by Joyce Trebilcot, 94–128. Totowa, NJ: Rowman and Allanheld, 1984.

———. "Rereading Freud on 'Femininity' or Why Not Womb Envy?" In *Hypatia Reborn: Essays in Feminist Philosophy*, edited by Azizah Y. Al-Hibri and Margaret A. Simons, 192–203. Bloomington and Indianapolis: Indiana University Press, 1990.

Klein, Melanie. "Early Stages in the Oedipal Conflict." In *The Selected Melanie Klein*, edited by Juliet Mitchell, 69–84. New York: Free Press, 1986.

———. "Envy and Gratitude" (1956). In Juliet Mitchell, *Melanie Klein*. Harmondsworth, UK: Penguin, 1986.

Klein, Renate, Gena Corea, and Ruth Hubbard. "German Women Say No to Gene and Reproductive Technology: Reflections on a Conference in Bonn, West Germany, April 19–21, 1985." *Women's Studies International Forum* 8, 3 (1985): 1–6.

Knight, Damon. "Mary" (1964). In *Clones*, edited by Jack Dann and Gardner Dozois, 31–53. New York: Ace Books, 1998.

Kochhar-Lindgren, Gray. *Narcissus Transformed: The Textual Subject in Psychoanalysis and Literature*. University Park: Pennsylvania State University Press, 1993.

Kolata, Gina. *The Road to Dolly and the Path Ahead*. New York: William Morrow, 1998.

Kress, Nancy. "Sex Education." In *Beaker's Dozen*, 209–226. New York: Tor, 1998.

———. "To Cuddle Amy." In *Year's Best SF 6*, edited by David G. Hartwell, 304–307. New York: EOS, 2001.

Kristeva, Julia. *La Révolution du langage poétique*. Paris: Seuil, 1974.

———. "Stabat Mater." In *Tales of Love*, translated by Leon S. Roudiez. New York: Columbia University Press, 1987.

———. *Tales of Love*, translated by Leon S. Roudiez. New York: Columbia University Press, 1987.

———. *Strangers to Ourselves*, translated by Leon S. Roudiez. New York: Harvester Wheatsheaf, 1991.

———. *Julia Kristeva Interviews*, edited by Ross Mitchell Guberman. New York: Columbia University Press, 1996.

———. "Extraterrestrials Suffering for Want of Love." In *The Portable Kristeva*, edited by Kelly Oliver, 170–179. New York: Columbia University Press, 1997.

———. *The Portable Kristeva*, edited by Kelly Oliver. New York: Columbia University Press, 1997.

———. "Women's Time." In *The Portable Kristeva*, edited by Kelly Oliver, 349–369. New York: Columbia University Press, 1997.

———. *Revolt, She Said*. An interview by Philippe Petit; translated by Brian O'Keefe; edited by Sylvère Lotringer. Los Angeles: Semiotext[e], 2002.

Lacan, Jacques. "Some Reflections on the Ego." *International Journal of Psychoanalysis* 34 (1953): 11–17.

———. "Aggressivity and Psychoanalysis." In *Écrits: A Selection*, translated by Alan Sheridan, 8–29. New York: W. W. Norton, 1977.

———. *Écrits: A Selection*, translated by Alan Sheridan. New York: W. W. Norton, 1977.

———. "The Mirror Stage as Formative of the Function of the I as Revealed in Psychoanalytic Experience." In *Écrits: A Selection*, translated by Alan Sheridan, 1–7. New York: W. W. Norton, 1977.

———. *The Four Fundamental Concepts of Psycho-Analysis*, edited by Jacques-Alain Miller; translated by Alan Sheridan. New York: W. W. Norton, 1978.

———. *Seminar, livre III: Les psychoses, 1955–1956*, translated by Russell Grigg. New York: W. W. Norton, 1993.

Lambspring. *Dyas chemica tripartita*. Frankfurt, 1625.

Lane, Mary E. Bradley. *Mizora* (1890), introduction by Joan Saberhagen. Lincoln and London: University of Nebraska Press, 1999.

Laplanche, J., and J. B. Pontalis. *The Language of Psychoanalysis*, translated by Donald Nicholson-Smith. London: Hogarth Press, 1973.

Larbalestier, Justine. *The Battle of the Sexes in Science Fiction*. Middletown, CT: Wesleyan University Press, 2002.

Lasch, Christopher. *The Culture of Narcissism: American Life in an Age of Diminishing Expectations* (1979). New York: W. W. Norton, 1991.

Lash, John. *Twins and the Double*. London: Thames and Hudson, 1993.
Lawrence, D. H. "The Undying Man." In *Phoenix*, edited with an introduction by Edward D. McDonald. Harmondsworth, UK: Penguin, 1978.
Lefanu, Sarah. "Difference and Sexual Politics in Naomi Mitchison's *Solution Three*." In *Utopian and Science Fiction by Women: Worlds of Difference*, edited by Jane L. Donawerth and Carol A. Kolmarten; foreword by Susan Gubar, 153–165. Liverpool: Liverpool University Press, 1994.
Le Guin, Ursula K. *The Left Hand of Darkness* (1969). New York: Ace Books, 1976.
———. "Nine Lives" (1969). In *Clones*, edited by Jack Dann and Gardner Dozois, 1–30. New York: Ace Books, 1998.
———. "On Theme." In *Those Who Can*, edited by Robin Scott Wilson, 204–205. New York: Signet, 1973.
Levin, Charles. *Jean Baudrillard: A Study in Cultural Metaphysics*. London: Prentice-Hall/Harvester Wheatsheaf, 1996.
Levin, Ira. *The Boys from Brazil*. London and Sydney: Pan Books, 1976.
Lévi-Strauss, Claude. *Structural Anthropology*. Garden City, NY: Doubleday, 1967.
Lief, Evelyn. *The Clone Rebellion*. New York: Pocket Books, 1980.
Livermore, Mary A. "Coöperative Womanhood in the State." *The North American Review* 153 (1891): 283–295.
Loewald, Hans. "The Waning of the Oedipus Complex." In *Papers on Psychoanalysis*. New Haven: Yale University Press, 1980.
Lomas, David. *The Haunted Self: Surrealism, Psychoanalysis, Subjectivity*. New Haven and London: Yale University Press, 2000.
Lomas, Peter. "Ritualistic Elements in the Management of Childbirth." *British Journal of Medical Psychology* 39 (1966): 207.
Loraux, Nicole. *The Children of Athens: Athenian Ideas about Citizenship and the Division between the Sexes*. Princeton: Princeton University Press, 1993.
Lublin, Nancy. *Pandora's Box: Feminism Confronts Reproductive Technology*. Lanham, MD: Rowman and Littlefield, 1998.
Ludovici, Anthony. *Lysistrata, or Woman's Future and Future Woman*. London: Kegan Paul, Trench Trubner & Co., 1924.
Lukacher, Ned. *Primal Scenes: Literature, Philosophy, Psychoanalysis*. Ithaca, NY, and London: Cornell University Press, 1986.
MacCannell, Juliet Flower. Introduction. In *Thinking Bodies*, edited by Juliet Flower MacCannell and Laura Zakarin, 1–15. Stanford, CA: Stanford University Press, 1994.
MacKenna, Erin. *The Task of Utopia: A Pragmatist and Feminist Perspective*. Lanham, MD: Rowman and Littlefield, 2001.
MacKinnon, Catherine. "Toward Feminist Jurisprudence." In *Feminist Jurisprudence*, edited by Patricia Smith, 610–620. New York: Oxford University Press, 1993.
MacLeod, Ian R. "Past Magic." In *Clones and Clones: Facts and Fantasies about Human Cloning*, edited by Martha Nussbaum and Cass R. Sustein, 159–175. New York: W. W. Norton, 1998.
Maier, Michael. *Atalanta fugiers*, translated and edited by Joscelyn Godwin. Grand Rapids, MI: Phanes Press, 1989.

Marcuse, Herbert. *One Dimensional Man*. London: Sphere Books, 1968.
———. *Counterrevolution and Revolt*. Boston: Beacon Press, 1972.
Marsh, Jan. *Pre-Raphaelite Women: Images of Femininity in Pre-Raphaelite Art*. London: Weidenfeld and Nicolson, 1987.
Martini, Virgillo. *Il Mondo senza Donne*. Guayaquil: Aliprandif, 1935.
———. *The World without Women* (1936). New York: Dial Press, 1971.
Marvell, Andrew. "The Garden." In *Complete Poems*, edited by Elizabeth Story Donno. Harmondsworth, UK: Penguin, 1972.
McElvaine, Robert S. *Eve's Seed: Biology, the Sexes, and the Course of History*. New York: McGraw-Hill, 2001.
McGee, Glenn. *The Perfect Baby: Parenthood in the New World of Cloning and Genetics*. Lanham, MD: Rowman and Littlefield, 2000.
Mears, A. Garland. *Mercia: The Astronomer Royal: A Romance*. London: Simpkin, Marshall, Hamilton, Kent and Company, 1895.
Melchior-Bonnet, Sabine. *The Mirror: A History*, translated by Katherine H. Jewett; preface by Jean Delumeau. New York and London: Routledge, 2002.
Mellor, Anne K. *Mary Shelley: Her Life, Her Fiction, Her Monsters*. New York: Methuen, 1988.
Merchant, Carolyn. *The Death of Nature: Women, Ecology, and the Scientific Revolution*. San Francisco: Harper & Row, 1983.
———. *Earthcare: Women and the Environment*. New York: Routledge, 1995.
Mérimé. *La Vénus d'Ille* (1837). Paris: Hachette, 2001.
Miller, Karl. *Doubles: Studies in Literary History*. New York: Oxford University Press, 1985.
Miller, Susan J., David J. Hickson, and David C. Wilson. "Decision-Making in Organizations." In *Managing Organizations: Current Issues*, 43–62.
Mills, Albert J. "Organizational Discourse and the Gendering of Identity." In *Postmodernism and Organization*, edited by John Hassard and Martin Parker, 132–147. London: Sage, 1994.
Mills, A. J., and S. J. Murgatroyd. *Organizational Rules: A Framework for Understanding Organizations*. Milton Keynes: Open University, 1991.
Mills, A. J., and P. Tancred, eds. *Gendering Organizational Analysis*. Newbury Park, CA: Sage, 1992.
Milton, John. *Paradise Lost: A Poem in Twelve Books*. New York: Odyssey Press, 1962.
Minsky, Rosalind. *Psychoanalysis and Culture: Contemporary States of Mind*. New Brunswick, NJ: Rutgers University Press, 1998.
Mitchison, Naomi. "Mary and Joe." In *Nova One*, edited by Harry Harrison, 161–173. New York: Dell, 1962.
———. "The Clone Mums." In *Haldane Papers*, November 1970.
———. *Not by Bread Alone*. London: Maryon Boyars, 1983.
———. *Memoirs of a Spacewoman* (1962). London: Women's Press, 1985.
———. *Solution Three* (1975), afterword by Susan M. Squier. New York: Feminist Press at the City University of New York, 1995.
Moers, Ellen. *Literary Women*. Garden City, NJ: Doubleday, 1976.
———. "Female Gothic." In *The Endurance of Frankenstein: Essays on Mary Shelley's Novel*, edited by George Levine and U. C. Knoepflmacher, 77–87. Berkeley and Los Angeles: University of California Press, 1982.

Moi, Toril, ed. *The Kristeva Reader*. New York: Columbia University Press, 1986.

Moraru, Christian. *Rewriting: Postmodern Narrative and Cultural Critique in the Age of Cloning*. Albany: State University of New York Press, 2001.

Morrissey, Thomas J. "Pamela Sargent's Science Fiction for Young Adults: Celebrations of Change." *Science-Fiction Studies* 16 (1989): 184–190.

Morrow, James. *Only Begotten Daughter*. San Diego: Harvest, 1990.

Moskowitz, Sam, ed. *When Women Rule*. New York: Walker, 1972.

Muller, Hermann. "Social Biology and Population Improvement" (The Geneticists' Manifesto). *Nature* 144 (1939): 521–522.

Murphy, Jane. "From Mice to Men? Implications of Progress in Cloning Research." In *Test-Tube Women: What Future for Motherhood?* edited by Rita Arditti, Renate Duelli Klein, and Shelley Minden, 76–91. London: Pandora, 1984.

Murphy, Julie. "Egg Farming and Women's Future." In *Test-Tube Women: What Future for Motherhood?* edited by Rita Arditti, Renate Duelli Klein, and Shelley Minden, 68–75. London: Pandora, 1984.

Murphy, Julien S. "Is Pregnancy Necessary? Feminist Concerns about Ectogenesis." In *Sex/Machine: Readings in Culture, Gender, and Technology*, edited by Patrick D. Hopkins, 185–200. Bloomington and Indianapolis: Indiana University Press, 1998.

National Bioethics Advisory Commission. "Religious Perspectives." In *Clones and Clones: Facts and Fantasies about Human Cloning*, edited by Martha C. Nussbaum and Cass R. Sunstein, 165–180. New York and London: W. W. Norton, 1998.

Nelkin, Dorothy, and M. Susan Lindee. *The DNA Mystique: The Gene as a Cultural Icon*. New York: W. H. Freeman, 1995.

Nesfield-Cookson, Bernard. *William Blake: Prophet of Universal Brotherhood*, foreword by George Trevelyan. London: Crucible, 1987.

Nietzsche, Friedrich. *The Gay Science*, translated by Walter Kaufmann. New York: Vintage, 1974.

———. *Thus Spake Zarathustra: A Book for All and None*, translated with a preface by Walter Kaufman. New York: Random House, 1995.

Neumann, Erich. *The Great Mother: An Analysis of the Archetype*, translated from the German by Ralph Manheim. London: Routledge & Kegan Paul, 1955.

Nussbaum, Martha. "Little C." In *Clones and Clones: Facts and Fantasies about Human Cloning*, edited by Martha C. Nussbaum and Cass R. Sustein. New York: W. W. Norton, 1998.

Nussbaum, Martha C., and Cass R. Sustein, eds. *Clones and Clones: Facts and Fantasies about Human Cloning*. New York: W. W. Norton, 1998.

Olson, Gary A., and Elizabeth Hirsch, eds. *Women Writing Culture*, foreword by Donna Haraway; afterword by Henry A. Giroux. Albany: State University of New York Press, 1995.

Ovid. *The Metamorphoses*, translated with an introduction by Mary M. Innes. Harmondswoth, UK: Penguin, 1955.

Pacteau, Francette. *The Symptom of Beauty*. London: Reaktion Books, 1994.

Paracelsus. *Hermetic Chemistry*, vol. 1 of *The Hermetic and Alchemical Writings of Aureolus Philippus Theophrastus Bombast, of Hohenheim, Called Paracelsus the Great*, edited by Arthur Edward Waite. Berkeley, CA: Shambala, 1976.

Parkin-Gounelas, Ruth. *Literature and Psychoanalysis: Intertextual Readings*. New York: Palgrave, 2001.

Pefanis, Julian. *Heterology and the Postmodern: Bataille, Baudrillard and Lyotard*. Durham, NC, and London: Duke University Press, 1991.

Pence, Gregory E. *Flesh of My Flesh: The Ethics of Human Cloning*, edited by Gregory E. Pence. Lanham, MD: Rowman and Littlefield, 1998.

———. *Who's Afraid of Human Cloning?* Lanham, MD: Rowman and Littlefield, 1998.

Petchesky, Rosalind Pollack. "Foetal Images: The Power of Visual Culture in the Politics of Reproduction." In Nancy Lublin, *Pandora's Box: Feminism Confronts Reproductive Technology*, 57–80. Lanham, MD: Rowman and Littlefield, 1998.

Pfaelzer, Jean. "Utopians Prefer Blondes—Mary Lane's *Mizora* and the Nineteenth-Century Utopian Imagination." In Mary E. Bradley Lane, *Mizora: A Prophecy*, edited with a critical introduction by Jean Pfaelzer, xi–xl. Syracuse, NY: Syracuse University Press, 2000.

Phillips, Adam. "Sameness Is All." In *Clones and Clones: Facts and Fantasies about Human Cloning*, edited by Martha C. Nussbaum and Cass R. Sunstein, 88–94. New York and London: W. W. Norton, 1998.

Piercy, Marge. *Woman on the Edge of Time*. New York: Fawcett, 1976.

Pinto Correia, Clara. *Clonai e Multiplicai-vos: Verdades e Mentiras*. Lisboa: Texto Editora, 1997.

———. *The Ovary of Eve: Egg and Sperm and Preformation*. Chicago and London: University of Chicago Press, 1997.

Plato. *Timaeus and Critias*, translated with an introduction and an appendix on *Atlantis* by Desmond Lee. Harmondsworth, UK: Penguin, 1981.

———. *Theaetetus*, edited by Seth Bernadete. Chicago: University of Chicago Press, 1986.

Plumwood, Val. "Women, Humanity and Nature." *Radical Philosophy* 48 (1988): 16–24.

———. *Feminism and the Mastery of Nature*. London and New York: Routledge, 1993.

Pringle, Rosemary. "Bureaucracy, Rationality and Sexuality: The Case of Secretaries." In *The Sexuality of Organization*, edited by Jeff Hearn, Deborah L. Sheppard, Pete Tancred-Sheriff, and Gibson Burrell, 158–177. London: Sage, 1989.

Ragland, Ellie. *Essays on the Pleasures of Death: From Freud to Lacan*. New York and London: Routledge 1995.

Raine, Kathleen. *Golgonooza: City of Imagination*. Ipswich, Golgonooza Press, 1991.

Ramsey, Paul. *Fabricated Man: The Ethics of Genetic Control*. New Haven: Yale University Press, 1970.

Rank, Otto. *The Double: A Psychoanalytic Study*, translated and edited with an introduction by Harry Tucker, Jr. London: Maresfield Library, 1989.

Raymond, Janice. "Feminist Ethics, Ecology, and Vision." In *Test-Tube Women: What Future for Motherhood?* edited by Rita Arditti, Renate Duelli Klein, and Shelley Minden, 427–437. London: Pandora, 1984.

Raymond, J. "Preface." In *Man-Made Women: How New Reproductive Technologies Affect Women*, edited by Gena Corea, Renate Duelli Klein, Jalna Hanmer, Helen B. Holmes, Betty Hoskins, Madhu Kishwar, Janice Raymond, Robyn Rowland, and Roberta Steinbacher. London: Hutch-inson, 1985.

Reynolds, Francis Joseph, ed. *Master Tales of Mystery by the World's Most Famous Authors of To-day*. 1915.

Rich, Adrienne. "The Theft of Childbirth." *New York Review of Books* 22, 15 (1975): 25–30.

———. *The Dream of a Common Language: Poems 1974–1977*. New York: W. W. Norton, 1978.

———. "Compulsory Heterosexuality and Lesbian Experience." *Signs* 5 (1983): 631–660.

———. *Of Woman Born: Motherhood as Experience and Institution*. London: Virago, 1992.

Roberts, Gareth. *The Mirror of Alchemy: Alchemical Ideas and Images in Manuscripts and Books*. London: The British Library, 1994.

Rorvik, David. *In His Image: The Cloning of a Man*. New York: Pocket Books, 1978.

Rose, Hilary. "Victorian Values in the Test-tube: The Politics of Reproductive Science and Technology." In Nancy Lublin, *Pandora's Box: Feminism Confronts Reproductive Technology*, 151–173. Lanham, MD: Rowman and Littlefield, 1998.

Rostand, Jean. *Can Man Be Modified? Predictions of Our Biological Future*, translated by Jonathan Griffin. New York: Basic Books, 1959.

Rowland, Robin. "Reproductive Technologies: The Final Solution to the Woman Question?" In *Test-Tube Women: What Future for Motherhood?* edited by Rita Arditti, Renate Duelli Klein, and Shelley Minden, 356–369. London: Pandora, 1984.

Roy, Ina. "Philosophical Perspectives." In *The Human Cloning Debate*, edited by Glenn McGee, 41–66. Berkeley, CA: Berkeley Hills Books, 1998.

Russ, Joanna. "Reflections on Science Fiction: An Interview with Joanna Russ." *Quest: A Feminist Quarterly* 2 (1975): 40–49.

———. *The Female Man* (1975). Boston: Beacon Press, 1986.

———. "*Amor Vincit Foeminam*: The Battle of the Sexes in Science Fiction." In *To Write like a Woman: Essays in Feminism and Science Fiction*, 41–59. Bloomington and Indianapolis: Indiana University Press, 1995.

———. "Recent Feminist Utopias." In *To Write like a Woman: Essays in Feminism and Science Fiction*, 133–148. Bloomington and Indianapolis: Indiana University Press, 1995.

———. *To Write like a Woman: Essays in Feminism and Science Fiction*. Bloomington and Indianapolis: Indiana University Press, 1995.

———. "For Women Only: Or, What Is That Man Doing under My Seat?" In *What Are We Fighting For? Sex, Race, Class, and the Future of Feminism*, 84–103. New York: St. Martin's Press, 1998.

Russell, Dora. *Hypatia, or, Women and Knowledge*. New York: E. P. Dutton & Company, 1925.

Ryman, Geoff. *The Child Garden, or, A Low Comedy*. London: Unwin Hyman, 1989.

Sallis, John. "Doublings." In *Derrida: A Critical Reader*, edited by David Wood, 120–136. Oxford: Blackwell, 1992.

Salvaggio, Ruth. "The Case of Two Cultures: Psychoanalytic Theories of Science and Literature." In *Discontented Discourse: Feminism/Textual Intervention/Psychoanalysis*, edited by Marleen S. Barr and Richard Feldstein, 54–65. Urbana and Chicago: University of Illinois Press, 1989.

Sargent, Pamela, ed. *Bio-Futures: Science Fiction Stories about Biological Metamorphosis*. New York: Vintage, 1976.

———. *Cloned Lives*. Greenwich, CT: Fawcett, 1976.

———. *The Shore of Women* (1987). London: Pan Books, 1988.

———. "Clone Sister." In *Clones*, edited by Jack Dann and Gardner Dozois, 177–201. New York: Ace Books, 1998.

Savage, M., and A. Witz. *Gender and Bureaucracy*. Oxford: Blackwell, 1992.

Saxton, Josephine. *Travels of Jane Saint and Other Stories*. London: The Women's Press, 1986.

Schwab, Gabriele. "Steps Toward a Millennial Imaginary: Reproductive Fantasies in Cussin's 'Confessions of a Bioterrorist.'" In *Playing Dolly: Technocultural Formations, Fantasies, and Fictions of Assisted Reproduction*, edited by E. Ann Kaplan and Susan Squier, 232–240. New Brunswick, NJ: Rutgers University Press, 1999.

Schwartz, Hillel. *The Culture of the Copy: Striking Likenesses, Unreasonable Facsimiles*. New York: Zone Books, 1996.

Scott, Jody. *I, Vampire*. London: The Women's Press, 1986.

Segel, Harold B. *Pinocchio's Progeny: Puppets, Marionettes, and Robots in Modernist and Avant-Garde Drama*. Baltimore and London: Johns Hopkins University Press, 1995.

Sendak, Maurice. *Outside over There*. New York: Harper and Row, 1981.

Senior, W. A. "Introduction." *Journal of the Fantastic in the Arts* 10, 3 (1999): 203–205.

Shakespeare, William. *Macbeth*, edited by K. Muir. London: Arden Shakespeare, 2001.

Shattuck, Roger. *Forbidden Knowledge: From Prometheus to Pornography*. San Diego and London: Harvest, 1996.

Shaw, Debra Benita. *Women, Science and Fiction: The "Frankenstein" Inheritance*. Houndmills, Basingstoke: Palgrave, 2000.

Shear, David. *Cloning*. New York: Pinnacle Books, 1972.

Sheffield, Charles. "Out of Copyright." In *Clones*, edited by Jack Dann and Gardner Dozois, 75–95. New York: Ace Books, 1998.

Shelley, Mary. *Frankenstein, or, The Modern Prometheus* (1818), edited by M. K. Joseph. Oxford: Oxford University Press, 1985.

———. *The Last Man* (1826), introduction by Brian Aldiss. London: Hogarth Press, 1985.

Shiva, Vandana. "The Seed and the Earth: Women, Ecology, and Biotechnology." *The Ecologist* 22, 1 (1992): 5.

Sidney, Philip. "Astrophel and Stella." In *Sir Philip Sidney: The Major Works*, edited by Katherine Duncan-Jones, 153–211. Oxford: Oxford University Press, 2002.

Silver, Lee M. *Remaking Eden: Cloning and Beyond in a Brave New World*. London: Weidenfeld & Nicolson, 1998.

Silverman, Kaja. *The Threshold of the Visible World*. New York and London: Routledge, 1996.

Singer, Peter, and Deane Wells. *Making Babies*. New York: Scribner's, 1984.

Singer, Rochelle. *The Demeter Flower*. New York: St. Martin's Press, 1980.

Slonzcewski, Joan. *A Door into Ocean* (1986). London: The Women's Press, 1987.

Smith, Michael Marshall. *Spares* (1996). London: HarperCollins, 1998.

Smith-Rosenberg, Carroll. "The Female World of Love and Ritual: Relations between Women in Nineteenth Century America." *Signs: Journal of Women in Culture and Society* 1 (1975): 1–29.

Songe-Moller, Vigdis. *Philosophy without Women: The Birth of Sexism in Western Thought*, translated by Peter Cripps. London and New York: Continuun, 2002.

Sourbut, Elizabeth. "Gynogenesis: A Lesbian Appropriation of Reproductive Technologies." In *Between Monsters, Goddesses and Cyborgs: Feminist Confrontations with Science, Medicine and Cyberspace*, edited by Nina Lykke and Rosi Braidotti, 227–241. London and Atlantic Highlands, NJ: Zed Books, 1996.

Spretnak, Charlene. "Naming the Cultural Forces That Push Us toward War." In *Exposing Nuclear Phallacies*, edited by Diana E. H. Russell, 53–62. New York: Pergamon Press, 1989.

Squier, Susan Merrill. *Babies in Bottles: Twentieth-Century Visions of Reproductive Technology*. New Brunswick, NJ: Rutgers University Press, 1994.

———. "Afterword." In Naomi Mitchison, *Solution Three*, 161–183. New York: The Feminist Press at the City University of New York, 1995.

———. "Reproducing the Posthuman Body: Ectogenetic Fetus, Surrogate Mother, Pregnant Man." In *Posthuman Bodies*, edited by Ira Livingston and Judith Halberstam, 113–132. Bloomington and Indianapolis: Indiana University Press, 1995.

Stableford, Brian. "Clones." In *The Encyclopedia of Science Fiction*, edited by John Clute and Peter Nicholls, 236–237. New York: St. Martin's Griffin, 1995.

———. "Biotechnology and Utopia." In *The Philosophy of Utopia*, edited by Barbara Goodwin, 189–204. London and Portland, OR: Frank Cass, 2001.

Stafford, Barbara Maria. *Visual Analogy: Consciousness as the Art of Connecting*. Cambridge, MA: MIT Press, 1999.

Stanworth, Michelle. "The Deconstruction of Motherhood." In *Reproductive Technologies: Gender, Motherhood, and Medicine*, edited by Michelle Stanworth, 1–35. Cambridge: Polity Press, 1987.

Steinbacher, Roberta. "Sex Preselection: From Here to Fraternity." In *Beyond Domination: New Perspectives on Women and Philosophy*, edited by C. Gould. Towota, NJ: Rowman and Allenheld, 1983.

Steinbacher, Roberta, and Helen B. Holmes. "Sex Choice: Survival and Sisterhood." In *Man-Made Women: How New Reproductive Technologies Affect Women*, edited by Gena Corea, Renate Duelli Klein, Jalna Hanmer, Helen B. Holmes, Betty Hoskins, Madhu Kishwar, Janice Raymond, Robyn Rowland, and Roberta Steinbacher. Bloomington and Indianapolis: Indiana University Press, 1985.

Sterne, Laurence. *Tristram Shandy*, edited by Graham Petree with an Introduction by Christopher Ricks. Harmondsworth, UK: Penguin, 1970.

Stewart, Susan. *On Longing: Narratives of the Miniature, the Gigantic, the Souvenir, the Collection*. Durham, NC, and London: Duke University Press, 1993.

Stock, Gregory. *Redesigning Humans: Choosing Our Children's Genes*. London: Profile Books, 2002.

Stonehouse, Julia. *From Idols to Incubators: Reproduction Theory through the Ages*. London: Scarlet Press, 1994.

Stoner, Dawn. *The Killing Clone*. Bloomington, IN: 1st Books Library, 2000.

Storch, Margaret. *Sons and Adversaries: Women in William Blake and D. H. Lawrence*. Knoxville: University of Tennessee Press, 1990.

Suleiman, Susan Rubin. *Subversive Intent: Gender, Politics, and the Avant-Garde.* Cambridge and London: Harvard University Press, 1990.

Swedenborg, Emanuel. *Heaven and Its Wonders and Hell: Drawn from Things Heard and Seen*, translated by George F. Dole; introduction by Bernhard Lang. New York: Chrysalis Books, 2001.

Taylor, Gordon Rattray. *The Biological Time Bomb.* New York: New American Library, 1968.

Tennant, Emma. *Two Women of London.* London: Faber and Faber, 1990.

———. *Faustine.* London: Faber and Faber, 1992.

Tepper, Sheri S. *The Gate to Women's Country.* New York: Bantam Books, 1988.

Teresi, Dick, and Kathleen Mcauliffe. "Male Pregnancy." In *Sex/Machine: Readings in Culture, Gender, and Technology*, edited by Patrick D. Hopkins, 175–183. Bloomington and Indianapolis: Indiana University Press, 1998.

Tiptree, James, Jr. "Houston, Houston, Do You Read?" In *Aurora: Beond Equality*, edited by Susan Janice Anderson and Vonda N. McIntyre, 36–98. Greenwich, CT: Fawcett, 1976.

———. *Up the Walls of the World.* New York: Berkley Publishing, 1978.

———. "Notes on 'Houston, Houston, Do You Read?'" In *Meet Me at Infinity*, 373–375. New York: Tor, 2000.

Toffler, Alvin. *Future Schock.* London: Pan Books, 1970.

Trotter, Wilfred. *Ihstincts of the Herd in Peace and War.* Oxford: Oxford University Press, 1919.

Tuana, Nancy. *Woman and the History of Philisophy.* New York: Paragon House, 1992.

———. "Aristotle and the Politics of Reproduction." In *Engendering Origins: Critical Feminist Readings in Plato and Aristotle*, edited by Bat-Ami Bar On, 189–206. Albany: State University of New York Press, 1994.

Turney, Jon. *Frankenstein's Footsteps: Science, Genetics and Popular Culture.* New Haven and London: Yale University Press, 1998.

Tuttle, Lisa. "World of Strangers." In *Clones and Clones: Facts and Fantasies about Human Cloning*, edited by Martha C. Nussbaum and Cass R. Sustein, 297–309. New York: W. W. Norton, 1998.

Van Scyoc, Sydney J. *Star Mother.* New York: Berkley Publishing, 1976.

Varley, John. "Lollipop and the Tar Baby" (1977). In *The Norton Book of Science Fiction*, edited by Ursula K. Le Guin and B. Atteberry, 357–375. New York: W. W. Norton, 1993.

———. *Steel Beach.* New York: Ace Books, 1993.

Villiers de L'Isle-Adam, Jean-Marie, Mathas-Phillipe-Auguste. *Eve of the Future Eden*, translated by Marilyn Gaddis Rose. Lawrence, KS: Coronado Press, 1981.

Vinge, Joan D. *Snow Queen.* New York: Popular Library, 1980.

Walker, Michelle Boulous. *Philosophy and the Maternal Body: Reading Silence.* London and New York: Routledge, 1998.

Walters, William A. W. "Transsexualism and Abdominal Pregnancy." *Developments in the Health Field with Bioethical Implications*, 2 (1990) [Australian] National Bioethics Consultive Committee, C-11–C-76.

Ward, Lester Frank. *Pure Sociology.* New York: Macmillan, 1903.

Warner, Marina. *Alone of All Her Sex: The Myth and the Cult of the Virgin Mary.* London: Vintage, 1976.

———. *Monuments and Maidens: The Allegory of the Female Form*. London: Picador, 1985.
———. *Managing Monsters: Six Myths of Our Time*. London: Vintage, 1994.
———. *No Go the Bogeyman: Scaring, Lulling, and Making Mock*. London: Vintage, 2000.
———. *Fantastic Metamorphoses, Other Worlds: Ways of Telling the Self*. Oxford: Oxford Universitry Press, 2002.
Warren, Mary Anne. "The Ethics of Sex Preselection." In *Sex/Machine: Readings in Culture, Gender, and Technology*, edited by Patrick D. Hopkins, 143–156. Bloomington and Indianapolis: Indiana University Press, 1998.
Watson, James D. "Moving toward the Clonal Man: Is This What We Want?" *Atlantic Monthly*, May 1971, 50–53.
Webb, Janeen. "Bone of My Bones, Flesh of My Flesh: A Brief History of the Clone in Science Fiction." In *Biotechnological and Medical Themes in Science Fiction*, edited by Domna Pastourmatzi, 152–164. Thessaloniki: University Studio Press, 2002.
Weil, Kari. *Androgyny and the Denial of Difference*. Charlottesville and London: University Press of Virginia, 1992.
Weldon, Fay. *The Cloning of Joanna May* (1989). London: Flamingo, 1993.
———. "Of Birth and Fiction." In *Fay Weldon's Wicked Fictions*, edited by Regina Barreca. Hanover and London: University Press of New England, 1994.
Wells, H. G. *The Island of Dr Moreau*, afterword by Brian Aldiss. New York: Signet, 1988.
———. *The Time Machine*, edited by Nicholas Ruddick. Peterborough, ON: Broadview Literary Texts, 2001.
Whitford, Margaret. *Luce Irigaray: Philosophy in the Feminine*. London and New York: Routledge, 1991.
Wilde, Oscar. *The Picture of Dorian Gray*, edited by Isabel Murray. Oxford: Oxford University Press, 1998.
Wilhelm, Kate. "Where Late the Sweet Birds Sang" (1974). In *Clones*, edited by Jack Dann and Gardner Dozois, 204–254. New York: Ace, 1998.
———. *Where Late the Sweet Birds Sang* (1976). New York: Pocket Books, 1977.
Williams, Linda Ruth. *Critical Desire: Psychoanalysis and the Literary Subject*. London: Edward Arnold, 1995.
Williams, Lynn. "Separatist Fantasies, 1690–1997: An Annotated Bibliography." *FEMSPEC* 1, 2 (2000): 30–42.
Wilmut, Ian, Keith Campbell, and Colin Tudge. *The Second Creation: The Age of Biological Control*. London: Headline, 2000.
Wilson, Anna. *Hatching Stones*. London: Onlywomen Press, 1991.
Wilson, Charles. *Embryo*. New York: St. Martin's Paperbacks, 1999.
Winston, Robert. *The IVF Revolution: The Definitive Guide to Assisted Reproductive Techniques*. London: Vermilion, 1999.
Wittig, Monique. *Les Guérillères*. Paris: Éditions de Minuit, 1969.
Wolfe, Gene. *The Fifth Head of Cerberus: Three Novellas* (1972). New York: Tom Doherty Associates, 1994.
Wolmark, Jenny. *Aliens and Others: Science Fiction, Feminism and Postmodernism*. New York: Harvester Wheatsheaf, 1994.

Woodward, Kathleen. *Aging and Its Discontents: Freud and Other Fictions.* Bloomington and Indianapolis: Indiana University Press, 1991.
Woolf, Virginia. *A Room of One's Own*. Harmondsworth, UK: Penguin, 1967.
———. "A Society." In *Selected Short Stories*, edited with an introduction by Sandra Kemp, 7–21. Harmondsworth, UK: Penguin, 1993.
Wylie, Philip. *The Disappearance* (1951). New York: Pocket Books, 1958.
Young, Donna J. *Retreat: As It Was*. Tallahassee: Naiad Press, 1979.
Young, Florence Ethell Mills. *The War of the Sexes*. London: John Long, 1905.
Yount, Lisa, ed. *Cloning*. San Diego: Greenhaven Press, 2000.
Žižek, Slavoj. *Enjoy Your Symptom! Jacques Lecan in Hollywood and Out*. New York and London: Routledge, 1992.
———. "Of Cells and Selves." In *The Zizek Reader*, edited by Elizabeth Wright and Edmond Wright, 302–320. Oxford: Blackwell, 2000.
———. "Bring Me My Philips Mental Jacket." *London Review of Books*, 22 May 2003, 3–5.
Zurbrugg, Nicholas, ed. *Jean Baudrillard: Art and Artefact*. London: Sage, 1997.

FILMOGRAPHY

Artificial Intelligence (Steven Spielberg, 2001, US)
Battlestar Galactica (Richard A. Colla, 1978, US)
Blade Runner (Ridley Scott, 1982, US)
The Boys from Brazil (Franklin J. Schaffner, 1977, US)
Gattaca (Andrew Nicol, 1997, US)
Junior (Ivan Reitman, 1994, US)
The Kid (Jan Turteltaub, 1999, US)
Metropolis (Fritz Lang, 1926, Germany)
Multiplicity (Harold Ramis, 1996, US)
The Sixth Day (Roger Spottiswoode, 2000, US)
Sleeper (Woody Allen, 1973, US)
Star Wars, Episode II, Attack of the Clones (George Lucas, 2002, US)
The Stepford Wives (Bryan Forbes, 1975, US)

Index

ABOUT THE AUTHOR

MARIA ALINE SALGUEIRO SEABRA FERREIRA is Associate Professor in the Department of Languages and Cultures, University of Aveiro, Portugal.

www.ingramcontent.com/pod-product-compliance
Lightning Source LLC
Chambersburg PA
CBHW060529310726
48982CB00002B/483

* 9 7 8 0 3 1 3 3 2 0 0 6 4 *